Engineering Hydrology and Earth Science

Engineering Hydrology and Earth Science

Edited by **Stacy Keach**

SYRAWOOD
PUBLISHING HOUSE

New York

Published by Syrawood Publishing House,
750 Third Avenue, 9th Floor,
New York, NY 10017, USA
www.syrawoodpublishinghouse.com

Engineering Hydrology and Earth Science
Edited by Stacy Keach

International Standard Book Number: 978-1-68286-042-7 (Hardback)

Contents

Preface   VII

Chapter 1   **Development of a large-sample watershed-scale hydrometeorological data set for the contiguous USA: data set characteristics and assessment of regional variability in hydrologic model performance**   1
A. J. Newman, M. P. Clark, K. Sampson, A. Wood, L. E. Hay, A. Bock, R. J. Viger, D. Blodgett, L. Brekke, J. R. Arnold, T. Hopson and Q. Duan

Chapter 2   **Assessment of precipitation and temperature data from CMIP3 global climate models for hydrologic simulation**   16
T. A. McMahon, M. C. Peel and D. J. Karoly

Chapter 3   **Precipitation variability within an urban monitoring network via microcanonical cascade generators**   33
P. Licznar, C. De Michele and W. Adamowski

Chapter 4   **What made the June 2013 flood in Germany an exceptional event? A hydro-meteorological evaluation**   55
K. Schröter, M. Kunz, F. Elmer, B. Mühr and B. Merz

Chapter 5   **Calibration approaches for distributed hydrologic models in poorly gaged basins: implication for streamflow projections under climate change**   74
S. Wi, Y. C. E. Yang, S. Steinschneider, A. Khalil and C. M. Brown

Chapter 6   **On the sensitivity of urban hydrodynamic modelling to rainfall spatial and temporal resolution**   94
G. Bruni, R. Reinoso, N. C. van de Giesen, F. H. L. R. Clemens and J. A. E. ten Veldhuis

Chapter 7   **Assessing the impact of different sources of topographic data on 1-D hydraulic modelling of floods**   113
A. Md Ali, D. P. Solomatine and G. Di Baldassarre

Chapter 8   **Estimates of global dew collection potential on artificial surfaces**   126
H. Vuollekoski, M. Vogt, V. A. Sinclair, J. Duplissy, H. Järvinen, E.-M. Kyrö, R. Makkonen, T. Petäjä, N. L. Prisle, P. Räisänen, M. Sipilä, J. Ylhäisi and M. Kulmala

Chapter 9   **Global trends in extreme precipitation: climate models versus observations**   139
B. Asadieh and N. Y. Krakauer

Chapter 10   **How does bias correction of regional climate model precipitation affect modelled runoff?**   **154**
J. Teng, N. J. Potter, F. H. S. Chiew, L. Zhang, B. Wang, J. Vaze and J. P. Evans

Chapter 11   **Evolving flood patterns in a Mediterranean region (1301–2012) and climatic factors – the case of Catalonia**   **172**
A. Barrera-Escoda and M. C. Llasat

Permissions

List of Contributors

Preface

Over the recent decade, advancements and applications have progressed exponentially. This has led to the increased interest in this field and projects are being conducted to enhance knowledge. The main objective of this book is to present some of the critical challenges and provide insights into possible solutions. This book will answer the varied questions that arise in the field and also provide an increased scope for furthering studies.

Hydrology is a significant discipline that aims to analyse the distribution and quality of water resources on earth. There has been an increasing emphasis on understanding the physico-chemical characteristics of global water reserves and hydrologic movement using computational modeling and measurement techniques. The chapters in this book discuss various topics like hydrometeorology, evaluation of hydrologic data from different parts of globe, climatology, water resource engineering, etc. It is an essential guide for both students and researchers seeking in-depth information of the field.

I hope that this book, with its visionary approach, will be a valuable addition and will promote interest among readers. Each of the authors has provided their extraordinary competence in their specific fields by providing different perspectives as they come from diverse nations and regions. I thank them for their contributions.

Editor

Development of a large-sample watershed-scale hydrometeorological data set for the contiguous USA: data set characteristics and assessment of regional variability in hydrologic model performance

A. J. Newman[1], M. P. Clark[1], K. Sampson[1], A. Wood[1], L. E. Hay[2], A. Bock[2], R. J. Viger[2], D. Blodgett[3], L. Brekke[4], J. R. Arnold[5], T. Hopson[1], and Q. Duan[6]

[1]National Center for Atmospheric Research, Boulder CO, USA
[2]United States Geological Survey, Modeling of Watershed Systems, Lakewood CO, USA
[3]United States Geological Survey, Center for Integrated Data Analytics, Middleton WI, USA
[4]US Department of Interior, Bureau of Reclamation, Denver CO, USA
[5]US Army Corps of Engineers, Institute for Water Resources, Seattle WA, USA
[6]Beijing Normal University, Beijing, China

Correspondence to: A. J. Newman (anewman@ucar.edu)

Abstract. We present a community data set of daily forcing and hydrologic response data for 671 small- to medium-sized basins across the contiguous United States (median basin size of $336 \, \text{km}^2$) that spans a very wide range of hydroclimatic conditions. Area-averaged forcing data for the period 1980–2010 was generated for three basin spatial configurations – basin mean, hydrologic response units (HRUs) and elevation bands – by mapping daily, gridded meteorological data sets to the subbasin (Daymet) and basin polygons (Daymet, Maurer and NLDAS). Daily streamflow data was compiled from the United States Geological Survey National Water Information System. The focus of this paper is to (1) present the data set for community use and (2) provide a model performance benchmark using the coupled Snow-17 snow model and the Sacramento Soil Moisture Accounting Model, calibrated using the shuffled complex evolution global optimization routine. After optimization minimizing daily root mean squared error, 90 % of the basins have Nash–Sutcliffe efficiency scores ≥ 0.55 for the calibration period and 34 % ≥ 0.8. This benchmark provides a reference level of hydrologic model performance for a commonly used model and calibration system, and highlights some regional variations in model performance. For example, basins with a more pronounced seasonal cycle generally have a negative low flow bias, while basins with a smaller seasonal cycle have a positive low flow bias. Finally, we find that data points with extreme error (defined as individual days with a high fraction of total error) are more common in arid basins with limited snow and, for a given aridity, fewer extreme error days are present as the basin snow water equivalent increases.

1 Introduction

With the increasing availability of gridded meteorological data sets, streamflow records and computing resources, large-sample hydrology studies have become more common in the last decade or more (i.e., Nathan and McMahon, 1990; Perrin et al., 2001; Maurer et al., 2002; Beldring et al., 2003; Merz and Bloschl, 2004; Andreassian et al., 2004; Lohmann et al., 2004; Duan et al., 2006; Oudin et al., 2006, 2010; Samaniego et al., 2010; Martinez and Gupta, 2010, 2011; Nester et al., 2011, 2012; Livneh and Lettenmaier, 2012, 2013; Kumar et al., 2013; Oubeidillah et al., 2013). Within the United States there have been several studies to produce large-sample hydrometeorological data sets (Maurer et al., 2002; Lohmann et al., 2004; Duan et al., 2006; Thornton et al., 2012; Xia et al., 2012; Livneh et al., 2013). Many of these data sets provide gridded data and may need to be further processed by the end user for their specific hydrologic model configuration.

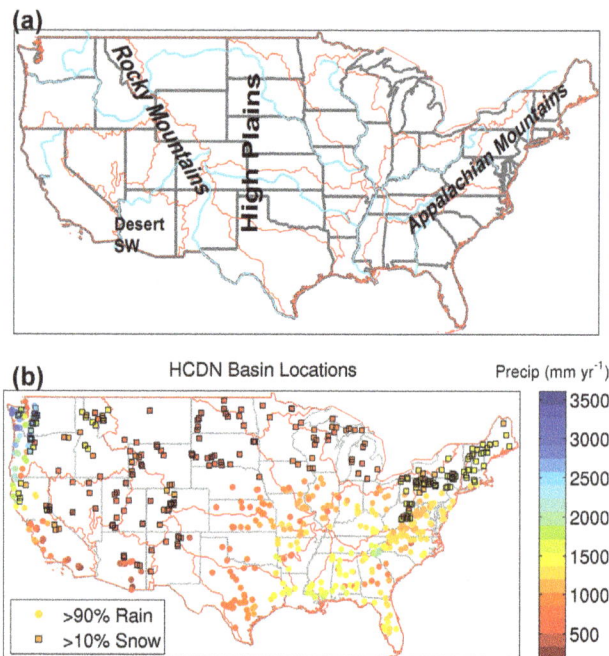

Figure 1. (**a**) Contiguous United States (CONUS) with states (gray), rivers (blue) and major hydrologic regions (red). Text indicates major geographic regions discussed in text. (**b**) Location of the 671 HCDN-2009 basins across the contiguous US used in the basin data set with precipitation shaded. Circles denote basins with > 90 % of their precipitation falling as rain, squares with black outlines denote basins with > 10 % of their precipitation falling as snow as determined by using a 0 °C daily mean Daymet temperature threshold. State outlines are in thin gray and hydrologic regions in thin red.

The Model Parameter Estimation Project (MOPEX) data set does provide basin mean hydrometeorological data and observed streamflow records for 438 basins across the contiguous United States (CONUS; Schaake et al., 2006) for over more than 30 years; making it one of the few, high-quality, freely available hydrometeorological data sets with immediate applicability to catchment-type hydrologic models.

Gupta et al. (2014) emphasize that more large-sample hydrologic studies are needed to "balance depth with breadth"; most hydrologic studies have traditionally focused on one or a small number of basins (depth), which hinders the ability to establish general hydrologic concepts applicable across regions (breadth). Gupta et al. (2014) go on to discuss practical considerations for large-sample hydrology studies, noting first and foremost that large data sets of quality basin data need to be available and shared in the community. In support of this philosophy, we present a large-sample hydrometeorological data set and modeling tools to understand regional variability in hydrologic model performance across the contiguous US (Fig. 1). The development of the basin data set presented herein takes advantage of high-quality, freely available data from various US government agencies and re-

search laboratories. It includes (1) daily forcing data for 671 basins for multiple spatial configurations over the 1980–2010 time period; (2) daily streamflow data; (3) basic metadata (e.g., location, elevation, size, and basin delineation shapefiles) and (4) benchmark model performance which contains the final calibrated model parameter sets, model output time series for all basins as well as summary graphics for each basin. This builds on the MOPEX data set by providing basin mean forcing data for 233 more basins along with two other spatial configurations and the benchmark model performance parameter sets and model output.

This data set and benchmark application is intended for the community to use as a test bed to facilitate the evaluation of hydrologic modeling and prediction questions. To this end, the benchmark consists of the calibrated, coupled Snow-17 snow model and the Sacramento Soil Moisture Accounting Model (SAC-SMA) for all 671 basins using the shuffled complex evolution (SCE) global optimization routine. Development of a large-sample hydrologic data set such as this will allow for exploration into many important scientific questions. We provide some basic analysis relating to questions such as (1) what is the model performance across a large sample of basins and how does model performance vary across basin hydroclimatic conditions? (2) How do error characteristics relate to basin calibration performance and hydroclimatic conditions? This basic analysis is intended to highlight some of the important questions that can be answered through large-sample hydrologic studies and provide example results for further exploration.

The next section describes the development of the basin data set from basin selection through forcing data generation. It then briefly describes the modeling system and calibration routine. Next, example results using the basin data set and modeling platform are presented. Finally, concluding thoughts and next steps are discussed.

2 Basin data set

The development of a freely available large-sample basin data set requires several choices and subsequent data acquisition. Three major decisions were made and are discussed in this section: (1) the selection process for the basins, (2) the various basin spatial configurations to be developed, and (3) selection of the underlying forcing data set used to develop forcing data time series. Additionally, aggregation of the necessary streamflow data is described.

2.1 Basin selection

The United States Geological Survey (USGS) developed an updated version of their Geospatial Attributes of Gages for Evaluating Streamflow (GAGES-II) in 2011 (Falcone et al., 2010; Falcone, 2011). This database contains geospatial information for over 9000 stream gages maintained by the

USGS. As a subset of the GAGES-II database, a portion of the basins with minimal human disturbance (i.e., minimal land use changes or disturbances, minimal human water withdrawals) are noted as "reference" gages. A further subsetting of the reference gages were made as a follow-on to the Hydro-Climatic Data Network (HCDN) 1988 data set (Slack and Landwehr, 1992). These gages, marked HCDN-2009 (Lins, 2012), meet the following criteria: (1) have at least 20 years of complete flow data between 1990 and 2009 and were active as of 2009, (2) are a GAGES-II reference gage, (c) have less than 5 % imperviousness as measured by the National Land Cover Database (NLCD-2011; Jin et al., 2013), and (d) passed a manual survey of human impacts in the basin by local Water Science Center evaluators (Falcone et al., 2010). There are 704 gages in the GAGES-II database that are considered HCDN-2009 across the CONUS. This study uses that portion of the HCDN-2009 basin set as the starting point since they should best represent natural flow conditions. After initial processing and data availability requirements, 671 basins are used for analysis in this study (Fig. 1b). Because these basins have minimal human influence they are almost exclusively smaller, headwater-type basins.

2.2　Forcing and streamflow data

Hydrologic models are run with a variety of spatial configurations, including entire watersheds (lumped), elevation bands, hydrologic response units (HRUs), or grids. For this data set, forcing data were calculated (via areal averaging) for watershed, HRU and elevation band spatial configurations. The basin spatial configurations were created from the base national geospatial fabric for hydrologic modeling developed by the USGS Modeling of Watershed Systems (MoWS) group (Viger, 2014; Viger and Bock, 2014). The geospatial fabric is a watershed-oriented analysis of the National Hydrography Data set that contains points of interest (e.g., USGS streamflow gauges), hydrologic response unit boundaries and simplified stream segments (not used in this study). This geospatial fabric contains points of interest that include USGS streamflow gauges and allowed for the determination of upstream total basin area and basin HRUs (Viger, 2014; Viger and Bock, 2014). A digital elevation model (DEM) was applied to the geospatial fabric data set to create elevation contour polygon shapefiles for each basin. The USGS Geo Data Portal (GDP) developed by the USGS Center for Integrated Data Analytics (CIDA) (Blodgett et al., 2011) was leveraged to produce area-weighted forcing data for the various basin spatial configurations over our time period. The GDP performs all necessary spatial subsetting and weighting calculations and returns the area-weighted time series for the specified inputs.

The Daymet data set was selected as the primary gridded meteorological data set to derive forcing data for our streamflow simulations (Thornton et al., 2012). Daymet was cho-

sen because of its high spatial resolution, a necessary requirement to more fully estimate spatial heterogeneity for basins in complex topography. Daymet is a daily, gridded (1 × 1 km) data set over the CONUS and southern Canada and is available from 1980 to present. It is derived solely from daily observations of temperature and precipitation. The Daymet variables used here are daily maximum and minimum temperature, precipitation, shortwave downward radiation, day length, and humidity; additionally, snow water equivalent is included (not used in this work). These daily values are estimated through the use of an iterative method dependent on local station density and the spatial convolution of a truncated Gaussian filter for station interpolation, and the Mountain Climate Simulator (MT-CLIM) to estimate shortwave radiation and humidity (Thornton et al., 1997; Thornton and Running, 1999; Thornton et al., 2000). Daymet does not include estimates of potential evapotranspiration (PET), a commonly needed input for conceptual hydrologic models or wind speed and direction. Therefore, PET was estimated using the Priestly–Taylor (P–T) method (Priestly and Taylor, 1972) and is discussed further in Sect. 3. Data quality is an ever-present issue in hydrologic modeling, and while the input data to Daymet are subject to rigorous quality control checks (Durre et al., 2008, 2010) potential errors may remain (Menne et al., 2009, 2010; Oubeidillah et al., 2013). Additionally, the Maurer et al. (2002) and National Land Data Assimilation System (NLDAS) (Xia et al., 2012) 12 km gridded data sets were processed to provide daily forcing data for the basin lumped configuration, resulting in three distinct data sets available for future forcing data impact studies.

Daily streamflow data for the HCDN-2009 gages were obtained from the USGS National Water Information System server (http://waterdata.usgs.gov/usa/nwis/sw) over the same forcing data time period, 1980–2010. While the period 1980–1990 is not covered by the HCDN-2009 review, it was assumed that these basins would have minimal human disturbances in this time period as well. For the portion of the basins that do not have streamflow records back to 1980, analysis is restricted to the available data records. The USGS provides streamflow data flags to identify periods of estimated flow and are included here. However, other data quality information is unavailable without further investigation and not available in this data set. For reference, 90 % (604) of the basins have 20 % or fewer flow days estimated and 75 % (503 basins) have 10 % or less flow values estimated.

The 671 basins span the entire CONUS and cover a wide range of hydroclimatic conditions. They range from wet, warm basins in the southeastern (SE) US to hot and dry basins in the southwestern (SW) US, to wet, cool basins in the northwestern (NW) and dry, cold basins in the intermountain (Rocky Mountains in Fig. 1a) western US. Figure 1b displays the basin annual precipitation (colored shading) along with symbols to denote rain- and snow-dominated basins. In terms of annual mean CDFs (cumulative density functions),

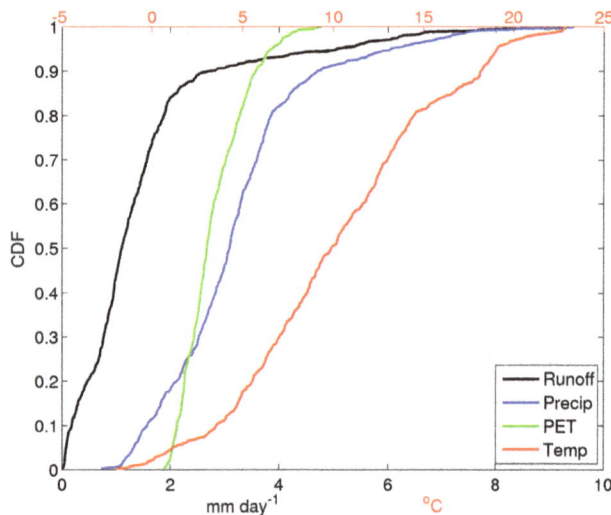

Figure 2. Annual CDFs of runoff (mm day^{-1}) (black, bottom x axis), precipitation (mm day^{-1}) (blue, bottom x axis), potential evapotranspiration (mm day^{-1}) (green, bottom x axis), and temperature ($^{\circ}$C) (red, top x axis).

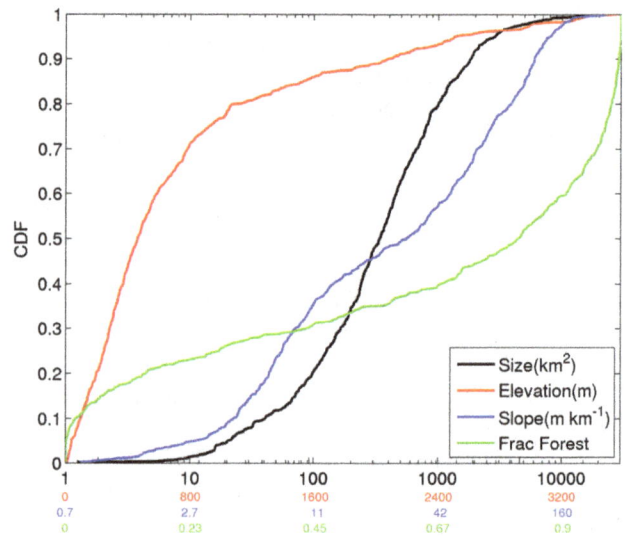

Figure 3. Cumulative density functions of basin size (km^2) (black), basin mean elevation (m) (red), mean slope (m km^{-1}) (blue), and fractional forest cover (green) for the basin set.

Daymet-estimated basin mean temperatures range from -2 to 23 $^{\circ}$C with precipitation amounts of 0.7–9.4 mm day^{-1} (Fig. 2). Annual observed mean runoff ranges from 0.01 to 9.3 mm day^{-1} with PET estimates ranging from 1.9 to 4.8 mm day^{-1}. Interestingly, this implies that Daymet precipitation itself is not enough to balance the observed runoff in some basins and is consistent with other recent large-sample hydrologic studies (Oubeidillah et al., 2013). Seasonal variations in these four variables are large as well, with some basins reaching mean winter time temperatures lower than $-10\,^{\circ}$C and summer time mean temperatures higher than 25 $^{\circ}$C (not shown). The seasonal water balance varies greatly with some basins experiencing much higher precipitation and runoff rates in one season versus another (e.g., spring runoff peaks in mountain snowmelt-dominated basins). As expected, PET varies seasonally with a minimum in winter and a maximum in summer.

Figure 3 gives CDFs for various physical descriptors of the basin set. The basins range in size from roughly 1 to 25 800 km^2 with the median basin size being about 335 km^2 and have mean elevations spanning from nearly sea level (10 m) to high alpine elevations (3570 m) with a median elevation of 462 m. Notably, 75 basins have mean elevations > 2000 m. Corresponding to the large range of elevations in the basin set, the mean slopes vary considerably, spanning over 2 orders of magnitude from near zero to over 200 m km^{-1}. The basin set covers a wide range of basin shapes with aspect ratios ranging from 0.08 to about 11. Finally, there is a large range of forest covers across the basin set which may have implications for hydrologic similarity (Oudin et al., 2010) with 20 % of the basins having less than

(more than) 14 % (98 %) forest cover and the median basin having about 80 % forest cover (NLCD-2011).

This basin set allows us to simulate a variety of energy- and water-limited basins with different snow storage, elevation, slope, and precipitation characteristics. Figure 4a shows runoff ratio (USGS streamflow / Daymet precipitation) versus the aridity index (Daymet Precipitation / PET). Immediately, it can be seen that some basins lie above the water limit line ($Y = 1$) indicating more runoff than precipitation, and many basins are near it ($Y > 0.9$). In these cases the model calibration process would struggle to produce an unbiased calibration, or never in basins above the water limit, because the basic water balance requires nearly zero evapotranspiration (ET) or is not satisfied. This requires a modification to incoming precipitation, which is discussed in the next section. Not coincidentally, the basins near and above the water limit are colder basins (mean annual $T < 10\,^{\circ}$C) with frozen precipitation during colder months. Additionally, two basins lie to the right of the curved line ($Y = 1-1/$aridity) indicating a surplus of water. These basins may also require modifications to input precipitation, but it is less clear in this case as observations of precipitation are generally underestimates, especially for snowfall (e.g., Yang et al., 1998). Examining the basin set using model output terms in the Budyko framework, there are many energy-limited basins with dryness ratios as small as 0.2 and many water-limited basins with model estimated dryness ratios as large as 4.5 (Fig. 4b). Note that now no basins lie above the water limit, indicating bulk precipitation corrections were applied as needed during the calibration process. Examination of hydrometeorological forcing data sets across a large spatial extent through the lens of water and energy balance draws attention to gross errors

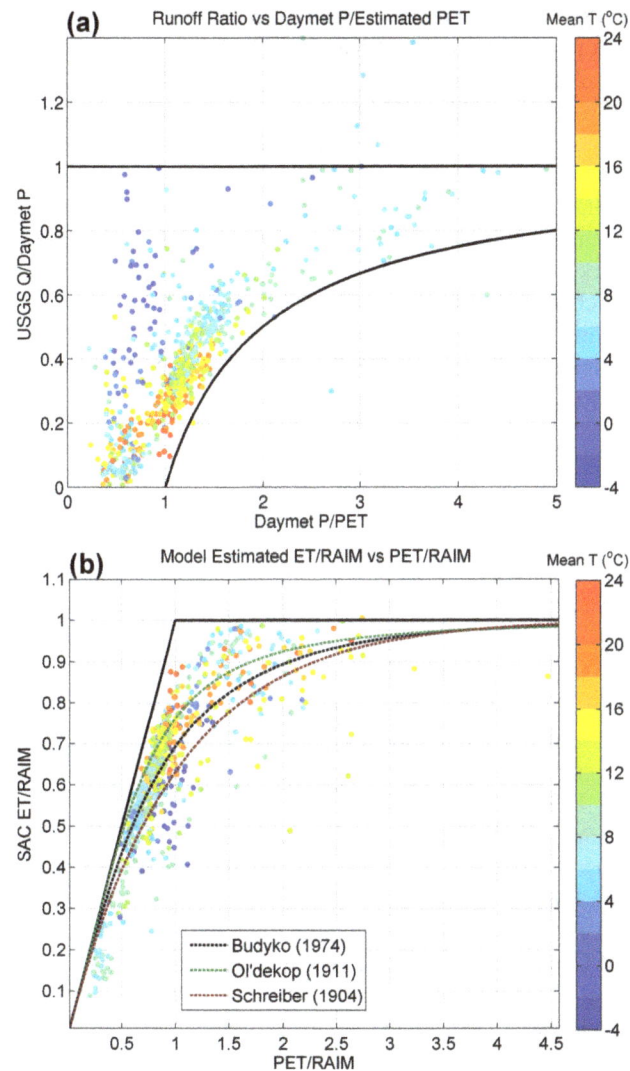

Figure 4. (a) Runoff ratio of observed runoff to Daymet-estimated precipitation versus ratio of Daymet-estimated precipitation to Priestly–Taylor-estimated PET. (b) Model-derived Budyko analysis using model ET, PET and total surface water input (rain plus melt, RAIM) for the 671 basins and three derivations of the Budyko curve (dashed lines). Basin mean temperature is shaded (coloring) in both panels.

in the forcing or streamflow data sets and permits any identified errors to be placed into spatial and temporal context, a benefit of large-sample studies.

As noted above, no additional quality control was performed on the candidate basins before calibration. For completeness and to more fully highlight some of the benefits and tradeoffs made when performing large-sample hydrologic studies, all basins are kept for analysis in this work.

3   Hydrologic modeling benchmark

As stated in the introduction, the intended purpose of this data set is a test bed to facilitate assessment of hydrologic modeling and prediction questions across broad hydroclimatic variations, and we focus here on providing a benchmark performance assessment for a widely used calibrated, conceptual hydrologic modeling system. This type of data set can be used for many applications including evaluation of new modeling systems against a well known benchmark system over wide ranging conditions, or as a base for comprehensive predictability experiments exploring the importance of meteorology or initial basin conditions. To this end, we have implemented and tested an initial model and calibration system described below, using the primary models and objective calibration approach that have been used by the US National Weather Service River Forecast Centers (NWSRFCs) in service of operational short-term and seasonal streamflow forecasting.

3.1   Models

The HCDN-2009 basins include those with substantial seasonal snow cover (Fig. 1b), necessitating a snow model in addition to a hydrologic model. Within the NWSRFCs, the coupled Snow-17 and SAC-SMA system is used. Snow-17 is a conceptual air-temperature-index-based snow accumulation and ablation model (Anderson, 1973). It uses near-surface air temperature to determine the energy exchange at the snow–air interface and the only time-varying inputs are typically air temperature and precipitation (Anderson, 1973, 2002). The SAC-SMA model is a conceptual hydrologic model that includes representation of physical processes such as evapotranspiration, percolation, surface flow, and subsurface lateral flow. Required inputs to SAC-SMA are potential evapotranspiration and water input to the soil surface (Burnash et al., 1973; Burnash, 1995). Snow-17 runs first and determines the partition of precipitation into rain and snow and the evolution of the snowpack. Any rain, snowmelt or rain passing unfrozen through the snowpack for a given time step becomes direct input to the SAC-SMA model. Finally, streamflow routing is accomplished through the use of a simple two-parameter, Nash-type instantaneous unit-hydrograph model (Nash, 1957).

3.2   Calibration

We employed a split-sample calibration approach following Klemes (1986): assigning the first 15 years of available streamflow data for calibration and the remainder for validation, then repeating the calibration using the last 15 years and the initial remaining period for validation; thus, approximately 5500 daily streamflow observations were used for each calibration. To initialize the model calibration moisture states on 1 October, we specified an initial wet SAC-SMA soil moisture state that was allowed to spin down to equilibrium for a given basin by running the first year of the calibration period repeatedly and assumed no initial snowpack. This was done until all SAC-SMA state variables had minimal year over year variations, which is a spin-up approach used by the Project for Intercomparison of Land-Surface Process Schemes (e.g., Schlosser et al., 2000). Determination of optimal calibration sampling and spin-up procedures is an area of active research. Spin-up was performed for every parameter set specified by the optimization algorithm, then the model was integrated for the calibration period and the RMSE (root mean square error) for that parameter set was calculated.

Objective calibration was done by minimizing the RMSE of daily modeled runoff versus observed streamflow using the SCE global search algorithm of Duan et al. (1992, 1993). The SCE algorithm uses a combination of probabilistic and deterministic optimization approaches that systematically spans the allowed parameter search space and also includes competitive evolution of the parameter sets (Duan et al., 1993). Prior applications to the SAC-SMA model have shown good results (Sorooshian et al., 1993; Duan et al., 1994). In the coupled Snow-17 and SAC-SMA modeling system, 35 potential parameters are available for calibration, of which we calibrated 20 parameters having either a priori estimates (Koren et al., 2000) or those found to be most sensitive following Anderson (2002) (Table 1). The SCE algorithm was run using 10 different random seed starts for the initial parameter sets for each basin, in part to evaluate the robustness of the optimum in each case, and the optimized parameter set with the minimum RMSE from the 10 different optimization runs was chosen for evaluation.

For Snow-17, six parameters were chosen for optimization (Table 1): the minimum and maximum melt factors (MFMIN, MFMAX), the wind adjustment for enhanced energy fluxes to the snowpack during rain on snow (UADJ), the rain/snow partition temperature, which may not be 0 °C (PXTEMP), the snow water equivalent for 100 % snow covered area (SI), and the gauge catch correction term for snowfall only (SCF). These six parameters were chosen because MFMIN, MFMAX, UADJ, SCF, and SI are defined as major model parameters by Anderson (2002). PXTEMP was also shown to be important in the Snow-17 model by Mizukami et al. (2013). The SCF is critical in many snow-dominated basins as precipitation is generally underestimated in these types of basins (e.g., Yang et al., 1998) and is certainly underestimated in some basins in Daymet as shown in Figs. 3 and 4.

The areal depletion curve (ADC) is considered a major parameter in Snow-17. However, to avoid expanding the parameter space by the number of ordinates on the curve (typically 10), we manually specified the ADC according to regional variations in latitude, topographic characteristics (e.g., plains, hills or mountains) and typical air mass characteristics (e.g., maritime polar, continental polar) (as suggested in Anderson, 2002). The remaining Snow-17 parameters were set in the same manner. Following the availability of a priori parameter estimates for SAC-SMA from a variety of data sets and various calibration studies with SAC-SMA (Koren et al., 2000; Anderson et al., 2006; Pokhrel and Gupta, 2010; Zhang et al., 2012), 11 parameters from SAC-SMA are included for calibration (Table 1). We use an instantaneous unit hydrograph, represented as a two-parameter gamma distribution for streamflow routing (Sherman, 1932; Clark, 1945; Nash, 1957; Dooge, 1959), the parameters of which were inferred as part of calibration. .

Finally, the scaling parameter in the Priestly–Taylor PET estimate is also calibrated. The P–T equation (Priestly and Taylor, 1972) can be written as

$$\text{PET} = \frac{a}{\lambda} \cdot \frac{s \cdot (R_n - G)}{s + \gamma}. \tag{1}$$

Where λ (MJ kg^{-1}) is the latent heat of vaporization, R_n (MJ m^{-2} day^{-1}) is the net radiation estimated using day of year, all Daymet variables and equations to estimate the various radiation terms (Allen et al., 1988; Zotarelli et al., 2009), G (MJ m^{-2} day^{-1}) is the soil heat flux (assumed to be zero in this case), s (kPa °C^{-1}) is the slope of the saturation vapor pressure–temperature relationship, γ (kPa °C^{-1}) is the psychrometric constant and a (unitless) is the P–T coefficient. The P–T coefficient replaces the aerodynamic term in the Penman–Monteith equation and varies by the typical conditions of the area where the P–T equation is being applied with humid forested basins typically having smaller values and exposed arid basins having larger values (Shuttleworth and Calder, 1979; Morton, 1983; Jensen et al., 1990). Thus, the P–T coefficient was included in the calibration since it should vary from basin to basin.

4 Benchmark results

4.1 Assessment objectives and metrics

Assessment of the models will focus on overall performance across the basin set, regional variations, and error characteristics. Nash–Sutcliffe efficiency (NSE) (Nash and Sutcliffe, 1970) and two of the decomposition components of NSE, variance bias (α) and total volume bias (β) (Gupta et al., 2009), are the first metrics examined in two variations. Because NSE scores model performance relative to the observed climatological mean, regions in which the model can track a strong seasonal cycle (large flow autocorrelation) perform relatively better when measured by NSE, and this seasonal enhancement may be imparted when using NSE as the objective function for both the calibration and validation phases (e.g., Schaefli and Gupta, 2007). Additionally, basins with higher streamflow variance and frequent precipitation events have better model performance. Therefore, to give a more standardized picture of model performance across varying hydroclimatologies, the NSE was recomputed using

Table 1. Table describing all parameters calibrated and their bounds for calibration.

Parameter	Description	Units	Calibration range
	Snow-17		
MFMAX	Maximum melt factor	$mm\,°C^{-1}\,6\,h^{-1}$	0.8–3.0
MFMIN	Minimum melt factor	$mm\,°C^{-1}\,6\,h^{-1}$	0.01–0.79
UADJ	Wind adjustment for enhanced flux during rain on snow	$km\,6\,h^{-1}$	0.01–0.40
SI	SWE for 100 % snow covered area	mm	1.0–3500.0
SCF	Snow gauge undercatch correction factor	–	0.1–5.0
PXTEMP	Temperature of rain/snow transition	$°C$	−1.0–3.0
	SAC-SMA		
UZTWM	Upper zone tension water maximum storage	mm	1.0–800.0
UZFWM	Upper zone free water maximum storage	mm	1.0–800.0
LZTWM	Lower zone tension water maximum storage	mm	1.0–800.0
LZFPM	Lower zone free water primary maximum storage	mm	1.0–1000.0
LZFSM	Lower zone free water secondary maximum storage	mm	1.0–1000.0
UZK	Upper zone free water lateral depletion rate	day^{-1}	0.1–0.7
LZPK	Lower zone primary free water depletion rate	day^{-1}	0.00001–0.025
LZSK	Lower zone secondary free water depletion rate	day^{-1}	0.001–0.25
ZPERC	Maximum percolation rate	–	1.0–250.0
REXP	Exponent of the percolation equation	–	0.0–6.0
PFREE	Fraction percolating from upper to lower zone free water storage	–	0.0–1.0
	Others		
USHAPE	Shape of unit hydrograph	–	1.0–5.0
USCALE	Scale of unit hydrograph	–	0.001–150.0
P–T	Priestly–Taylor coefficient	–	1.26–1.74

the long-term monthly mean flow instead of mean flow (denoted MNSE hereafter), thus preventing climatological seasonality from inflating the NSE and more accurately ranking basins by the degree to which the model added value over climatology in response to weather events (Garrick et al., 1978; Martinec and Rango, 1989; Schaefli et al., 2005). MNSE in this context is defined for each day of year (DOY) via a 31-day window centered on a given DOY. The long-term flow for that 31-day "month" is computed giving rise to a "monthly" mean flow. Using this type of climatology as the base for an NSE-type analysis provides improved standardization in basins with large flow autocorrelations. This definition is similar to the one proposed by Garrick et al. (1978) but with the addition of the 31-day smoother, which is done to provide a smoother reference climatology.

Also, several other advanced, more physically based, metrics of model performance are provided. First, three diagnostic signatures based on the flow duration curve (FDC) from Yilmaz et al. (2008) are computed: (1) the top 2 % flow bias, (2) the bottom 30 % flow bias, and (3) the bias of the slope of the middle portion (20–70 percentile) of the FDC. Second, examination of the time series of squared error contribution to the RMSE statistic was performed to highlight events in which the model performs poorly following Clark et al. (2008). This analysis was performed to gauge the representativeness of performance metrics over the model record by using the sorted (highest to lowest) time series of squared error to identify the N number of the largest error days and determine their fractional error contribution to the total. Finally, we extend this analysis to introduce a simple, normalized general error index for application and comparison across varying modeling and calibration studies. We coin the index, E50, the fraction of calibration points contributing 50 % of the error. This captures the number of points determining the majority of the error and thus the optimal parameter set.

4.2 Spatial variability

It is informative to examine spatial patterns of the aforementioned metrics to elucidate factors leading to weak (and strong) model performance. This also allows for identification of outlier basins and characterization of contributing factors (i.e., forcing or streamflow data issues or poor calibration). Poorly performing basins are most common along the high plains and desert southwest (Fig. 5a, Sect. 3c). When examining MNSE (Fig. 5b), basins with high nonseasonal streamflow variance and frequent precipitation events (SE and NW US) have the highest model MNSE, while most of the snowmelt-dominated basins see MNSE scores reduced

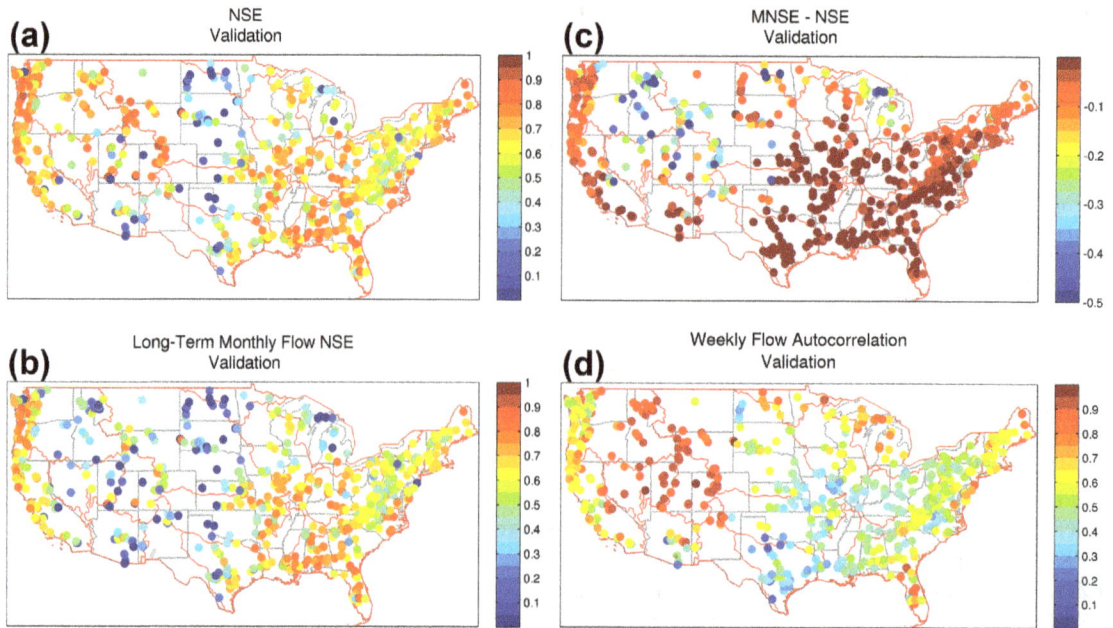

Figure 5. (a) Spatial distribution of NSE, **(b)** NSE using MNSEs rather than the long-term mean flow, **(c)** MNSE – NSE for the validation period, and **(d)** weekly flow autocorrelation.

relative to NSE, particularly in the validation phase (Fig. 5c). This indicates that RMSE as an objective function may not be well suited for model calibration in basins with high flow autocorrelation (Kavetski and Fenicia, 2011; Evin et al., 2014). This is confirmed by comparing Fig. 5d to Fig. 5c, basins with large flow autocorrelations (1 week mean flow for example) generally have lower MNSE scores.

Areas with low-validation NSE and MNSE scores have generally large biases when looking at FDC metrics as well (Fig. 6). Focusing on the high plains, high flow biases of $\pm 50\%$ are common. Extreme negative low flow biases are also present along the high plains and desert SW along with a general model trend to have large negative FDC slope biases, consistent with a poorly calibrated model. For the 72 % of basins with validation NSE > 0.55 (basins with yellow-green to dark red colors in Fig. 6a), there is no noticeable spatial pattern across the CONUS in regard to high flow periods. However, basins with a more pronounced seasonal cycle (e.g., snowpack-dominated watersheds, central west coast) generally have a negative low flow bias, while basins with a smaller seasonal cycle have a positive low flow bias (Fig. 6b). Correspondingly, basins with a pronounced seasonal cycle generally have a near zero or positive slope of the FDC bias, while basins with a smaller seasonal cycle have a negative slope bias (Fig. 6c).

Past applications with similar conceptual snow and hydrologic modeling systems across the CONUS have shown comparable spatial performance patterns. Clark et al. (2008) applied many conceptual models to a subset of the MOPEX basin set and found poor performance in arid regions. Mar-

tinez and Gupta (2010), using a monthly water balance model, found the best performance generally along the east coast, most of SE CONUS, and along the west coast with scattered good performance in the Rocky Mountains. They found that many basins along the high plains and north side of the Appalachian Mountains perform poorly. They also note that arid regions have high variability error (variability bias term in KGE – Kling–Gupta efficiency).

4.3 Cumulative performance

Two basic cumulative thresholds for model performance are highlighted here, NSE values of 0.55 and 0.8. An NSE of 0.55 indicates some model skill, and an NSE of 0.8 suggests reasonably good model performance. For the calibration period, 90 % (604) of the basins have a NSE greater than 0.55, while 72 % (484) of the basins had a validation period NSE > 0.55 (Fig. 7a). At the NSE > 0.8 level, 34 % (225) of the basin models perform better during calibration and 12 % (78) of the basin models meet that criteria during the validation phase. When using MNSE, 85 and 57 % (568 and 385) of the basins lie above 0.55, and 1 and 4 % (114 and 29) of the basins lie above 0.8 during the calibration and validation phases. The decomposition of the NSE (Gupta et al., 2009) shows that 90 % of the basins have a calibration (validation) model–observation flow correlation > 0.75 (0.68) and 30 % (12 %) of the basins have a model–observation flow correlation > 0.9 (Fig. 7b). However, nearly all basins have too little modeled variance (values less than one) for both the calibration and validation phases (Fig. 7c). The total volume biases are generally small with 94 % (79 %) of the basins having a

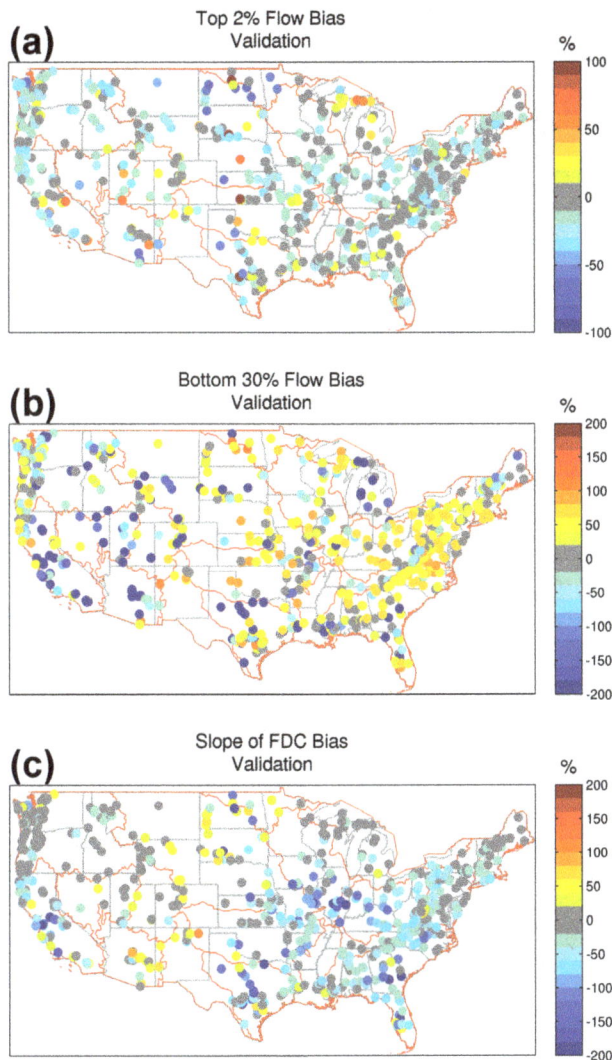

(a) Top 2% Flow Bias Validation

(b) Bottom 30% Flow Bias Validation

(c) Slope of FDC Bias Validation

Figure 6. (a) Spatial distribution of the high flow bias, **(b)** low flow bias, and **(c)** flow duration curve bias for the validation period.

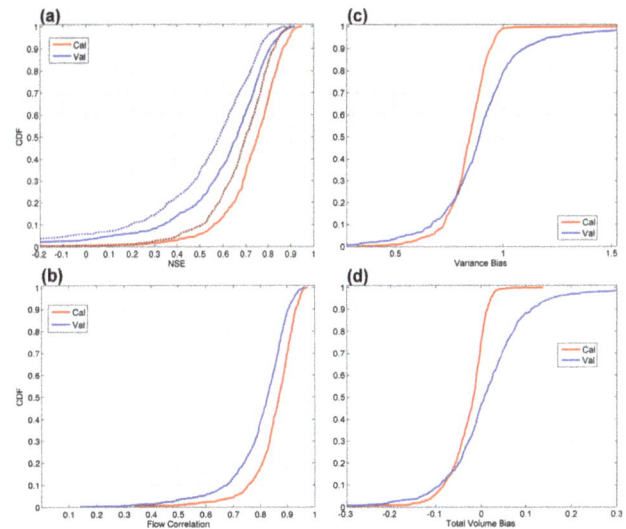

Figure 7. (a) CDFs of the model NSE (solid) for the calibration (red) and validation periods (blue) and NSE using the MNSEs (dark shaded and dashed). CDFs for **(b)** simulated–observed flow correlation in the decomposition of the NSE, **(c)** for the variance bias in the decomposition of the NSE, and **(d)** total volume bias in the decomposition of the NSE.

calibration (validation) period total flow bias within 10 % of the observed flow (Fig. 7d). These are expected results when using RMSE for the objective function (Gupta et al., 2009) and reaffirm that our implementation of the SCE calibrates the model properly.

Figure 8 highlights the full split sample approach for calibration following Klemes (1986). It is seen that the calibration and validation statistics give quite similar results regardless of which time period is used for calibration and validation using the Daymet data. This could indicate that both halves of the data are equally challenging to model with this modeling system. We have also included basin calibrations using only the first 15 years for the Maurer et al. (2002) and NLDAS-II (Xia et al., 2012) data sets. It can be seen that the Daymet forcing provides better model performance overall than both Maurer et al. and NLDAS forcing data. This likely relates to the coarser resolution of the Maurer et

al. (2002) and NLDAS data (12 km) and the somewhat small basin sizes in this basin set. More importantly, the inclusion of the Klemes (1986) split-sample approach provides users of this data set two parameter estimates for each basin using different calibration periods, while the inclusion of three total forcing data sets begins to allow for ensemble-type forcing data impact studies across a large basin sample size. In the remaining discussion, only model performance results using the first half of the split sample for calibration are presented.

With respect to advanced diagnostics, the model underpredicts high flow events in nearly all basins during calibration and slightly less so for the validation period (Fig. 9a). This is an expected result when using RMSE as the objective function because the optimal calibration underestimates flow variability (Gupta et al., 2009). Low flow periods are more evenly over- and underpredicted (Fig. 9b) for both the calibration and validation time frames with 58 and 61 % of basins having more modeled low flow. Finally, the bias in the slope of the FDC is generally underpredicted with about 75 % of the basins having a negative model bias (FDC slope is negative, thus a negative bias indicates the model slope is more positive and that the modeled flow variability is too compressed). The slope of the FDC indicates the variance of daily flows, which primarily relate to the seasonal cycle or the "flashiness" of a basin. Again this indicates model variability is less than that observed, at both short and longer timescales. In aggregate, these results agree with Fig. 5 and are expected based on the analysis of Gupta et al. (2009). Optimization using RMSE or NSE as the objective function generally results in underprediction of flow variance and

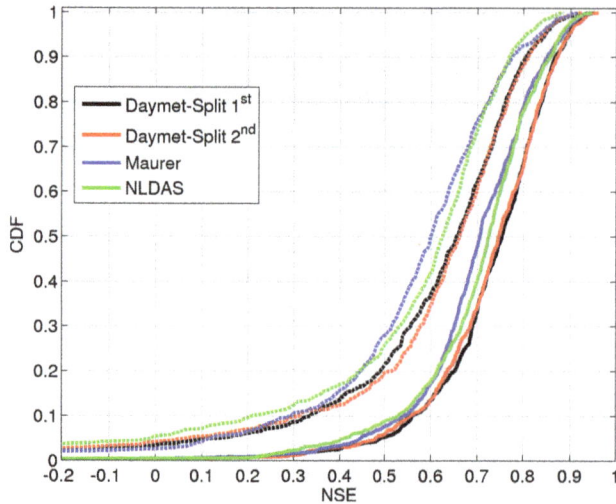

Figure 8. Cumulative density functions for model Nash–Sutcliffe efficiency for the calibration (solid) and validation (dashed) periods using three different forcing data sets (Daymet, Maurer, NLDAS). The Daymet data set was calibrated using the first 15 years (Split 1st) and validated against the remaining data and also calibrated using the last 15 years (Split 2nd) and validated against the initial streamflow data. Maurer and NLDAS calibrations performed using the first 15 years of observed streamflow only.

near-zero total flow bias (Fig. 7). This manifests itself in the simulated hydrograph as underpredicted high flows, generally overpredicted low flows and a more positive slope to the middle portion of the FDC (Fig. 9). It is worth repeating that the goal of this initial application is to provide to community with a benchmark of model performance using well known models, calibration systems and widely used, simple objective functions, thus the use of RMSE.

4.4   Error characteristics

When examining fractional error statistics for the basin set, 15 basins have single days that contribute at least half the total squared error (potential outlier basins), whereas at the median, the largest error day contributes 8.3 % of the total squared error for the median basin (Fig. 10). The fractional error contribution for the 10, 100 and 1000 largest error days for the median basin are 33, 70 and 96 % of the total squared error respectively. This indicates that for nearly all basins, there are 100 or fewer points that drive the RMSE and therefore optimal model parameters. This type of analysis can be undertaken for any objective function to identify the most influential points and allow for more in-depth examination of forcing data, streamflow records, and calibration strategies (i.e., Kavetski et al., 2006; Vrugt et al., 2008; Beven and Westerberg, 2011; Beven et al., 2011; Kauffeldt et al., 2013), or if different model physics are warranted.

The spatial distribution of fractional error contributions show that the issue of model performance being explained

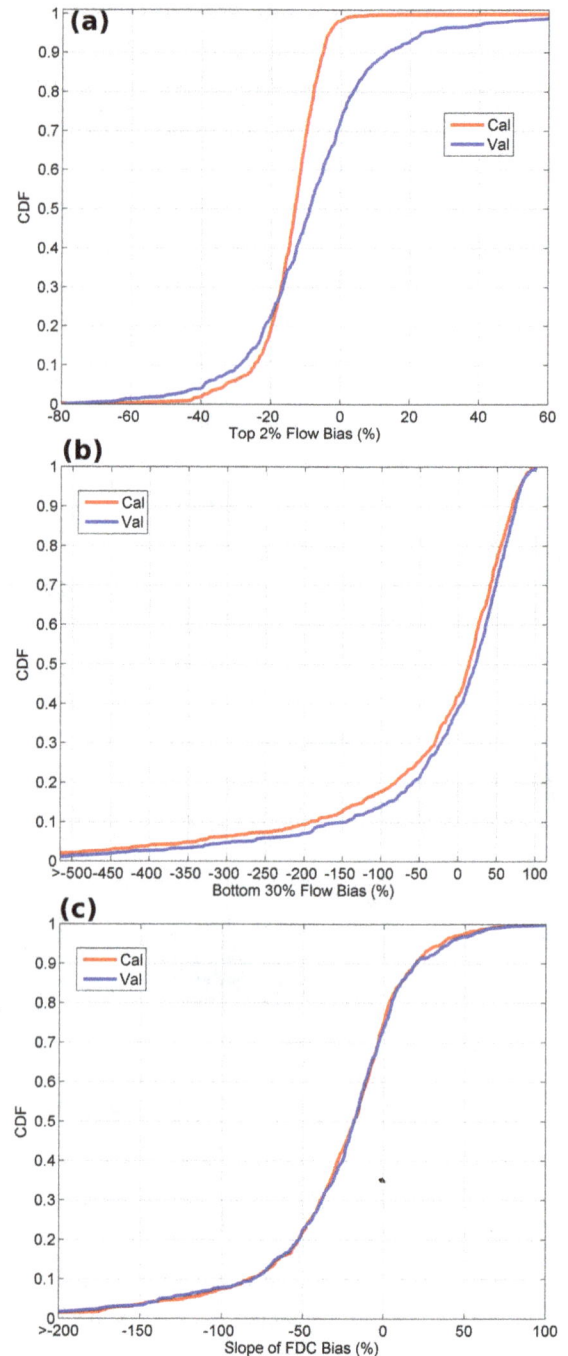

Figure 9. (a) CDFs for model high flow bias for the calibration (red) and validation periods (blue), (b) model low flow bias, and (c) model flow duration curve slope bias.

by a relatively small set of days is more prevalent in arid regions of the CONUS (desert SW US and high plains) as well as basins slightly inland from the east coast of the CONUS (Fig. 11a, b). The arid basins are generally dry with sporadic high precipitation (and flow) events, while the Appalachian basins are wetter (Fig. 1b) with extreme precipitation events

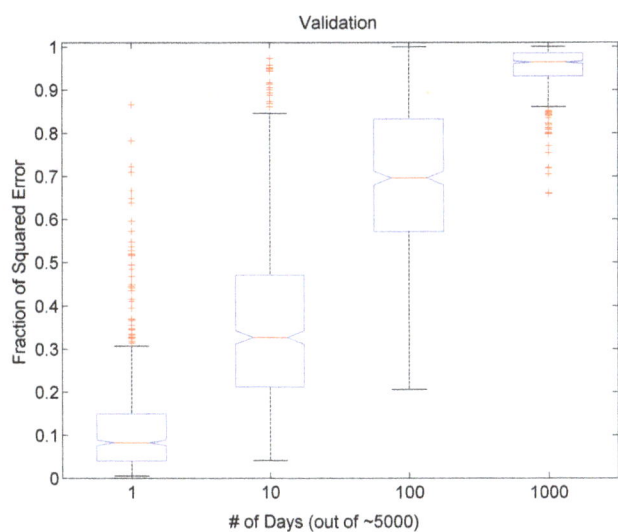

Figure 10. Fractional contribution of the total squared error for the 1, 10, 100, 1000 largest error days. The box plots represent the 671 basins with the blue area defining the interquartile range, the whiskers representing reasonable values and the red crosses denoting outliers. The median is given by the red horizontal line with the notch in the box denoting the 95 % confidence interval of the median value.

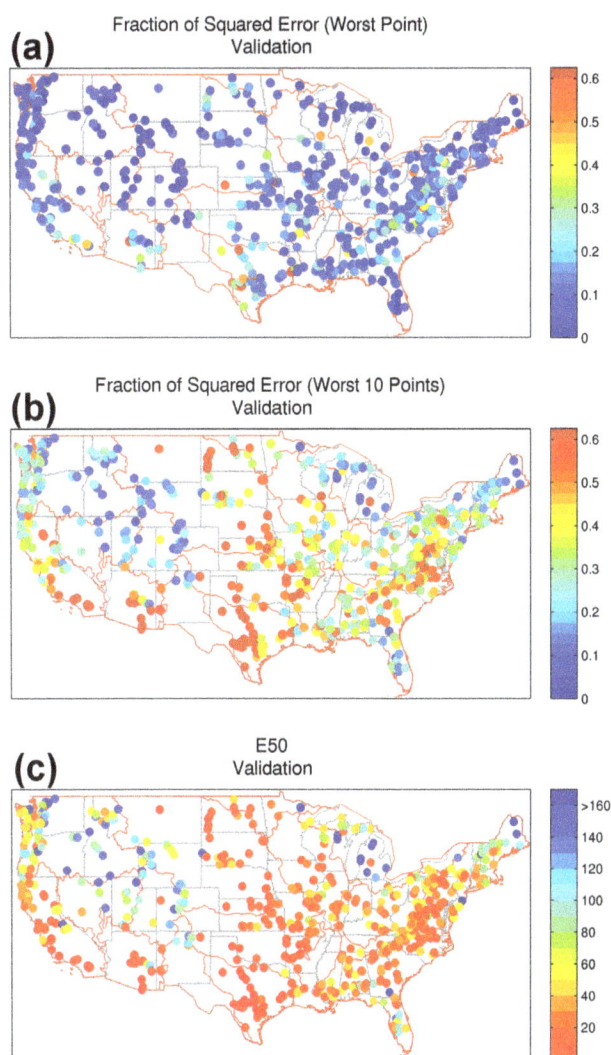

Figure 11. (**a**) Spatial distribution of the fractional contribution of total squared error for the largest day during the validation period, the (**b**) 10 largest error days, and (**c**) the number of days contributing 50 % of the total objective function error, E50.

interspersed throughout the record. Basins with significant snowpack tend to have lower error contributions from the largest error days (Fig. 11a, b). The E50 metric highlights mean peak snow water equivalent (SWE) and frequent precipitation basins as well. These regions contain and order of magnitude more days than the high plains and desert SW, giving insight into how representative of the entire streamflow time series the optimal model parameter set really is.

Additionally, ranking the basins using their fractional error characteristics provides a similar insight. As the aridity index increases, the fractional error contribution increases for basins with little to no mean peak SWE. For basins with significant SWE, the fractional error contribution decreases with increasing aridity (Fig. 12). Alternatively, for a given aridity index the fractional error contribution for N days will decrease with increasing SWE. This dynamic arises because more arid basins with SWE produce a relatively greater proportion of their runoff from snowmelt, without intervening rainfall. This implies that the optimized model produces a more uniform error distribution with less heteroscedacity in basins with more SWE. Moreover, as the fractional error contribution for the 10 largest error days increases, model NSE generally decreases in the validation phase (Fig. 13). This indicates fractional error metrics are related to overall model performance and that calibration methods to reduce extreme error days should improve model performance. This is not unexpected due to the fact that the residuals from an RMSE-type calibration are heteroscedastic. Arid basins typically have few high flow events, which are generally subject

to larger errors when minimizing the RMSE. Using advanced calibration methodologies that account for heteroscedasticity (Kavetski and Fenicia, 2011; Evin et al., 2014) may produce improved calibrations for arid basins in this basin set and provide different insights into model behavior using this type of analysis.

4.5 Limitations and uncertainties

One interesting example of the usefulness (and a potential limitation) of large-sample hydrology stemming from this work lies in the identification of issues with forcing data sets. Figures 3 and 4 show Daymet has too little precipitation in certain regions, which is also seen in Oubeidillah et al. (2013). When examining calibrated model performance in

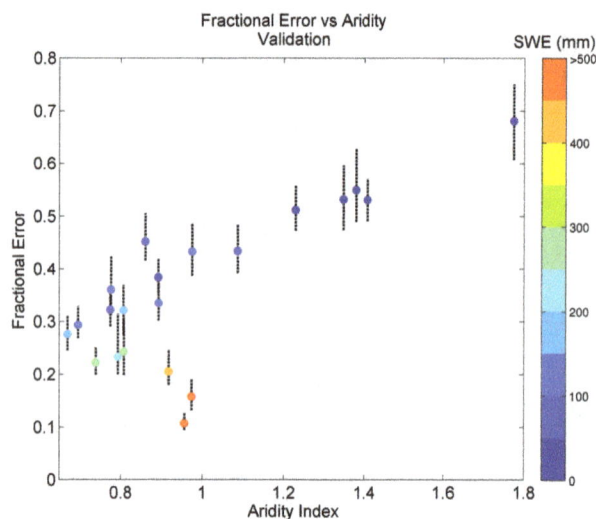

Figure 12. Ranked fractional squared error contribution for the 100 largest error days for the 671 basins versus the aridity index with mean maximum SWE shaded. Each dot represents a 32-basin bin defined by the rank of the fractional error contribution for the 100 largest error days for all basins. The dashed vertical black lines denote the 95 % confidence interval for the mean of the fractional error contribution for a given bin.

the Pacific northwest, it is seen that several basins along the west coast have low outlier NSE scores. Tracing this unexpected result, we find the Daymet forcing data available for those basins has a negative temperature bias, preventing mid-winter rain and melt episodes in the modeling system, identifying scope for improvement in the Daymet forcing. Moreover, winter periods of observed precipitation and streamflow rises coincide with subzero T_{max} in the Daymet data set, also suggesting areas to improve the Daymet forcing. The large-sample of basins in this region (91) allowed for identification of the outlier basins and the underlying causes.

This may also limit interpretation of these results and other large-sample hydrologic studies. As noted by Gupta et al. (2014), large-sample hydrology requires a tradeoff between breadth and depth. The lack of depth may inhibit discovery and identification of all data quality issues and the underlying causes of outliers in any analysis (e.g., Fig. 13). Explanation of these outliers is sometimes difficult and not complete in the initial development and analysis due to the lack of familiarity with specific basins and any forcing or validation data peculiarities. However, providing forcing data, model parameters and model output permits additional focused studies and helps reduce these limitations. Additional prescreening using the methods of Martinez and Gupta (2011) can also help identify outliers due to data quality issues and help identify basins and regions where model physics errors are present.

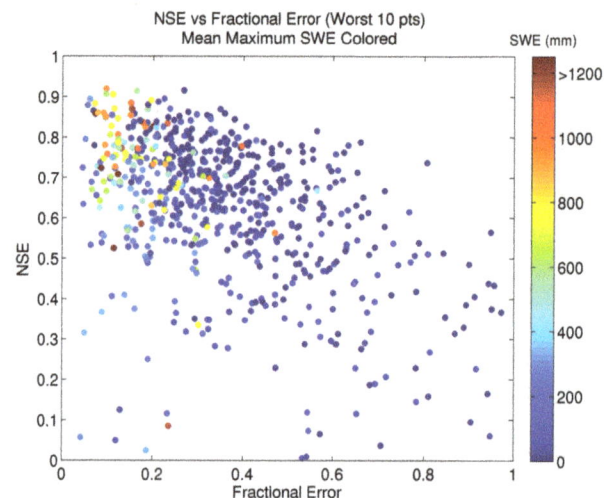

Figure 13. Nash–Sutcliffe efficiency versus the fractional error of the 10 largest error days for the validation period for all basins with basin mean peak snow water equivalent (mm) colored.

5   Summary and discussion

Most hydrologic studies focus in detail on a small number of watersheds, providing comprehensive but highly local insights, and may be limited in their ability to inform general hydrologic concepts applicable across regions (Gupta et al., 2014). To facilitate large-sample hydrologic studies, large-sample basin data sets and corresponding benchmarks of model performance using standard methodology across all basins need to be freely available to the community. To that end, we have compiled a community data set of daily forcing and streamflow data for 671 basins and provide a benchmark of performance using a widely used conceptual a hydrologic modeling and calibration scheme over a wide range of conditions.

Overall, application of the basin set to assessing an objectively calibrated conceptual hydrologic model representation of the 671 watersheds yielded calibration NSE scores of > 0.55 (0.8) for 90 % (34 %) of the basins. Performance of the models varied regionally, and the main factors influencing this variation were found to be aridity and precipitation intermittency, contribution of snowmelt, and runoff seasonality. Analysis of the cumulative fractional error contributions from the largest error days showed that the presence of significant SWE offset the negative impact of increasing aridity on simulation performance. This study has identified potential outlier basins for this modeling system and has provided insights into potential forcing data limitations. Although this modeling application utilized a conceptual hydrologic model with a single-objective calibration strategy, the findings provide a baseline for assessing more complex strategies in each area, including multiobjective calibration of more highly distributed hydrologic models (e.g., in Shi et al., 2008). The unusually broad variation of hydroclimatologies represented by the data set, which contains forcing and streamflow data obtained by consistent methodology and retains outlier basins, makes it a notable resource for these and other future large-sample watershed-scale hydrologic analysis efforts.

This data set and the applications presented are made available to the community (see http://ral.ucar.edu/projects/hap/flowpredict/subpages/modelvar.php or http://dx.doi.org/10.5065/D6MW2F4D).

Acknowledgements. This work is funded by the US Army Corps of Engineers Climate Preparedness and Resilience Programs and the US Department of the Interior Bureau of Reclamation. The authors would like to thank the USGS Modeling of Watershed Systems (MoWS) group, specifically for providing technical support and the national geospatial fabric data to generate all the basin spatial configurations. We would also like to thank Jordan Read and Tom Kunicki of the USGS Center for Integrated Data Analytics for their help with the USGS Geodata Portal.

Edited by: S. Archfield

References

Allen, R. G., Pereira, L. S., Raes, D., and Smith, M.: Crop evapotranspiration: guidelines for computing crop water requirements. Food and Agriculture Organization of the United Nations, Rome, 15 pp., 1988.

Anderson, E. A.: National Weather Service River Forecast System – Snow accumulation and ablation model. NOAA Technical Memorandum, NWS, HYDRO-17, US Department of Commerce, Silver Spring, MD, 217 pp., 1973.

Anderson, E. A.: Calibration of conceptual hydrologic models for use in river forecasting. NOAA Technical Report, NWS 45, Hydrology Laboratory, Silver Spring, MD, 2002.

Anderson, R. M., Koren, V. I., and Reed, S. M.: Using SSURGO data to improve Sacramento Model a priori parameter estimates, J. Hydrol., 320, 103–116, 2006.

Andreassian, V., Oddos, A., Michel, C., Anctil, F., Perrin, C., and Loumange, C.: Impact of spatial aggregation of inputs and parameters on the efficiency of rainfall-runoff models: A theoretical study using chimera watersheds, Water Resour. Res., 40, W05209, doi:10.1029/2003WR002854, 2004.

Beldring, S., Engeland, K., Roald, L. A., Sælthun, N. R., and Voksø, A.: Estimation of parameters in a distributed precipitation-runoff model for Norway, Hydrol. Earth Syst. Sci., 7, 304–316, doi:10.5194/hess-7-304-2003, 2003.

Beven, K. and Westerberg, I.: On red herrings and real herrings: disinformation and information in hydrological inference, Hydrol. Process., 25, 1676–1680, 2011.

Beven, K., Smith, P. J., and Wood, A.: On the colour and spin of epistemic error (and what we might do about it), Hydrol. Earth Syst. Sci., 15, 3123–3133, doi:10.5194/hess-15-3123-2011, 2011.

Blodgett, D. L., Booth, N. L., Kunicki, T. C., Walker, J. L., and Viger, R. J.: Description and testing of the geo data portal: A data integration framework and web processing services for environ-

mental science collaboration. US Geological Survey, Open-File Report 2011-1157, 9 pp., Middleton WI, USA, 2011.

Burnash, R. J. C.: The NWS River Forecast System – Catchment model, in: Computer Models of Watershed Hydrology, edited by: Singh, V. P., 311–366, Water Resources Publications, Highlands Ranch, Colo, 1995.

Burnash, R. J. C., Ferral, R. L., McGuire, R. A.: A generalized streamflow simulation system conceptual modeling for digital computers, US Department of Commerce National Weather Service and State of California Department of Water Resources, 1973.

Clark, C. O.: Storage and the unit hydrograph. Proc. Am. Soc. Civ. Eng., 9, 1333–1360, 1945.

Clark, M. P., Slater, A. G., Rupp, D. E., Woods, R. A., Vrugt, J. A., Gupta, H. V., Wagener, T., and Hay, L. E.: Framework for Understanding Structural Errors (FUSE): A modular framework to diagnose differences between hydrologic models, Water Resour. Res., 44, W00B02, doi:10.1029/2007WR006735, 2008.

Dooge, J. C. I.: A general theory of the unit hydrograph, J. Geophys. Res., 64, 241–256, 1959.

Duan, Q., Sorooshian, S., and Gupta, V. K.: Effective and efficient global optimization for conceptual rainfall-runoff models, Water Resour. Res., 28, 1015–1031, 1992.

Duan, Q., Gupta, V. K., and Sorooshian, S.: A shuffled complex evolution approach for effective and efficient optimization, J. Optimiz. Theor. Appl., 76, 501–521, 1993.

Duan, Q., Sorooshian, S., and Gupta, V. K.: Optimal use of the SCE-UA global optimization method for calibrating watershed models, J. Hydrol., 158, 265–284, 1994.

Duan, Q., Schaake, J., Andreassian, V., Franks, S., Goteti, G., Gupta, H. V., Gusev, Y. M., Habets, F., Hall, A., Hay, L., Houge, T., Huang, M., Leavesley, G., Liang, X., Nasonova, O. N., Noilhan, J., Oudin, L., Sorooshian, S., Wagener, T., and Wood, E. F.: Model Parameter Estimation Experiment (MOPEX): An overview of science strategy and major results from the second and third workshops, J. Hydrol., 320, 3–17, 2006.

Durre, I., Menne, M. J., and Vose, R. S.: Strategies for evaluating quality assurance procedures, J. Appl. Meteor. Climatol., 47, 1785–1791, doi:10.1175/2007JAMC1706.1, 2008.

Durre, I., Menne, M. J., Gleason, B. E., Houston, T. G., and Vose, R. S.: Comprehensive Automated Quality Assurance of Daily Surface Observations, J. Appl. Meteor. Climatol., 49, 1615–1633, doi:10.1175/2010JAMC2375.1, 2010.

Evin, G., Thyer, M., Kavetski, D., McInerney, D., and Kuczera, G.: Comparison of joint versus postprocessor approaches for hydrological uncertainty estimation accounting for error autocorrelation and heteroscedasticity, Water Resour. Res., 50, 2350–2375, doi:10.1002/2013WR014185, 2014.

Falcone, J. A.: GAGES-II: Geospatial Attributes of Gages for Evaluating Streamflow. Digital spatial data set 2011, available at: http://water.usgs.gov/GIS/metadata/usgswrd/XML/gagesII_Sept2011.xml (last access: 10 October 2013), 2011.

Falcone. J. A., Carlisle, D. M., Wolock, D. M., and Meador, M. R.: GAGES: A stream gage database for evaluating natural and altered flow conditions in the conterminous United States. Ecology, 91, p. 621, A data paper in Ecological Archives E091-045-D1, available at: http://esapubs.org/Archive/ecol/E091/045/metadata.htm (last access: 5 April 2014), 2010.

Garrick, M., Cunnane, C., and Nash, J. E.: A criterion of efficiency for rainfall-runoff models, J. Hydrology, 36, 375–381, 1978.

Gupta, H. V., Kling, H., Yilmaz, K. K., and Martinez-Barquero, G. F.: Decomposition of the mean squared error and NSE performance criteria: Implications for improving hydrological modeling, J. Hydrol., 377, 80–91, doi:10.1016/j.jhydrol.2009.08.003, 2009.

Gupta, H. V., Perrin, C., Blöschl, G., Montanari, A., Kumar, R., Clark, M., and Andréassian, V.: Large-sample hydrology: a need to balance depth with breadth, Hydrol. Earth Syst. Sci., 18, 463–477, doi:10.5194/hess-18-463-2014, 2014.

Jensen, M. E., Burman, R. D., and Allen, R. G.: Evapotranspiration and irrigation water requirements. American Society of Civil Engineers, ASCE Manual and Reports on Engineering Practice, 332 p., New York, NY, 1990.

Jin, S., Yang, L., Danielson, P., Homer, C., Fry, J., and Xian, G.: A comprehensive change detection method for updating the National Land Cover Database to circa 2011, Remote Sens. Environ., 132, 159–175, 2013.

Kauffeldt, A., Halldin, S., Rodhe, A., Xu, C.-Y., and Westerberg, I. K.: Disinformative data in large-scale hydrological modelling, Hydrol. Earth Syst. Sci., 17, 2845–2857, doi:10.5194/hess-17-2845-2013, 2013.

Kavetski, D. and Fenicia, F.: Elements of a flexible approach for conceptual hydrological modeling: 2. Application and experimental insights, Water Resour. Res., 47, W11511, doi:10.1029/2011WR010748, 2011.

Kavetski, D., Kuczera, G., and Franks, S. W.: Bayesian analysis of input uncertainty in hydrological modeling: 2. Application, Water Resour. Res., 42, W03407, doi:10.1029/2005WR004376, 2006.

Klemes, V.: Operational testing of hydrological simulation models, Hydrol. Sci. J., 31, 13–24, 1986.

Koren, V. I., Smith, M., Wang, D., and Zhang, Z.: Use of soil property data in the derivation of conceptual rainfall-runoff model parameters. American Meteorological Society 15th Conference on Hydrology, Long Beach, CA, 103–106, 2000.

Kumar, R., Samaniego, L., and Attinger, S.: Implications of distributed hydrologic model parameterization on water fluxes at multiple scales and locations, Water Resour. Res., 49, 360–379, doi:10.1029/2012WR012195, 2013.

Lins, H. F.: USGS Hydro-Climatic Data Network 2009 (HCDN-2009), US Geological Survey, Fact Sheet 2012-3047, Reston VA, USA, 2012.

Livneh, B. and Lettenmaier, D. P.: Multi-criteria parameter estimation for the Unified Land Model, Hydrol. Earth Syst. Sci., 16, 3029–3048, doi:10.5194/hess-16-3029-2012, 2012.

Livneh, B. and Lettenmaier, D. P.: Regional parameter estimation for the Unified Land Model, Water Resour. Res., 49, 100–114, doi:10.1029/2012WR012220, 2013.

Livneh, B., Rosenberg, E. A., Lin, C., Nijssen, B., Mishra, V., Andreadis, K. M., Maurer, E. P., and Lettenmaier, D. P.: A Long-Term Hydrologically Based Dataset of Land Surface Fluxes and States for the Conterminous United States: Update and Extensions, J. Climate, 26, 9384–9392, doi:10.1175/JCLI-D-12-00508.1, 2013.

Lohmann, D., Mitchell, K. E., Houser, P. R., Wood, E. F., Schaake, J. C., Robock, A., Cosgrove, B. A., Sheffield, J., Duan, Q., Luo, L., Higgins, R. W., Pinker, R. T., and Tarpley, J. D.: Streamflow and water balance intercomparisons of four land surface models in the North American Land Data Assimilation System project, J. Geophys. Res., 109, D07S91, doi:10.1029/2003JD003517, 2004.

Martinec, J. and Rango, A.: Merits of statistical criteria for the performance of hydrological models, Water Resour. B., 25, 421–432, 1989.

Martinez, G. and Gupta, H. V.: Toward improved identification of hydrologic models: A diagnostic evaluation of the "abcd" monthly water balance model for the conterminous United States, Water Resour. Res., 46, W08507, doi:10.1029/2009WR008294, 2010.

Martinez, G. and Gupta, H. V.: Hydrologic consistency as a basis for assessing complexity of monthly water balance models for the continental United States, Water Resour. Res., 47, W12540, doi:10.1029/2011WR011229, 2011.

Maurer, E. P., Wood, A. W., Adam, J. C., Lettenmaier, D. P., and Nijssen, B.: A long-term hydrologically-based data set of land surface fluxes and states for the conterminous United States, J. Climate, 15, 3237–3251, 2002.

Menne, M. J., Williams Jr., C. N., and Vose, R. S.: The U.S. Historical Climatology Network monthly temperature data, version 2, Bull. Am. Meteor. Soc., 90, 993–1007, doi:10.1175/2008BAMS2613.1, 2009

Menne, M. J., Williams, C. N., and Palecki, M. A.: On the reliability of the U.S. surface temperature record, J. Geophys. Res., 115, D11108, doi:10.1029/2009JD013094, 2010.

Merz, R. and Bloschl, G.: Regionalization of catchment model parameters, J. Hydrol., 287, 95–123, 2004.

Mizukami, N., Koren, V., Smith, M., Kingsmill, D., Zhang, Z., Cosgrove, B., and Cui, Z.: The impact of precipitation type discrimination on hydrologic simulation: Rain-snow partitioning derived from HMT-West radar-detected brightband height versus surface temperature data, J. Hydrometeorol., 14, 1139–1158, doi:10.1175/JHM-D-12-035.1, 2013.

Morton, F. I.: Operational estimates of actual evapotranspiration and their significance to the science and practice of hydrology, J. Hydrol., 66, 1–76, 1983.

Nash, J. E.: The form of the instantaneous unit hydrograph, International Association of Scientific Hydrology Publication, 45, 114–121, Toronto ON, CA, 1957.

Nash, J. E. and Sutcliffe, J. V.: River flow forecasting through conceptual models. Part I: A discussion of principles, J. Hydrol., 10, 282–290, doi:10.1016/0022-1694(70)90255-6, 1970.

Nathan, R. J. and McMahon, T. A.: The SFB model, Part I – Validation of fixed model parameters, Civil Eng. Trans., CE32, 157–161, 1990.

Nester, T., Kirnbauer, R., Gutknecht, D., and Bloschl, G.: Climate and catchment controls on the performance of regional flood simulations, J. Hydrol., 402, 340–356, 2011.

Nester, T., Kirnbauer, R., Parajka, J., and Bloschl, G.: Evaluating the snow component of a flood forecasting model, Hydrol. Res., 43, 762–779, 2012.

Oubeidillah, A. A., Kao, S.-C., Ashfaq, M., Naz, B. S., and Tootle, G.: A large-scale, high-resolution hydrological model parameter data set for climate change impact assessment for the conterminous US, Hydrol. Earth Syst. Sci., 18, 67–84, doi:10.5194/hess-18-67-2014, 2014.

Oudin, L., Andreassian, V., Mathevet, T., Perrin, C., and Michel, C.: Dynamic averaging of rainfall-runoff model simulations from complementary model parameterizations, Water Resour. Res., 42, W07410, doi:10.1029/2005WR004636, 2006.

Oudin, L., Kay, A. L., Andreassian, V., and Perrin, C.: Are seemingly physically similar catchments truly hydrologically similar?, Water Resour. Res., 46, W11558, doi:10.1029/2009WR008887, 2010.

Perrin, C., Michel, C., and Andreassian, V.: Does a large number of parameters enhance model performance? Comparative assessment of common catchment model structures on 429 catchments, J. Hydrol., 242, 275–301, doi:210.1016/S0022-1694(1000)00393-00390, 2001.

Pokhrel, P. and Gupta, H. V.: On the use of spatial regularization strategies to improve calibration of distributed watershed models, Water Resour. Res., 46, W01505, doi:10.1029/2009WR008066, 2010.

Priestly, C. H. B. and Taylor, R. J.: On the assessment of surface heat flux and evaporation using large-scale parameters, Mon. Weather Rev., 100, 81–82, 1972.

Samaniego, L., Bardossy, A., and Lumar, R.: Streamflow prediction in ungauged catchments using copula-based dissimilarity measures, Water Resour. Res., 46, W02506, doi:10.1029/2008WR007695, 2010.

Schaake, J., Cong, S., Duan, Q.: U.S. MOPEX data set. Report UCRL-JRNL-221228, Lawrence Livermore National Laboratory, Livermore CA, USA, available at: https://e-reports-ext.llnl.gov/pdf/333681.pdf (last access: 10 September 2014), 2006.

Schaefli, B., Hingray, B., Niggli, M., and Musy, A.: A conceptual glacio-hydrological model for high mountainous catchments, Hydrol. Earth Syst. Sci., 9, 95–109, doi:10.5194/hess-9-95-2005, 2005.

Schaefli, B. and Gupta, H. V.: Do Nash values have value?, Hydrol. Process., 21, 2075–2080, doi:10.1002/hyp.6825, 2007.

Schlosser, C. A., Slater, A. G., Robock, A., Pitman, A. J., Vinnikov, K. Y., Henderson-Sellers, A., Speranskaya, N. A., Mitchell, K., and the PILPS 2(d) contributors: Simulations of a boreal grassland hydrology at Valdai, Russia: PILPS phase 2(d), Mon. Weather Rev., 128, 301–321, 2000.

Sherman, L. K.: Streamflow from rainfall by the unit graph method, Eng. News Rec., 108, 501–505, 1932.

Shi, X., Wood, A. W., and Letenmaier, D. P.: How essential is hydrologic model calibration to seasonal streamflow forecasting?, J. Hydrometeorol., 9, 1350–1363, 2008.

Shuttleworth, W. J. and Calder, I. R.: Has the Priestly-Taylor equation any relevance to forest evaporation?, J. Appl. Meteorol., 18, 639–646, 1979.

Slack, J. R. and Landwehr, J. M.: Hydro-Climatic Data Network (HCDN): A US Geological Survey streamflow data set for the United States for the study of climate variations, 1874–1988, US Geological Survey, Open-File Report 92-129, Reston VA, USA, 1992.

Sorooshian, S., Duan, Q., and Gupta, V. K.: Calibration of conceptual rainfall-runoff models using global optimization: application to the Sacramento soil moisture accounting model, Water Resour. Res., 29, 1185–1194, 1993.

Thornton, P. E. and Running, S. W.: An improved algorithm for estimating incident daily solar radiation from measurements of temperature, humidity and precipitation, Agr. Forest Meteorol., 93, 211–228, 1999.

Thornton, P. E., Running, S. W., and White, M. A.: Generating surfaces of daily meteorological variables over large regions of complex terrain, J. Hydrol., 190, 214–251, doi:10.1016/S0022-1694(96)03128-9, 1997.

Thornton, P. E., Hasenauer, H., and White, M. A.: Simultaneous estimation of daily solar radiation and humidity from observed temperature and precipitation: An application over complex terrain in Austria, Agr. Forest Meteorol., 104, 255–271, 2000.

Thornton, P. E., Thornton, M. M., Mayer, B. W., Wilhelmi, N., Wei, Y., and Cook, R. B.: Daymet: Daily surface weather on a 1 km grid for North America, 1980–2012, available at: http://daymet.ornl.gov/ (last access: 15 July 2013) from Oak Ridge National Laboratory Distributed Active Archive Center, Oak Ridge, Tennessee, USA, 2012.

Viger, R. J.: Preliminary spatial parameters for PRMS based on the Geospatial Fabric, NLCD2001 and SSURGO, US Geological Survey, doi:10.5066/F7WM1BF7, 2014.

Viger, R. J. and Bock, A.: GIS Features of the Geospatial Fabric for National Hydrologic Modeling, US Geological Survey, doi:10.5066/F7542KMD, 2014.

Vrugt, J. A., ter Braak, C. J. F., Clark, M. P., Hyman, J. M., and Robinson, B. A.: Treatment of input uncertainty in hydrologic modeling: Doing hydrology backward with Markov chain Monte Carlo simulation, Water Resour. Res., 44, W00B09, doi:10.1029/2007WR006720, 2008.

Xia, Y., Mitchell, K., Ek, M., Sheffield, J., Cosgrove, B., Wood, E., Luo, L., Alonge, C., Wei, H., Meng, J., Livneh, B., Lettenmaier, D., Koren, V., Duan, Q., Mo, K., Fan, Y., and Mocko, D.: Continental-scale water and energy flux analysis and validation for the North American Land Data Assimilation System project phase 2 (NLDAS-2): 1. Intercomparison and application of model products, J. Geophys. Res., 117, D03109, doi:10.1029/2011JD016048, 2012.

Yang, D., Goodison, B. E., Metcalfe, J. R., Golubev, V. S., Bates, R., Pangburn, T., and Hanson, C. L.: Accuracy of NWS 8" standard nonrecording precipitation gauge: Results and application of WMO intercomparison, J. Atmos. Ocean. Technol., 15, 54–68, 1998.

Yilmaz, K. K., Gupta, H. V., and Wagener, T.: A process-based diagnostic approach to model evaluation: Application to the NWS distributed hydrologic model, Water Resour. Res., 44, W09417, doi:10.1029/2007WR006716, 2008.

Zhang, Z., Koren, V., Reed, S., Smith, M., Zhang, Y., Moreda, F., and Cosgrove, B.: SAC-SMA a priori parameter differences and their impact on distributed hydrologic model simulations, J. Hydrol., 420–421, 216–227, 2012.

Zotarelli, L., Dukes, M. D., Romero, C. C., Migliaccio, K. W., and Morgan, K. T.: Step by step calculation of the Penman-Monteith Evapotranspiration (FAO-56 Method). University of Florida Extension, AE459, available at: http://edis.ifas.ufl.edu (last access: 1 April 2014), 10 pp., 2009.

2

Assessment of precipitation and temperature data from CMIP3 global climate models for hydrologic simulation

T. A. McMahon[1]**, M. C. Peel**[1]**, and D. J. Karoly**[2]

[1]Department of Infrastructure Engineering, University of Melbourne, Victoria, 3010, Australia
[2]School of Earth Sciences and ARC Centre of Excellence for Climate System Science, University of Melbourne, Victoria, 3010, Australia

Correspondence to: M. C. Peel (mpeel@unimelb.edu.au)

Abstract. The objective of this paper is to identify better performing Coupled Model Intercomparison Project phase 3 (CMIP3) global climate models (GCMs) that reproduce grid-scale climatological statistics of observed precipitation and temperature for input to hydrologic simulation over global land regions. Current assessments are aimed mainly at examining the performance of GCMs from a climatology perspective and not from a hydrology standpoint. The performance of each GCM in reproducing the precipitation and temperature statistics was ranked and better performing GCMs identified for later analyses. Observed global land surface precipitation and temperature data were drawn from the Climatic Research Unit (CRU) 3.10 gridded data set and re-sampled to the resolution of each GCM for comparison. Observed and GCM-based estimates of mean and standard deviation of annual precipitation, mean annual temperature, mean monthly precipitation and temperature and Köppen–Geiger climate type were compared. The main metrics for assessing GCM performance were the Nash–Sutcliffe efficiency (NSE) index and root mean square error (RMSE) between modelled and observed long-term statistics. This information combined with a literature review of the performance of the CMIP3 models identified the following better performing GCMs from a hydrologic perspective: HadCM3 (Hadley Centre for Climate Prediction and Research), MIROCm (Model for Interdisciplinary Research on Climate) (Center for Climate System Research (The University of Tokyo), National Institute for Environmental Studies, and Frontier Research Center for Global Change), MIUB (Meteorological Institute of the University of Bonn, Meteorological Research Institute of KMA, and Model and Data group), MPI (Max Planck Institute for Meteorology) and MRI (Japan Meteorological Research Institute). The future response of these GCMs was found to be representative of the 44 GCM ensemble members which confirms that the selected GCMs are reasonably representative of the range of future GCM projections.

1 Introduction

Our primary objective in this paper is to identify better performing GCMs from a hydrologic perspective. To do this we assess how well 22 global climate models (GCMs) from the World Climate Research Programme's (WCRP) Coupled Model Intercomparison Project phase 3 (CMIP3) multi-model data set (Meehl et al., 2007) are able to reproduce GCM grid-scale climatological statistics of observed precipitation and temperature over global land regions. We recognise that GCMs model different variables with a range of success and that no single model is best for all variables and/or for all regions (Lambert and Boer, 2001; Gleckler et al., 2008). The approach adopted here is not inconsistent with Dessai et al. (2005) who regarded the first step in evaluating GCM projection skill is to assess how well observed climatology is simulated. We also recognise there have been assessments published in peer-reviewed journals, but all appear to be assessed from a climate science perspective. This review concentrates on GCM variables and statistical techniques that are relevant to engineering hydrologic practice.

GCM runs for the observed period do not seek to replicate the observed monthly record at any point in time and space. Rather a better performing GCM is expected to produce long-term mean annual statistics that are broadly similar to observed conditions across a wide range of locations. Here, the assessment of CMIP3 GCMs is made by comparing their long-term mean annual precipitation (MAP), standard deviation of annual precipitation (SDP), mean annual temperature (MAT), mean monthly patterns of precipitation and temperature and Köppen–Geiger climate type (Peel et al., 2007) with concurrent observed data for 616 to 11 886 terrestrial grid cells worldwide (the number of grid cells depends on the resolution of the GCM under consideration). These variables were chosen to assess GCM performance because they provide insight into the mean annual, interannual variability and seasonality of precipitation and temperature, which are sufficient to estimate the mean and variability of annual runoff from a traditional monthly rainfall–runoff model (Chiew and McMahon, 2002) or from a top-down annual rainfall–runoff model (McMahon et al., 2011) for hydrologic simulation purposes.

The GCMs included in this assessment are detailed in Table 1 (model acronyms adopted are listed in the table). Although no quantitative assessment of the BCCR (Bjerknes Centre for Climate Research) model is made, this model is included in Table 1 as details of its performance are available in the literature which is discussed in Sect. 2. Other details in the table include the originating group for model development, country of origin, model name given in the CMIP3 documentation (Meehl et al., 2007), the number of 20C3M runs available for analysis, the model resolution and the number of terrestrial grid cells used in the precipitation and temperature comparisons.

Readers should note that when this project began as a component of a larger study in 2010, runs from the CMIP5 were not available. We are of the view that the approach adopted here is equally applicable to evaluating CMIP5 runs for hydrologic simulations. Conclusions about better performing models drawn from this analysis may prove similar to a comparable analysis of CMIP5 runs since most models in CMIP5 are, according to Knutti et al. (2013), "strongly tied to their predecessors". Analysis of the CMIP5 models indicates that the CMIP3 simulations are of comparable quality to the CMIP5 simulations for temperature and precipitation at regional scales (Flato et al., 2013).

This study is part of a larger research project that seeks to enhance our understanding of the uncertainty of future annual river flows worldwide through catchment-scale hydrologic simulation, leading to more informed decision-making for the sustainable management of scarce water resources, nationally and internationally. To achieve this, it is necessary to determine, as a minimum, how the mean and variability of annual streamflows will be affected by climate change. Other factors of less importance are changes in the autocorrelation of annual streamflow, changes in net evaporation from reservoir water surfaces and changes in monthly flow patterns, with the latter being more important for relatively small reservoirs. In this paper we deal with the key drivers of streamflow production, namely the mean and the standard deviation of annual precipitation and mean annual temperature, the latter is adopted here as a surrogate for potential evapotranspiration (PET), along with secondary factors, the mean monthly patterns of precipitation and temperature. Adopting temperature as a surrogate for PET is contentious. We provide a detailed discussion of this issue in the Supplementary Material associated with this paper. Suffice to say that a more complex PET formulation requires additional GCM variables other than temperature which are less reliable. This simplicity comes at the expense of potentially inadequate representation of future changes in PET, which may have important negative consequences when modelling streamflow in energy limited catchments. Nevertheless, in the following discussion we concentrate on mean annual temperature as the GCM variable representing PET.

Computer models of most water resource systems that rely on surface reservoirs to offset streamflow variability adopt a monthly time step to ensure that seasonal patterns in demand and reservoir inflows are adequately accounted for. However, in a climate change scenario it is more likely that an absolute change in streamflow will have a greater impact on system yield than shifts in the monthly inflow or demand patterns. This will certainly be the case for reservoirs that operate as carryover systems rather than as within-year systems (for an explanation see McMahon and Adeloye, 2005). Therefore, in this paper we assess the GCMs in terms of annual precipitation and annual temperature, and patterns of mean monthly precipitation and temperature.

Following this introduction we describe, and summarise in the next section, several previous assessments of CMIP3 GCM performance. We also include some general comments on GCM assessment procedures. In Sect. 3, data (observed and GCM based) used in the analysis are described. Details and results of the subsequent analyses comparing GCM estimates of present climate mean and standard deviation of annual precipitation, mean annual temperature, mean monthly precipitation and temperature patterns and Köppen–Geiger climate type against observed data are set out in Sect. 4. In Sect. 5, we review the results and compare the literature information with our assessments of the GCMs. The final section of the paper presents several conclusions.

2 Literature

As noted above, to assess the impact of climate change on surface water resources of a region through hydrologic simulation, it is necessary to assess, as a minimum, the performance of the mean and the standard deviation of annual precipitation and mean annual temperature, and the mean monthly patterns of precipitation and temperature. Noting this background we describe in the next section procedures that have been adopted in the literature to assess GCM performance.

Table 1. Details of 23 GCMs considered in this paper.

Acronym	Originating group	Country	Model name in CMIP3	Number of 20C3M runs available	Resolution		Number of prec. grid cells[c]	Number of temp. grid cells[b]
					Lat (°)	Long (°)		
BCCR	Bjerknes Centre for Climate Research	Norway	bccr-bcm2.0	na[a]	1.9	1.9	na	na
CCCMA-t47	Canadian Centre for Climate Modeling and Analysis	Canada	cccma_cgm3_1_t47	1	∼3.75	3.75	631	916
CCCMA-t63	Canadian Centre for Climate Modeling and Analysis	Canada	cccma_cgm3_1_t63	1	∼2.8	2.8125	1169	1706
CCSM	National Centre for Atmospheric Research	USA	ccsm	8	∼1.4	1.40625	5184	7453
CNRM	Météo-France/Centre National de Recherches Météorologiques	France	cnrm	1	∼2.8	2.8125	1169	1706
CSIRO	Australia CSIRO	Australia	csiro_mk3_0	1	∼1.87	1.875	2820	4068
GFDL2.0	NOAA Geophysical Fluid Dynamics Laboratory	USA	gfdl2_cm2_0	1	2	2.5	1937	2828
GFDL2.1	NOAA Geophysical Fluid Dynamics Laboratory	USA	gfdl2_cm2_1	1	∼2	2.5	1911	2758
GISS-AOM	NASA Goddard Institute of Space Studies	USA	giss_aom_r1, 2	2	3	4	754	1076
GISS-EH	NASA Goddard Institute of Space Studies	USA	giss_eh1, 2,3	3	3 and 4	5	425	616
GISS-ER	NASA Goddard Institute of Space Studies	USA	giss_model_e_r	3	3 and 4	5	425	616
HadCM3	Hadley Centre for Climate Prediction and Research	UK	hadcm3	1	2.5	3.75	982	1421
HadGEM	Hadley Centre for Climate Prediction and Research	UK	HadGem	1	1.25	1.875	4316	6239
IAP	Institute of Atmospheric Physics, Chinese Acad. Sciences	China	iap_fgoals1.0_g	3	6.1∼2.8	2.8125	1159	1664
INGV	National Institute of Geophysics and Vulcanology, Italy	Italy	ingv20c ECHAM4.6	1	∼1.1	1.125	8291	11 886
INM	Institute for Numerical Mathematics, Russia	Russia	inmcm3.0	1	4	5	420	620
IPSL	Institut Pierre Simon Laplace	France	ipsl_cm4	1	∼2.5	3.75	980	1403
MIROCh	Model for Interdisciplinary Research on Climate, Center for Climate System Research (The University of Tokyo), National Institute for Environmental Studies, and Frontier Research Center for Global Change	Japan	miroc3_2_hires (mirochi)	1	∼1.1	1.125	8291	11 886
MIROCm	Model for Interdisciplinary Research on Climate, Center for Climate System Research (The University of Tokyo), National Institute for Environmental Studies, and Frontier Research Center for Global Change	Japan	miroc3_2_medres (mirocmedr)	3	∼2.8	2.8125	1169	1706
MIUB	Meteorological Institute of the University of Bonn, Meteorological Research Institute of KMA, and Model and Data group	Germany South Korea	miub_echo_g	3	∼3.7	3.75	631	916
MPI	Max Planck Institute for Meteorology	Germany	mpi_echam5 (mpi)	3	∼1.8	1.875	2820	4068
MRI	Japan Meteorological Research Institute	Japan	mri_cgcm2_3_2a (mri)	5	∼2.8	2.8125	1169	1706
PCM	National Center for Atmospheric Research	USA	pcm	1	∼2.8	2.8125	1169	1706

[a] na: not available. [b] Based on mean annual temperature comparison between GCM and CRU. [c] Based on mean annual precipitation comparison between GCM and CRU.

2.1 Procedures to assess GCM performance

Ever since the first GCM was developed by Phillips (1956) (see Xu, 1999), attempts have been made to assess the adequacy of GCM modelling. Initially, these evaluations were simple side-by-side comparisons of individual monthly or seasonal means or multi-year averages (Chervin, 1981). To assess model performance, Chervin (1981) extended the evaluation procedure by examining statistically the agreement or otherwise of the ensemble average and standard deviation between the GCM modelled climate and the observed data using the vertical transient heat flux in an example application. Legates and Willmott (1992) compared observed with

simulated average precipitation rates by 10° latitude bands. On a two-dimensional plot, Taylor (2001) developed a diagram in which each point consisted of the spatial correlation coefficient and the spatial root mean square (RMS) along with the ratio of the variances of the modelled and the observed variables. Recently, some authors have used the Taylor diagram (Covey et al., 2003; Bonsal and Prowse, 2006) or a similar approach (Lambert and Boer, 2001; Boer and Lambert, 2001). Murphy et al. (2004) introduced a climate prediction index (CPI) which is based on a broad range of present-day climates. This index was later used by Johns et al. (2006) for a different set of climate variables than those used by

Murphy et al. (2004). Whetton et al. (2005) introduced a demerit point system in which GCMs were rejected when a specified threshold was exceeded. Min and Hense (2006) introduced a Bayesian approach to evaluate GCMs and argued that a skill-weighted average with Bayes factors is more informative than moments estimated by conventional statistics. Shukla et al. (2006) suggested that differences in observed and GCM simulated variables should be examined in terms of their probability distributions rather than individual moments. They proposed the differences could be examined using relative entropy. Perkins et al. (2007) also claimed that assessing the performance of a GCM through a probability density function (PDF) rather than using the first or a second moment would provide more confidence in model assessment. To compare the reliability of variables (in time and space) rather than individual models, Johnson and Sharma (2009a, b) developed the variable convergence score which is used to rank a variable based on the ensemble coefficient of variation. They observed the variables with the highest scores were pressure, temperature and humidity. Reichler and Kim (2008) introduced a model performance index by first estimating a normalised error variance based on the square of the grid-point differences between simulated (interpolated to the observational grid) and the observed annual climate weighted and standardised with respect to the variance of the annual observations. The error variance was scaled by the average error found in the reference models and, finally, averaged over all climates.

It is clear from this brief review that no one procedure has been universally accepted to assess GCM performance, which is consistent with the observations of Räisänen (2007). We also note the comments of Smith and Chandler (2010, p. 379) who said "It is fair to say that any measure of performance can be subjective, simply because it will tend to reflect the priorities of the person conducting the assessment. When different studies yield different measures of performance, this can be a problem when deciding on how to interpret a range of results in a different context. On the other hand, there is evidence that some models consistently perform poorly, irrespective of the type of assessment. This would tend to indicate that these model results suffer from fundamental errors which render them inappropriate."

In 1992, Legates and Willmott (1992) assessed the adequacy of GCMs based mainly on January and July precipitation fields. Although a number of GCM assessments were carried out during the following one and a half decades, it was not until 2008 that mean precipitation, either absolute or bias, was included in GCM published assessments. In that year, Reichler and Kim (2008, p. 303) argued that the mean bias is an important component of model error.

In Table 2a and b we summarize the application of the numerical metrics and the ranking metrics of precipitation and temperature respectively applied to CMIP3 data sets at the global or country scales. These references cover the period from 2006 to 2014. Across these 15 papers, we observe

that for precipitation and temperature the spatial root mean square error, either using raw data (root mean square error – RMSE) or normalised data as a percentage of the mean value (RRMSE), is adopted in 7 of the 15 studies. (The data are normalised by the corresponding standard deviation of the reference or observed data.) This spatial root mean square metric, as well as the bias in the mean of the data, is relevant to hydrologists as it provides an indication of the uncertainty in the climate variables of interest to them. Of more relevance to hydrologists is the uncertainty in temporal mean and variance of climatic variables, which for precipitation are only reported in 4 of the 15 studies. Although spatial correlation is not used directly in general hydrologic investigations, in GCM assessments it is often combined with the variance and spatial RMSE through the Taylor diagram (Taylor, 2001) which is an excellent summary of the performance of a GCM projected variable. As noted in Table 2, three papers utilise this approach. Lambert and Boer (2001, p. 89) extended the Taylor diagram to display the relative mean square differences, the pattern correlations and the ratio of variances for modelled and observed data. This approach to displaying the second-order statistics appears not to have been widely adopted. It is noted in Table 2a that only four papers include the mean or bias of the raw precipitation data in the GCM assessments which is important from a hydrologic perspective. The second set of metrics listed in Table 2b is used essentially for ranking GCMs by performance. Several other assessment tools not included in Table 2b are the climate prediction index (Murphy et al., 2004) and Bayesian approaches (Min and Hense, 2006).

Specific climate features like the preservation of the ENSO (El Niño–Southern Oscillation) signal (van Oldenborgh et al., 2005) would also be considered to be a non-numerical measure of GCM performance, but in some regions to be no less important to hydrologists than the numerical measures. Most of these ranking metrics have been developed for specific purposes with respect to GCMs and several have little utility for the practicing hydrologist who is primarily interested in bias, variance and uncertainty in projected estimates of precipitation and temperature (plus net radiation, wind speed and humidity to derive potential ET) as input to drive stand-alone global and catchment hydrologic models.

2.2 Results of CMIP3 GCMs assessments

Table 2a indicates that only two papers (Räisänen, 2007; Gleckler et al., 2008) detail numerical measures for both mean annual precipitation and temperature for 21 and 22 CMIP3 GCMs, respectively, at a global scale. Reifen and Toumi (2009) (17 GCMs) and Knutti et al. (2010) (23 GCMs) address, inter alia only mean annual temperature. Hagemann et al. (2011) used three GCMs to estimate precipitation and temperature characteristics, but the paper includes only precipitation results.

Table 2a. Numerical measures of performance assessment of CMIP3 GCMs.

Reference	Global, country, large region	GCMs	Precipitation							Temperature					
			Reference data sets	Mean of raw data	Bias in mean of raw data	Variance of raw data	RMS or similar metric	Spat. correl.	Taylor plots	Reference data sets	Mean of raw data	Bias in mean of raw data	RMS or similar metric	Spat. correl.	Taylor plots
Bonsal and Prowse (2006)	Northern Canada	7 GCMs	CRU and other data	yes as figure	yes				yes (abs)		yes				yes (abs)
Suppiah et al. (2007)	Australia	23 GCMs	Bureau of Met., Australia				yes (abs)∧	yes		Bureau of Met., Australia			yes	yes	
Räisänen (2007)	Global	21 GCMs	CRU, GPCPv2		yes as figure	yes as figure	yes	yes		CRU TS2.0, NCEP-NCAR		yes as figure	yes (abs)	yes	
Gleckler et al. (2008)	Global	22 GCMs	GPCP/CMAP				yes (norm)#		yes (norm)	ERA40/NCEP-NCAR			yes (norm)		yes (norm)
Reifen and Toumi (2009)	Global	17 GCMs								HadCRUT3 5°×5°			yes (abs)		
Knutti et al. (2010)	Global	23 GCMs								ERA40		yes	yes (abs)	yes	
Macadam et al. (2010)	Global	17 GCMs	HadCRUT3 data set								yes as figure*				
Hagemann et al. (2011)	Global	MPI CNRM IPSL	WFD (ERA-40)	yes as figure	yes as figure	yes as figure									
Heo et al. (2014)	East Asia	21 GCMs	CMAP				yes (norm)	yes	yes (norm)	NCEP-1			yes (norm)	yes	yes (norm)
Raju and Kumar (2014)	India	11 GCMs	NCAP/NCAR 2.5°×2.5°		yes (norm)		yes (abs)			NCAP/NCAR 2.5°×2.5°		yes (norm)	yes (norm)		
Number of references	10			1	4	2	5	3	3		2	2	7	4	3

∧ (abs): based on absolute data; # (norm): based on normalized data; * also as an anomaly.

Table 2b. Ranking measures of performance assessment of CMIP3 GCMs.

Reference	Global, country, large region	GCMs	PDF and related measures	Performance index based on variance	Entropy	Skill score	Variance convergence score	Signal noise ratio
Shukla et al. (2006)	Global	13 GCMs			yes			
Perkins et al. (2007)	Australia	16 GCMs	yes			yes		
Gleckler et al. (2008)	Global	22 GCMs				yes		
Reichler and Kim (2008)	Global	21 GCMs		yes				
Watterson (2008)	Australia	23 GCMs	yes					
Johnson and Sharma (2009b)	Australia	9 GCMs					yes	
Knutti et al. (2010)	Global	23 GCMs	yes					
Heo et al. (2014)	East Asia	21 GCMs			yes			yes
Number of references		8	3	1	2	2	1	1

Räisänen (2007) results illustrate the wide range of model performances that exist: for precipitation, RMSE $= 1.35\,\mathrm{mm\,day^{-1}}$ with a range of 0.97–1.86 and for temperature, RMSE $= 2.32\,°C$ with a range of 1.58–4.56. Reichler and Kim (2008) considered 14 variables covering mainly the period 1979–1999 to assess the performance of CMIP3 models using their model performance index. They concluded that there was a continuous improvement in model performance from the CMIP1 models compared to those available in CMIP3 but there are still large differences in the CMIP3 models' ability to match observed climates. Gleckler et al. (2008) normalised the data in Taylor diagrams for a range of climate variables and concluded that some models performed substantially better than others. However, they also concluded that it is not yet possible to answer the question: what is the best model?

Reifen and Toumi (2009) (Table 2b) using temperature anomalies observed that "...there is no evidence that any subset of models delivers significant improvement in prediction accuracy compared to the total ensemble". On the other hand, Macadam et al. (2010) (Table 2a) assessed the performance of 17 CMIP3 GCMs comparing the observed and modelled temperatures over five 20-year periods and concluded that GCM rankings based on anomalies can be inconsistent over time, whereas rankings based on actual temperatures can be consistent over time.

In summary, Gleckler et al. (2008) stated that the best GCM will depend on the intended application. In the overarching project of which this study is a component, we are interested in the uncertainty in annual streamflow estimated through hydrologic simulation using GCM precipitation and temperature and how that uncertainty will affect estimates of future yield from surface water reservoir systems. Consequently, we are interested in which GCMs reproduce precipitation and temperature satisfactory. Based on the references of Reichler and Kim (2008), Gleckler et al. (2008) and Macadam et al. (2010), the performance of 23 CMIP3 GCMs assessed at a global scale are ranked in Table 3. In Ta-

ble 3 eight models that meet the Reichler and Kim (2008) criterion are also ranked in the upper 50 % based on the Macadam et al. (2010) and Gleckler et al. (2008) references. These models are CCCMA-t47 (Canadian Centre for Climate Modeling and Analysis), CCSM (Community Climate System Model), GFDL2.0 (Geophysical Fluid Dynamics Laboratory), GFDL2.1, HadCM3 (Hadley Centre for Climate Prediction and Research), MIROCm (Model for Interdisciplinary Research on Climate), MPI (Max Planck Institute for Meteorology) and MRI (Japan Meteorological Research Institute).

3 Data

Two data sets are used in the GCM assessment that follows in Sect. 4. One is based on observed data and the other on GCM simulations of present climate (20C3M). It should be noted that of the 22 GCMs examined herein, multiple runs or projections were available for nine models. The resulting 46 runs are identified in the tables summarising the results.

The first data set is based on monthly observed precipitation and temperature gridded at $0.5° \times 0.5°$ resolution over the global land surface from Climatic Research Unit (CRU) 3.10 (New et al., 2002) for the period January 1950 to December 1999. For grid cells where monthly observations are not available, the CRU 3.10 data set is based on interpolation of observed values within a correlation decay distance of 450 km for precipitation and 1200 km for temperature. The CRU 3.10 data set provides information about the number of observations within the correlation decay distance of each grid cell for each month. In this analysis we defined a grid cell as observed if $\geq 90\,\%$ of months at that grid cell has at least one observation within the correlation decay distance for the period January 1950 to December 1999. Only observed grid cells are used to compute summary statistics in the following analysis.

The second data set is monthly precipitation and temperature data for the present climate (20C3M) from 22 of the

Table 3. Summary of performance of 23 CMIP3 GCMs in simulating present climate based on literature review.

GCM	Source	Macadam et al. (2010)	Gleckler et al. (2008)		Reichler and Kim (2008)
	Variables method	Temperature ranking[b]	Precipitation relative error ranking	Overall rank	Many performance index*
BCCR		15	$=13^{\#}$	14	No
CCCMA-t47		9	$=1$	$=3$	Yes
CCCMA-t63		na[a]	$=3$	na	Yes
CCSM		6	$=13$	$=10$	Yes
CNRM		8	$=19$	13	No
CSIRO		12	$=3$	7	No
GFDL2.0		16	$=3$	$=10$	Yes
GFDL2.1		5	$=3$	2	Yes
GISS-AOM		na	$=13$	na	No
GISS-EH		na	$=19$	na	No
GISS-ER		1	$=17$	9	No
HadCM3		2	$=9$	5	Yes
HadGEM		14	$=17$	15	Yes
IAP		na	$=9$	na	No
INGV		na	na	na	Yes
INM		13	$=19$	16	No
IPSL		10	$=9$	$=10$	No
MIROCh		na	$=9$	na	Yes
MIROCm		7	$=3$	$=3$	Yes
MIUB		4	$=1$	1	na
MPI		3	$=13$	8	Yes
MRI		11	$=3$	6	Yes
PCM		17	22	17	No

* As summarised in Smith and Chandler (2010) (The performance index is based on the error variance between modelled and observed climate for 14 climate and ocean variables. "Yes" indicates the variance error is less than the median across the GCMs.) [a] na: not available or not applicable. [b] Rank 1 is best rank. [#] more than one GCM with this rank.

23 GCMs listed in Table 1 and consists of 46 GCM runs. The 20C3M monthly data for precipitation and temperature were extracted from the CMIP3 data set. As shown in Table 1 the GCMs have a wide range of spatial resolutions, all of which are coarser than the observed CRU data. In order to make comparisons between observed and GCM data either the CRU and/or GCM data must be re-sampled to the same resolution. To avoid re-sampling coarse resolution data to a finer resolution we only re-sampled the CRU data here. Thus, in the following analysis the performance of each GCM is assessed at the resolution of the GCM and the CRU data are re-sampled to match the GCM resolution. Therefore, the number of grid cells in each comparison varies with the GCM resolution and ranged from 616 to 11 886 for the temperature comparisons and 425 to 8291 for the precipitation comparisons. The difference in number of grid cells between temperature and precipitation is due to more terrestrial grid cells having observed temperature data than precipitation data over the period 1950–1999.

In the following analysis comparisons are made between observed and GCM values of mean and standard deviation of annual precipitation and mean annual temperature. The GCM values are based on *concurrent raw* (that is, not downscaled nor bias corrected) data from the 20C3M simulation. For example, if a grid cell has observed calendar-year data from 1953 to 1994, then the comparison will be made with GCM values from the 20C3M run for the concurrent calendar years 1953–1994. Although the aim of a 20C3M run from a given GCM is not to strictly replicate the observed monthly record, we expect better performing GCMs to reproduce mean annual statistics that are broadly similar to observed conditions. Average monthly precipitation and temperature patterns are also compared to assess how well GCM runs reproduce observed seasonality. Finally, we assess how well the Köppen–Geiger climate classification (Peel et al., 2007) estimated from the CMIP3 data compares with present-day gridded observed climate classification.

4 Comparison of present climate GCM data with observed data

In the analyses that follow, GCM estimates of mean annual precipitation and temperature and the standard deviation of annual precipitation are compared against observed estimates for terrestrial grid cells with $\geq 90\%$ observed data during the period 1950–1999.

Eight standard statistics – Nash–Sutcliffe efficiency (NSE) (Nash and Sutcliffe, 1970), product moment coefficient of determination (R^2) (MacLean, 2005), standard error of regression (Maidment, 1992), bias (MacLean, 2005), percentage bias (Maidment, 1992), absolute percentage bias (MacLean, 2005), root mean square error (RMSE) (MacLean, 2005) and mean absolute error (MacLean, 2005) – were computed as the basis of comparison, but we report only the NSE, R^2 and RMSE in the following discussion. For our analysis, the NSE is the most useful statistic as it shows the proportion of explained variance relative to the 1 : 1 line in a comparison of two estimates of the same variable. R^2 is included because many analysts are familiar with its interpretation. Both NSE and R^2 were computed in arithmetic (untransformed) and natural log space. We have also included RMSE values (computed from the untransformed values) as many GCM analyses include this measure.

In the following sub-sections comparisons between the concurrent raw GCM data and observed values for MAP, SDP, MAT, long-term average monthly precipitation and temperature patterns and Köppen–Geiger climate classification at the grid cell scale are presented and discussed. Although we rank the models by each selection criteria and combine the ranks by addition, we note the warning of Stainforth et al. (2007) who argue that model response should not be weighted but ruled in or out. We follow this approach in this paper by identifying better performing GCMs to be used for hydrologic simulations reported in a companion paper (Peel et al., 2015). This approach is consistent with the concept recognised by Randall et al. (2007, p. 608) that "…for models to predict future climatic conditions reliably, they must simulate the current climatic state with some as yet unknown degree of fidelity. Poor model skill in simulating present climate could indicate that certain physical or dynamical processes have been misrepresented". It is noted that our comparisons are conducted over the global terrestrial land surface rather than focussing on a single catchment, region or continent. This allows us to assess whether a GCM performs consistently well across a large area and reduces the chance of a GCM being selected due to a random high performance over a small area.

4.1 Mean annual precipitation

Comparisons of mean annual precipitation and the standard deviation of annual precipitation between GCM estimates and observed data for the grid cells across the 46 runs are presented in Table 4. For MAP, the NSE varied from a maximum of 0.68 ($R^2 = 0.69$) with a RMSE value of 335 mm year^{-1} for model MIUB(3) (Meteorological Institute of the University of Bonn) to -0.54 for GISS-EH(3) (NASA Goddard Institute of Space Studies). (GCM run number is enclosed by parenthesis, for example MIUB(3) is run 3 for the GCM

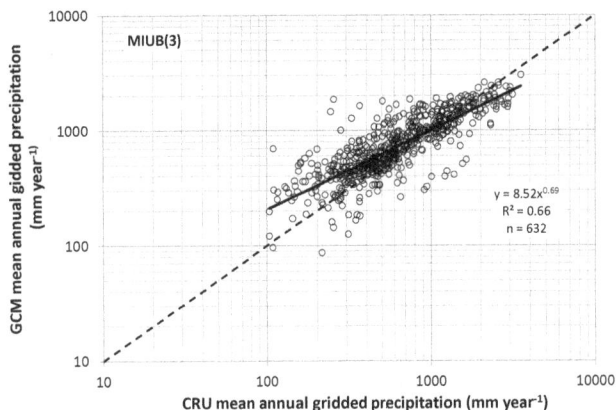

Figure 1. Comparison of MIUB(3) model estimates of observed mean annual precipitation with CRU estimates. (Based on untransformed precipitation NSE = 0.678, rank 1 of 46 runs, and $R^2 = 0.691$.)

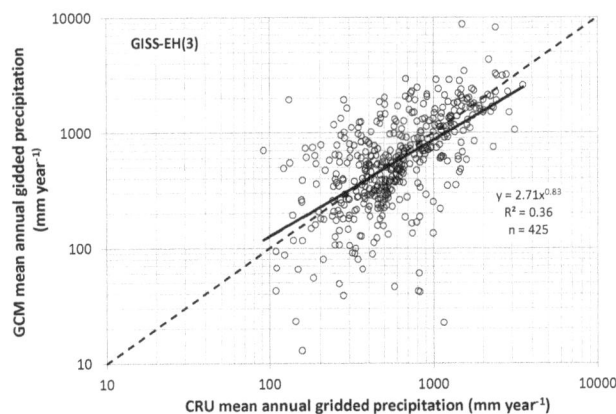

Figure 2. Comparison of GISS-EH(3) model estimates of observed mean annual precipitation with CRU estimates. (Based on untransformed precipitation NSE = -0.535, rank 46 of 46 runs, and $R^2 = 0.368$.)

MIUB.) The MAP values for MIUB(3) are compared with the observed CRU MAP values in Fig. 1. Each data point in this figure represents a MAP comparison at one of the 632 MIUB(3) terrestrial grid cells where observed CRU 3.10 data were available for the period January 1950 to December 1999. The relationship between GCM and observed MAP shown in this figure is representative of the other GCMs where high MAP is underestimated and low MAP is overestimated. GISS-EH(3), shown in Fig. 2, is an example of a poorly performing GCM in terms of mean annual precipitation. Here, based on untransformed data, the NSE is -0.54 ($R^2 = 0.37$) with a RMSE value of 697 mm year^{-1}.

The range of NSE values for the MAP comparisons across the 46 GCM runs is plotted in Fig. 3. The results may be classified into four groups: 5 runs exhibiting NSE > 0.6, 27 runs $0.4 < \text{NSE} \leq 0.6$, 6 runs $0 < \text{NSE} \leq 0.4$ and 8 runs ≤ 0, where the predictive power of the GCM is less than using

Table 4. Performance statistics comparing CMIP3 GCM mean and standard deviation of annual precipitation, mean annual temperature, and mean monthly patterns of precipitation and temperature with concurrent observed data. (Analysis based on untransformed data.)

GCM Name	MAP			SDP			MAT			Monthly pattern	
	R^2	NSE	RMSE	R^2	NSE	RMSE	R^2	NSE	RMSE	NSE Prec	NSE Temp
CCCMA-t47	0.498	0.457	435	0.342	0.252	63	0.984	0.953	3.14	0.409	0.838
CCCMA-t63	0.519	0.458	447	0.397	0.328	65	0.984	0.940	3.59	0.364	0.797
CCSM(1)*	0.496	0.483	460	0.426	0.413	71	0.982	0.981	2.06	−0.178	0.910
CCSM(2)	0.488	0.473	464	0.423	0.411	71	0.982	0.981	2.03	−0.210	0.912
CCSM(3)	0.493	0.479	462	0.418	0.403	71	0.981	0.980	2.08	−0.195	0.908
CCSM(4)	0.500	0.488	457	0.426	0.410	71	0.982	0.980	2.08	−0.174	0.911
CCSM(5)	0.493	0.480	461	0.423	0.410	71	0.983	0.981	2.02	−0.210	0.909
CCSM(6)	0.494	0.480	461	0.437	0.426	70	0.982	0.981	2.04	−0.181	0.909
CCSM(7)	0.496	0.483	460	0.429	0.420	71	0.982	0.981	2.06	−0.173	0.907
CCSM(9)	0.500	0.488	457	0.400	0.393	72	0.982	0.980	2.08	−0.157	0.910
CNRM	0.445	0.246	527	0.479	0.321	65	0.979	0.967	2.67	−0.631	0.879
CSIRO	0.387	0.363	503	0.462	0.452	65	0.971	0.959	2.99	0.034	0.825
GFDL2.0	0.544	0.528	434	0.588	0.460	63	0.980	0.934	3.79	−0.092	0.760
GFDL2.1	0.534	0.518	436	0.570	0.196	77	0.979	0.970	2.54	0.071	0.884
GISS-AOM(1)	0.330	−0.093	624	0.142	0.039	73	0.972	0.969	2.55	−0.325	0.873
GISS-AOM(2)	0.330	−0.087	623	0.132	0.027	74	0.972	0.970	2.54	−0.306	0.876
GISS-EH(1)	0.373	−0.510	692	0.210	−0.397	78	0.963	0.956	3.03	−0.856	0.858
GISS-EH(2)	0.375	−0.502	690	0.176	−0.589	83	0.962	0.955	3.07	−0.920	0.852
GISS-EH(3)	0.368	−0.535	697	0.181	−0.521	81	0.962	0.955	3.06	−0.858	0.856
GISS-ER(1)	0.386	−0.347	653	0.254	−0.115	70	0.970	0.960	2.87	−0.819	0.854
GISS-ER(2)	0.381	−0.357	656	0.203	−0.372	77	0.970	0.959	2.90	−0.739	0.850
GISS-ER(4)	0.386	−0.340	652	0.223	−0.214	72	0.970	0.960	2.88	−0.742	0.854
HadCM3	0.662	0.630	363	0.618	0.572	51	0.988	0.973	2.43	0.227	0.893
HadGEM	0.571	0.302	531	0.457	0.178	82	0.977	0.953	3.22	0.046	0.824
IAP(1)	0.496	0.438	456	0.191	0.096	75	0.963	0.894	4.64	−0.910	0.777
IAP(2)	0.493	0.433	458	0.188	0.041	77	0.962	0.895	4.61	−0.989	0.779
IAP(3)	0.499	0.440	455	0.186	0.048	77	0.963	0.896	4.60	−0.922	0.781
INGV	0.681	0.672	371	0.492	0.468	70	0.983	0.973	2.45	−0.263	0.882
INM	0.450	0.439	431	0.287	0.099	65	0.969	0.952	3.21	−0.247	0.833
IPSL	0.394	0.116	563	0.421	0.223	68	0.967	0.957	3.05	−0.147	0.846
MIROCh	0.588	0.370	514	0.583	0.570	63	0.974	0.971	2.54	0.107	0.906
MIROCm(1)	0.555	0.512	424	0.477	0.454	58	0.970	0.969	2.58	0.061	0.899
MIROCm(2)	0.552	0.508	425	0.525	0.501	56	0.970	0.969	2.58	0.054	0.900
MIROCm(3)	0.549	0.505	427	0.459	0.428	60	0.971	0.970	2.52	0.041	0.902
MIUB(1)	0.689	0.676	336	0.527	0.510	51	0.979	0.960	2.92	0.166	0.870
MIUB(2)	0.684	0.671	338	0.529	0.513	51	0.979	0.962	2.85	0.155	0.867
MIUB(3)	0.691	0.678	335	0.524	0.515	51	0.979	0.958	2.99	0.167	0.860
MPI(1)	0.543	0.538	429	0.464	0.437	66	0.985	0.984	1.88	0.014	0.939
MPI(2)	0.541	0.536	430	0.462	0.415	67	0.985	0.983	1.90	−0.002	0.939
MPI(3)	0.542	0.536	430	0.507	0.479	63	0.986	0.984	1.87	0.007	0.940
MRI(1)	0.617	0.535	414	0.507	0.499	56	0.977	0.969	2.57	0.217	0.912
MRI(2)	0.615	0.537	413	0.513	0.491	56	0.976	0.968	2.64	0.216	0.907
MRI(3)	0.617	0.541	411	0.523	0.505	55	0.977	0.969	2.57	0.222	0.911
MRI(4)	0.619	0.539	412	0.532	0.523	54	0.977	0.969	2.60	0.195	0.911
MRI(5)	0.615	0.538	412	0.503	0.487	56	0.977	0.968	2.62	0.211	0.907
PCM	0.360	0.190	546	0.336	0.135	73	0.975	0.943	3.49	−0.415	0.798

* In parentheses after a GCM name, throughout this paper, indicates the run number.

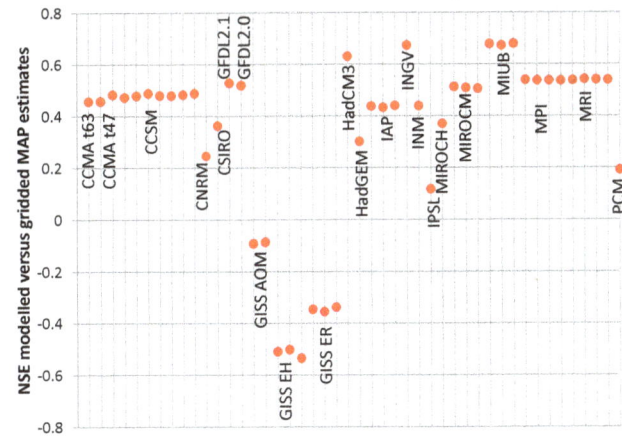

Figure 3. Nash–Sutcliffe efficiency (NSE) values for modelled versus observed MAP untransformed estimates for 46 CMIP3 GCM runs.

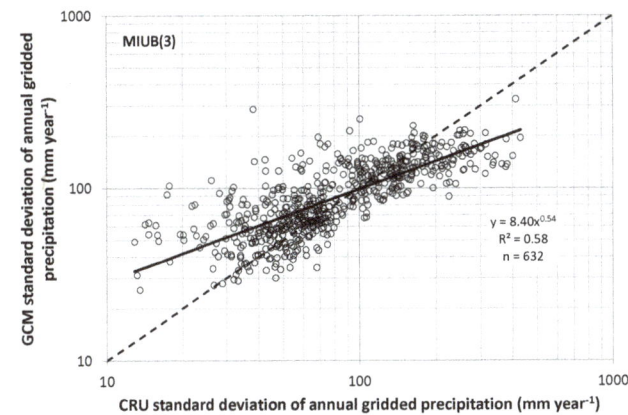

Figure 4. Comparison of MIUB(3) model estimates of the standard deviation of annual precipitation with CRU observed estimates. (Based on untransformed precipitation NSE = 0.515, rank 4 of 46 runs, and $R^2 = 0.524$.)

the average observed MAP across all grid cells (Gupta et al., 2009).

4.2 Standard deviation of annual precipitation

For the standard deviation of annual precipitation, HadCM3 was the best performing model with a NSE of 0.57, R^2 of 0.62 and a RMSE of 51 mm year^{-1}. MIROCh also yielded a NSE of 0.57 and an R^2 of 0.58 but with a RMSE of 63 mm year^{-1}. These results along with other standard deviation values are listed in Table 4. Figure 4 is a plot for MIUB(3), which is representative (rank 4, that is the fourth best performance of the 46 runs) of the relationship between GCM and observed SDP, and shows the model underestimates the standard deviation of annual precipitation for high values and overestimates at low values of standard deviation compared with observed values.

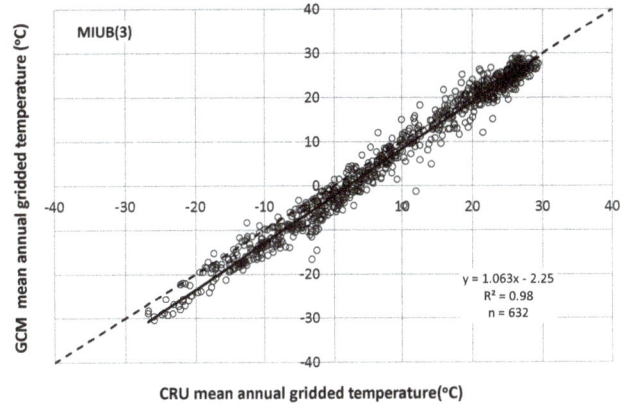

Figure 5. Comparison of MIUB(3) model estimates of mean annual temperature with CRU estimates. (Based on untransformed temperature NSE = 0.958, rank 33 of 46 runs, and $R^2 = 0.979$.)

4.3 Mean annual temperature

The comparison of the GCM mean annual temperatures with concurrent observed data for the grid cells are listed for each model run in Table 4. In contrast to the precipitation modelling, the mean annual temperatures are simulated satisfactorily by most of the GCMs. Except for the IAP (Institute of Atmospheric Physics, Chinese Acad. Sciences) and the GFDL2.0 models (NSE = ~ 0.90 and 0.93, respectively), all model runs exhibit NSE values ≥ 0.94 with 17 of the 46 GCM runs having a NSE value ≥ 0.97. A comparison between MIUB(3) estimates of mean annual temperature (NSE = 0.96, rank 33) and observed values from the CRU data set is presented in Fig. 5. Also shown in Fig. 5 is a linear fit between GCM and observed MAT. The average fit for the 46 GCM runs (not shown) exhibited a small negative bias of −1.03 °C and a slope of 1.01.

4.4 Average monthly precipitation and temperature patterns

Because a monthly rainfall–runoff model is applied in the next phase of our analysis (reported in a companion paper) it is considered appropriate to assess how well the GCMs simulate the observed mean monthly patterns of precipitation and temperature (see also the argument of Charles et al., 2007). The NSE was used for the assessment by comparing the 12 long-term average monthly values. For each GCM run the average precipitation and temperature values for each month were calculated for each grid cell. NSEs were computed between the equivalent 12 GCM-based and 12 CRU-based monthly averages. The median NSE values across terrestrial grid cells where observed CRU 3.10 data were available for the period January 1950 to December 1999 for each GCM run are summarised in Table 4. As shown in Table 4 average monthly patterns of precipitation are poorly modelled. In fact, 57 % of the 46 model runs have a median

Table 5. Köppen–Geiger climate classification (adapted from Peel et al., 2007).

Köppen–Geiger class	Description of climate
Af	Tropical, rainforest
Am	Tropical, monsoon
Aw	Tropical, savannah
BWh	Arid, desert hot
BWk	Arid, desert cold
BSh	Arid, steppe hot
BSk	Arid, steppe cold
Csa	Temperate, dry and hot summer
Csb	Temperate, dry and warm summer
Csc	Temperate, dry and cold summer
Cwa	Temperate, dry winter and hot summer
Cwb	Temperate, dry winter and warm summer
Cwc	Temperate, dry winter and cold summer
Cfa	Temperate, without dry season and hot summer
Cfb	Temperate, without dry season and warm summer
Cfc	Temperate, without dry season and cold summer
Dsa	Cold, dry and hot summer
Dsb	Cold, dry and warm summer
Dsc	Cold, dry and cool summer
Dsd	Cold, dry summer and very cold winter
Dwa	Cold, dry winter and hot summer
Dwb	Cold, dry winter and warm summer
Dwc	Cold, dry winter and cool summer
Dwd	Cold, dry winter and very cold winter
Dfa	Cold, without dry season and hot summer
Dfb	Cold, without dry season and warm summer
Dfc	Cold, without dry season and cool summer
Dfd	Cold, without dry season and very cold winter
ET	Polar, tundra
EF	Polar, frost

NSE value of < 0. For these GCMs their predictive power for the monthly precipitation pattern is less than using the average of the 12 monthly values at each of the terrestrial grid cells. Only two GCMs have NSE values > 0.25. In contrast, the median NSEs of all monthly temperature patterns are > 0.75, with 41 % > 0.90. The NSE metric reflects how well the GCM replicates both the monthly pattern and the overall average monthly value (bias). Thus, the monthly pattern of temperature is generally well reproduced by the GCMs, whereas the monthly pattern of precipitation is not, which is mainly due to the bias in the GCM average monthly precipitation.

4.5 Köppen–Geiger classification

The Köppen–Geiger climate classification (Peel et al., 2007) (see Table 5) provides an alternate way to assess the adequacy of how well a GCM represents climate because the classification is based on a combination of annual and monthly precipitation and temperature data. Two comparisons between the MPI(3) model and CRU observed data are presented in Table 6. The MPI(3) was chosen as an example here as over the three levels of climate classes it estimated the

observed climate correctly more often than the other model runs. In Table 6a a comparison at the first letter level of the Köppen–Geiger climate classification is shown. This comparison reveals how well the GCM reproduces the distribution of broad climate types: tropical, arid, temperate, cold and polar over the terrestrial surface. In Table 6b the comparison shown is for the second letter level of the Köppen–Geiger climate classification, which assesses how well the GCM reproduces finer detail within the broad climate types; for example, the seasonal distribution of precipitation or whether a region is semi-arid or arid. The bold diagonal values shown in Table 6a and b represent the number of grid cells correctly classified by the GCM, whereas the off-diagonal values are the number of grid cells incorrectly classified by the GCM for the one- and two-letter level. At the first letter level MPI(3) reproduces the correct climate type at 81 % of the terrestrial grid cells. Within this good performance the MPI(3) produces more polar climate and fewer tropical and cold grids cells than observed. At the second letter level, MPI(3) reproduces the correct climate type at 67 % of the terrestrial grid cells. The model produces fewer grid cells of tropical rainforest, cold with a dry winter and cold without a dry season than expected and more cold with a dry summer and polar tundra than expected.

Table 7 summarises the overall proportion of GCM grid cells that were classified correctly for each GCM run across the three levels of classification. As we wish to have a ranking of the comparisons we adopted this simple measure as it is regarded as "…one of the most basic and widely used measures of accuracy…" for comparing thematic maps (Foody, 2004, p. 632). From Table 7 we observe that GCM accuracy in reproducing the climate classification decreases as one moves from coarse to fine detail climate classification. The average accuracy (and range) for the three classes are 0.48 (0.36–0.60) for the three-letter classification, 0.57 (0.47–0.68) for the two-letter classification and for one-letter 0.77 (0.66–0.82). In other words, at the three-letter scale nearly 50 % of GCM Köppen–Geiger estimates are correct, increasing to nearly 60 % at the two-letter level and, finally, at the one-letter aggregation more than 75 % are correct across the 46 GCM runs. Using these average values across the three classes, the following seven models performed satisfactorily in identifying Köppen–Geiger climate class correctly: CNRM (Météo-France/Centre National de Recherches Météorologiques), CSIRO (Commonwealth Scientific and Industrial Research Organisation), HadCM3, HadGEM, MIUB, MPI and MRI. Of these models the least successful run was for CSIRO with the percentage correct for each class as follows: three-letter 51 %, two-letter 60 % and one-letter 78 %.

Table 6. Köppen–Geiger climate estimated by MPI(3) compared with the observed Köppen–Geiger climate for (a) the one-letter and (b) the two-letter climate classification. Bold values are correctly classified grid cells.

(a)				CRU			
	Land surface	A	B	C	D	E	Sum
GCM	A	**414**	19	8	0	0	441
	B	68	**339**	52	17	0	476
	C	24	62	**319**	27	0	432
	D	0	76	16	**1085**	17	1194
	E	0	6	7	143	**121**	277
	Sum	506	502	402	1272	138	2820

(b)							CRU							
	Land surface	Af	Am	Aw	BW	BS	Cs	Cw	Cf	Ds	Dw	Df	ET	EF	Sum
GCM	Af	**57**	0	2	0	0	0	0	0	0	0	0	0	0	59
	Am	24	**19**	13	0	0	0	0	0	0	0	0	0	0	56
	Aw	25	49	**225**	0	19	0	4	4	0	0	0	0	0	326
	BW	2	1	2	**134**	50	3	4	0	0	0	2	0	0	198
	BS	4	11	48	50	**105**	13	19	13	4	0	11	0	0	278
	Cs	0	0	0	10	18	**35**	9	20	1	0	6	0	0	99
	Cw	0	1	17	0	5	0	**62**	1	0	1	0	0	0	87
	Cf	2	2	2	3	26	1	35	**156**	0	0	19	0	0	246
	Ds	0	0	0	0	33	2	1	1	**38**	1	40	0	0	116
	Dw	0	0	0	0	5	0	1	0	0	**102**	2	0	0	110
	Df	0	0	0	3	35	0	4	7	2	57	**843**	17	0	968
	ET	0	0	0	0	6	2	2	3	8	22	113	**93**	0	249
	EF	0	0	0	0	0	0	0	0	0	0	0	11	**17**	28
	Sum	114	83	309	200	302	56	141	205	53	183	1036	121	17	2820

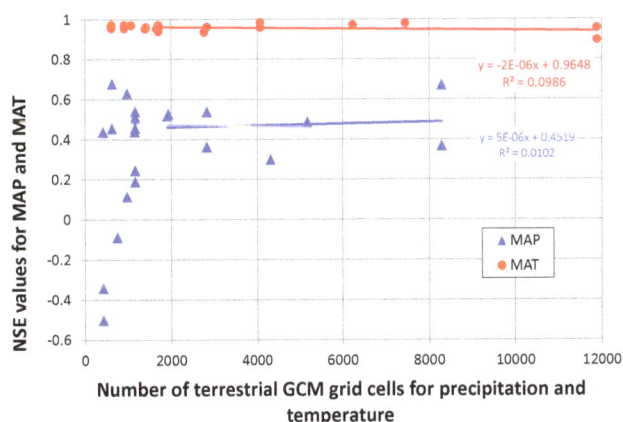

Figure 6. Relating 22 CMIP3 GCM resolutions (as the number of terrestrial grid cells for MAP) to model performance based on Nash–Sutcliffe efficiency (NSE) for mean annual precipitation and mean annual temperature. (The trend lines are fitted to data with > 1500 grid cells.)

5 Discussion

5.1 Relating GCM resolution to performance

In the analysis presented in the previous section each GCM's performance in reproducing observed climatological statistics was assessed at the resolution of the individual GCM. The question of whether GCMs with a finer resolution outperform GCMs with a coarser resolution is addressed in Fig. 6, where GCM performance in reproducing observed terrestrial MAP and MAT, based on the NSE, is related to GCM resolution, defined as the number of grid cells used in the comparison. The plot suggests there is no significant relationship between GCM resolution and GCM performance beyond 1500 grid cells for either MAP or MAT. Interestingly, some lower resolution GCMs, < 1500 grid cells, perform as well as higher resolution GCMs for MAP and MAT, yet for others, they perform poorly. While it is sometimes assumed that higher resolution should normally lead to improved performance, there are many other factors that affect performance. These include the sophistication of the parameterisation schemes for different sub-grid-scale processes, the time spent in developing and testing the individual schemes and their interactions. Our purpose here is to report this observation rather than speculate what it might mean for GCM

Table 7. Proportion of CMIP3 GCM grid cells (20C3M) that reproduce observed CRU Köppen–Geiger climate classification over the period January 1950–December 1999.

GCM Name	Köppen–Geiger climate class*		
	Three-letter	Two-letter	One-letter
CCCMA-t47	0.498	0.620	0.753
CCCMA-t63	0.429	0.558	0.709
CCSM(1)	0.488	0.558	0.749
CCSM(2)	0.489	0.563	0.748
CCSM(3)	0.424	0.545	0.744
CCSM(4)	0.466	0.549	0.749
CCSM(5)	0.444	0.519	0.727
CCSM(6)	0.490	0.563	0.757
CCSM(7)	0.488	0.556	0.749
CCSM(9)	0.489	0.560	0.755
CNRM	0.539	0.602	0.775
CSIRO	0.506	0.601	0.775
GFDL2.0	0.430	0.563	0.726
GFDL2.1	0.508	0.590	0.781
GISS-AOM(1)	0.460	0.559	0.773
GISS-AOM(2)	0.456	0.561	0.773
GISS-EH(1)	0.407	0.487	0.751
GISS-EH(2)	0.402	0.482	0.741
GISS-EH(3)	0.400	0.473	0.744
GISS-ER(1)	0.426	0.478	0.732
GISS-ER(2)	0.424	0.468	0.722
GISS-ER(4)	0.426	0.478	0.732
HadCM3	0.549	0.624	0.797
HadGEM	0.563	0.676	0.818
IAP(1)	0.362	0.484	0.790
IAP(2)	0.368	0.480	0.784
IAP(3)	0.369	0.490	0.784
INGV	0.495	0.616	0.815
INM	0.452	0.526	0.731
IPSL	0.459	0.544	0.749
MIROCh	0.496	0.631	0.806
MIROCm(1)	0.477	0.597	0.749
MIROCm(2)	0.477	0.594	0.759
MIROCm(3)	0.469	0.583	0.748
MIUB(1)	0.528	0.604	0.783
MIUB(2)	0.528	0.604	0.783
MIUB(3)	0.520	0.610	0.778
MPI(1)	0.599	0.666	0.801
MPI(2)	0.593	0.657	0.805
MPI(3)	0.602	0.669	0.808
MRI(1)	0.534	0.644	0.808
MRI(2)	0.521	0.625	0.798
MRI(3)	0.527	0.632	0.798
MRI(4)	0.528	0.634	0.799
MRI(5)	0.532	0.641	0.803
PCM	0.397	0.481	0.660

* The three-, two- and one-letter climate classes are listed in Table 5.

Table 8. CMIP3 GCM run rank (rank 1 = best) based on Nash–Sutcliffe efficiency (NSE) values from comparison of 20C3M and concurrent observed grid cell data.

GCM Name	MAP rank	SDP rank	MAT rank	Monthly pattern rank*	Rank sum	Overall GCM rank
CCCMA-t47	28	30	38	19	115	12
CCCMA-t63	27	28	42	22	119	13
CCSM(1)	21	22	7	18	68	8
CCSM(2)	26	23	5	17	71	
CCSM(3)	25	26	10	21	82	
CCSM(4)	20	25	11	16	72	
CCSM(5)	24	24	4	21	73	
CCSM(6)	23	19	6	20	68	
CCSM(7)	22	20	8	19.5	69.5	
CCSM(9)	19	27	9	17	72	
CNRM	36	29	26	30.5	121.5	14
CSIRO	34	16	32	28.5	110.5	11
GFDL2.0	14	14	43	34	105	10
GFDL2.1	15	32	15	17.5	79.5	9
GISS-AOM(1)	40	39	20	30.5	129.5	
GISS-AOM(2)	39	40	17	29.5	125.5	15
GISS-EH(1)	45	44	35	35.5	159.5	22
GISS-EH(2)	44	46	37	39	166	
GISS-EH(3)	46	45	36	36.5	163.5	
GISS-ER(1)	42	41	28	36	147	19
GISS-ER(2)	43	43	31	36.5	153.5	
GISS-ER(4)	41	42	29	36	148	
HadCM3	5	1	13	12	31	1
HadGEM	35	33	39	28	135	17
IAP(1)	31	36	46	44	157	
IAP(2)	32	38	45	45	160	
IAP(3)	29	37	44	44	154	21
INGV	3	13	12	28	56	5
INM	30	35	40	35	140	18
IPSL	38	31	34	29.5	132.5	16
MIROCh	33	2	14	14.5	63.5	7
MIROCm(1)	16	15	22	17	70	
MIROCm(2)	17	8	21	17	63	6
MIROCm(3)	18	18	16	17.5	69.5	
MIUB(1)	2	6	30	18.5	56.5	
MIUB(2)	4	5	27	19.5	55.5	4
MIUB(3)	1	4	33	19	57	
MPI(1)	9	17	2	10.5	38.5	
MPI(2)	12	21	3	12	48	
MPI(3)	11	12	1	10.5	34.5	2
MRI(1)	13	9	18	5	45	
MRI(2)	10	10	25	10.5	55.5	
MRI(3)	6	7	19	5.5	37.5	3
MRI(4)	7	3	23	8	41	
MRI(5)	8	11	24	11.5	54.5	
PCM	37	34	41	38.5	150.5	20

* Monthly pattern rank is the rank of the average of the monthly pattern NSEs for precipitation and temperature.

model development. Our observation is consistent with Masson and Knutti (2011) who comment that "...model resolution in CMIP3 seems to only affect performance in simulating present-day temperature for small scales over land" (p. 2691) and for precipitation they comment that "...no clear relation seems to exist at least within the relatively narrow range of resolutions covered by CMIP3" (p. 2686).

5.2   Joint comparison of precipitation and temperature

In using GCM climate scenarios in a water resources study, it is appropriate to ensure consistency between precipitation and temperature by adopting projections of these variables

Table 9. Better performing CMIP3 GCMs identified from the literature and our analyses.

Grid cells (Tables 4 and 8) (Col. 1)	Literature (Table 3) (Col. 2)	Better performing GCMs (Col. 3)
	CCCMA-t47	
	CCSM	
	GFDL2.0	
	GFDL2.1	
HadCM3	HadCM3	HadCM3
INGV		
	MIROCh*	
MIROCm	MIROCm	MIROCm
MIUB	MIUB*	MIUB
MPI	MPI	MPI
MRI	MRI	MRI

* Added to list – see Section 5.3 for explanation.

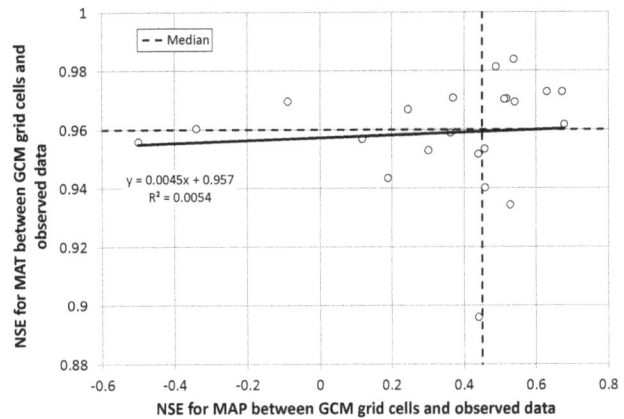

Figure 7. Comparison of Nash–Sutcliffe efficiency (NSE) values between CMIP3 GCM and observed mean annual temperatures with NSE values between CMIP3 GCM and observed mean annual precipitation.

from the same GCM run. Grid cell based NSEs for mean annual temperature and mean annual precipitation from each GCM are compared in Fig. 7, which illustrates the performance of each GCM for both variables. Models that have relatively high NSEs for precipitation do not necessarily have relatively high values for temperature. It is interesting to note that the rank of the models based on NSE of the MAP is unrelated to the ranking of the models based on MAT. Fortunately, however, most of the NSEs for MAT are relatively high and the acceptance or rejection of a GCM as a better performing model is largely dependent on its precipitation characteristics.

5.3 Identifying better performing GCMs

To identify the better performing GCMs across the different variables assessed, the results in Table 4 are ranked by NSE and summarised in Table 8. The monthly patterns of precipitation and temperature are combined by ranking the average of their respective NSE values. The overall rank for each GCM run is based on combining, by addition, the ranks for the individual variables and, finally, identifying the best performing run from each GCM. Selection of the better performing GCMs using these rankings is not inconsistent with Stainforth et al. (2007) who argued that model response should not be weighted but ruled in or out. From Table 8 we identify several GCMs, listed in Table 9, as better performing models. These selected GCMs were based on the assumption that performance across the four variables (MAP, SDP, MAT and combined monthly pattern) is equally weighted. GCMs that achieved MAP NSE > 0.50, SDP NSE > 0.45, MAT NSE > 0.95 and mean monthly pattern of precipitation NSE > 0.0 (Table 4) were identified as better performing. (Because nearly all the GCM runs modelled mean monthly patterns of temperature satisfactorily,

this measure was not considered in the selection of models listed in column 1, Table 9.) The following GCMs were selected (Table 9): HadCM3, INGV (National Institute of Geophysics and Vulcanology, Italy), MIROCm, MIUB, MPI and MRI. INGV was included although it failed the monthly precipitation pattern criterion. The above criteria were selected to identify a small number of GCMs that would require less bias correction to produce annual precipitation and temperature consistent with observations.

In Table 9, we summarise our observations from the literature review in Sect. 2 and the results from our analyses in Tables 4 and 8, where we identified six GCMs that satisfied our selection criteria (Table 9, column 1). From the literature review (Table 3), eight GCMs were identified as being satisfactory. We have added MIUB because in the literature review it ranked first overall, although no guidance was available from Reichler and Kim (2008). We also added MIROCh to this list as it performed better according to Gleckler et al. (2008) than several models in the above list and met the performance index of Reichler and Kim (2008). Columns 1 and 2 of Table 9 suggest there is some consistency between our analyses from a hydrologic perspective and that reported in the literature from a climatological perspective. From the table, we identify that, in terms of our objective to assess how well the CMIP3 GCMs are able to reproduce observed annual precipitation and temperature statistics and the mean monthly patterns of precipitation and temperature, the following models are deemed acceptable for the next phase of our project: HadCM3, MIROCm, MIUB, MPI and MRI. Although not used in the selection criteria we observe our selected GCMs performed well in the Köppen–Geiger climate assessment. We note here that INGV also performed satisfactorily but it was not included in our adopted GCMs as it was not reviewed in the papers of Gleckler et al. (2008), Reichler and Kim (2008) and Macadam et al. (2010).

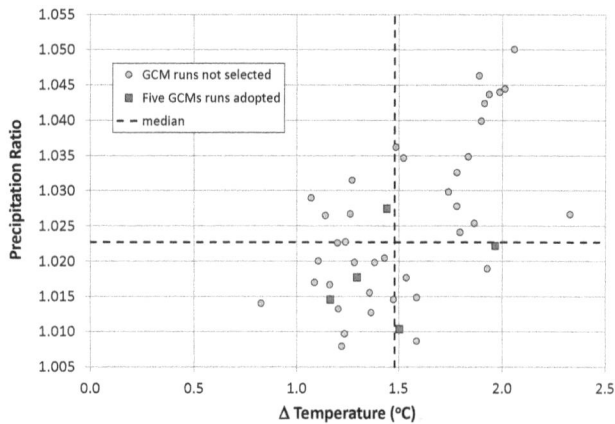

Figure 8. Ratio of 2015–2034 to 1965–1994 mean annual precipitation compared with the change in mean annual temperature (2015–2034 to 1965–1994) for the selected five CMIP3 GCMs runs compared with the 23 CMIP3 GCMs including all ensemble members for the global land surface.

5.4   Comparing future responses of selected GCMs

In order to confirm that the selected GCM runs are representative of the range of future responses to climate change in the CMIP3 ensemble, we plot in Fig. 8 the ratio of mean annual precipitation for the period 2015–2034 (from the A1B scenario) to 1965–1994 against the mean annual temperature difference between 2015–2034 and 1965–1994 for the global land surface. The five selected GCM runs are well distributed amongst the 44 GCM ensemble members, which indicates that the selected GCMs are reasonably representative of the range of future GCM projections if all the runs were considered. We observe that most GCM runs are clustered around the median response, except for the seven CCSM runs in the top right quadrant with a precipitation ratio $> \sim 1.04$.

6   Conclusions

Our primary objective in this paper is to identify better performing GCMs from a hydrologic perspective over global land regions. The better performing GCMs were identified by their ability to reproduce observed climatological statistics (mean and the standard deviation of annual precipitation and mean annual temperature, and the mean monthly patterns of precipitation and temperature) for hydrologic simulation. The GCM selection process was informed by our results presented here and by a literature review of CMIP3 GCM performance. In terms of the NSE there was a large spread in values for mean annual precipitation and the standard deviation of annual precipitation over concurrent periods. The highest NSE for mean annual precipitation was 0.68 and 0.57 for the standard deviation of annual precipitation. On the other hand, for mean annual temperatures, the NSEs between modelled and observed data were very high, with median NSE being

0.97. Overall, all GCMs reproduced the Köppen–Geiger climate satisfactorily at the broad first letter level. From the literature, the following GCMs were identified as being suitable to simulate annual precipitation and temperature statistics: CCCMA-T47, CCSM, GFDL2.0, GFDL2.1, HadCM3, MIROCh, MIROCm, MIUB, MPI and MRI. After combining our results with the literature the following GCMs were considered the better performing models from a hydrologic perspective: HadCM3, MIROCm, MIUB, MPI and MRI. The future response of the better performing GCMs was found to be representative of the 44 GCM ensemble members which confirms that the selected GCMs are reasonably representative of the range of future GCM projections. Our approach for evaluating GCM performance for hydrologic simulation could be applied to CMIP5 runs.

The Supplement related to this article is available online at doi:10.5194/hess-12-361-2015-supplement.

Acknowledgements. This research was financially supported by Australian Research Council grant LP100100756 and FT120100130, Melbourne Water and the Australian Bureau of Meteorology. Lionel Siriwardena, Sugata Narsey and Dr Ian Smith assisted with extraction and analysis of CMIP3 GCM data. Lionel Siriwardena also assisted with extraction and analysis of the CRU 3.10 data. We acknowledge the modelling groups, the Program for Climate Model Diagnosis and Intercomparison (PCMDI) and the WCRP's Working Group on Coupled Modelling (WGCM) for their roles in making available the WCRP CMIP3 multi-model data set. Support of this data set is provided by the Office of Science, U.S. Department of Energy. The authors thank two anonymous reviewers who provided stimulating comments on the discussion paper.

Edited by: A. Loew

References

Boer, G. J. and Lambert, S. J.: Second order space–time climate difference statistics, Clim. Dynam., 17, 213–218, 2001.

Bonsal, B. T. and Prowse, T. D.: Regional assessment of GCM-simulated current climate over Northern Canada, Arctic, 59, 115–128, 2006.

Charles, S. P., Bari, M. A., Kitsios, A., and Bates, B. C.: Effect of GCM bias on downscaled precipitation and runoff projections for the Serpentine catchment, Western Australia, Int. J. Climatol., 27, 1673–1690, 2007.

Chervin, R. M.: On the Comparison of Observed and GCM Simulated Climate Ensembles, J. Atmos. Sci., 38, 885–901, 1981.

Chiew, F. H. S. and McMahon, T. A.: Modelling the impacts of climate change on Australian streamflow, Hydrol. Process., 16, 1235–1245, 2002.

Covey, C., Achutarao, K. M., Cubasch, U., Jones, P., Lambert S. J., Mann, M. E., Phillips, T. J., and Taylor, K. E.: An overview of results from the Coupled Model Intercomparison Project, Global Planet. Change, 37, 103–133, 2003.

Dessai, S., Lu, X., and Hulme, M.: Limited sensitivity analysis of regional climate change probabilities for the 21st century, J. Geophys. Res., 110, D19108, doi:10.1029/2005JD005919, 2005.

Flato, G., Marotzke, J., Abiodun, B., Braconnot, P., Chou, S. C., Collins, W., Cox, P., Driouech, F., Emori, S., Eyring, V., Forest, C., Gleckler, P., Guilyardi, E., Jakob, C., Kattsov, V., Reason, C., and Rummukainen, M.: Evaluation of Climate Models, in: Climate Change 2013: The Physical Science Basis. Contribution of Working Group I to the Fifth Assessment Report of the Intergovernmental Panel on Climate Change, edited by: Stocker, T. F., Qin, D., Plattner, G.-K., Tignor, M., Allen, S. K., Boschung, J., Nauels, A., Xia, Y., Bex, V., and Midgley, P. M., Cambridge University Press, Cambridge, United Kingdom and New York, NY, USA, 2013.

Foody, G. M.: Thematic map comparison: Evaluating the statistical significance of differences in classification accuracy, Photogramm. Eng. Remote S., 70, 627–633, 2004.

Gleckler, P. J., Taylor, K. E., and Doutriaux, C.: Performance metrics for climate models, J. Geophys. Res.-Atmos., 113, D06104, doi:10.1029/2007JD008972, 2008.

Gupta, H. V., Kling, H., Yilmaz, K. K., and Martinez, G. F.: Decomposition of the mean squared error and NSE performance criteria: Implications for improving hydrological modelling, J. Hydrol., 377, 80–91, 2009.

Hagemann, S., Chen, C., Haerter, J. O., Heinke, J., Gerten, D., and Piani, C.: Impact of a statistical bias correction on the projected hydrological changes obtained from three GCMs and two hydrology models, J. Hydrometeorol. 12, 556–578, 2011.

Heo, K.-Y., Ha, K.-J., Yun, K.-S., Lee, S.-S., Kim, H.-J., and Wang, B.: Methods for uncertainty assessment of climate models and model predictions over East Asia, Int. J. Climatol., 34, 377–390, doi:10.1002/joc.2014.34.issue-2, 2014.

Johns, T. C., Durman, C. F., Banks, H. T., Roberts, M. J., Mclaren, A. J., Ridley, J. K., Senior, C. A., Williams, K. D., Jones, A., Rickard, G. J., Cusack, S., Ingram, W. J., Crucifix, M., Sexton, D. M. H., Joshi, M. M., Dong, B.-W., Spencer, H., Hill, R. S. R., Gregory, J. M., Keen, A. B., Pardaens, A. K., Lowe, J. A., Bodas-Salcedo, A., Stark, S., and Searl, Y.: The new Hadley Centre climate model (HadGEM1): evaluation of coupled simulations, J. Climate, 19, 1327–1353, 2006.

Johnson, F. M. and Sharma, A.: GCM simulations of a future climate: How does the skill of GCM precipitation simulations compare to temperature simulations, 18th World IMACS/MODSIM Congress, Cairns, Australia, 2009a.

Johnson, F. and Sharma, A.: Measurement of GCM skill in predicting variables relevant for hydroclimatological assessments, J. Climate, 22, 4373–4382, 2009b.

Knutti, R., Furrer, R., Tebaldi, C., Cermak, J., and Meehl, G. A.: Challenges in combining projections from multiple climate models, J. Climate, 23, 2739–2758, 2010.

Knutti, R., Masson, D., and Gettelman, A.: Climate model genealogy: Generation CMIP5 and how we got there, Geophys. Res. Lett., 40, 1194–1199, 2013.

Lambert, S. J. and Boer, G. J.: CMIP1 evaluation and intercomparison of coupled climate models, Clim. Dynam., 17, 83–106, 2001.

Legates, D. R. and Willmott, C. J.: A comparison of GCM-simulated and observed mean January and July precipitation, Global Planet. Change, 5, 345–363, 1992.

Macadam, I., Pitman, A. J., Whetton, P. H., and Abramowitz, G.: Ranking climate models by performance using actual values and anomalies: Implications for climate change impact assessments, Geophys. Res. Lett., 37, L16704, doi:10.1029/2010GL043877, 2010.

MacLean, A.: Statistical evaluation of WATFLOOD (Ms), University of Waterloo, Ontario, Canada, 2005.

Maidment, D. R.: Handbook of Hydrology, McGraw-Hill Inc., New York, 1992.

Masson, D. and Knutti, R.: Spatial-scale dependence of climate model performance in the CMIP3 ensemble, J. Climate, 24, 2680-2692, 2011.

McMahon, T. A. and Adeloye, A. J.: Water Resources Yield, Water Resources Publications, CO, USA, 220 pp., 2005.

McMahon, T. A., Peel, M. C., Pegram, G. G. S., and Smith, I. N.: A simple methodology for estimating mean and variability of annual runoff and reservoir yield under present and future climates, J. Hydrometeorol., 12, 135–146, 2011.

Meehl, G. A., Covey, C., Delworth, T., Latif, M., McAvaney, B., Mitchell, J. F. B., Stouffer, R. J., and Taylor, K. E.: The WCRP CMIP3 multi-model dataset: A new era in climate change research, B. Am. Meteorol. Soc., 88, 1383–1394, 2007.

Min, S.-K. and Hense, A.: A Bayesian approach to climate model evaluation and multi-model averaging with an application to global mean surface temperatures from IPCC AR4 coupled climate models, Geophys. Res. Lett., 33, L08708, doi:10.1029/2006GL025779, 2006.

Murphy, J. M., Sexton, D. M. H., Barnett, D. N., Jones, G. S., Webb, M. J., Collins, M. J., and Stainforth, D. A.: Quantification of modelling uncertainties in a large ensemble of climate change simulations, Nature, 430, 768–772, 2004.

Nash, J. E. and Sutcliffe, J. V.: River flow forecasting through conceptual models Part 1 – A discussion of principles, J. Hydrol., 10, 282–290, 1970.

New, M., Lister, D., Hulme, M., and Makin, I.: A high-resolution data set of surface climate over global land areas, Clim. Res., 21, 1–25, 2002.

Peel, M. C., Finlayson, B. L., and McMahon, T. A.: Updated world map of the Köppen–Geiger climate classification, Hydrol. Earth Syst. Sci., 11, 1633-1644, doi:10.5194/hess-11-1633-2007, 2007.

Peel, M. C., Srikanthan, R., McMahon, T. A., and Karoly, D. J.: Approximating uncertainty of annual runoff and reservoir yield using stochastic replicates of Global Climate Model data, Hydrol. Earth Syst. Sci. Discuss., under review, 2015.

Perkins, S. E., Pitman, A. J., Holbrook, N. J., and McAneney, J.: Evaluation of the AR4 climate models simulated daily maximum temperature, minimum temperature and precipitation over Australia using probability density functions, J. Climate, 20, 4356–4376, 2007.

Phillips, N. A.: The general circulation of atmosphere: a numerical experiment, Q. J. Roy. Meteorol. Soc., 82, 123–164, 1956.

Räisänen, J.: How reliable are climate models?, Tellus A, 59, 2–29, 2007.

Raju, K. S. and Kumar, D. N.: Ranking of global climate models for India using multicriterion analysis, Clim. Res., 60, 103–117, 2014.

Randall, R. A. and Wood, R. A. (Coordinating lead authors): Climate models and their evaluation. Contribution of Working Group I to the Fourth Assessment Report of the Intergovernmental Panel on Climate Change AR4, Chap. 8, 589–662, 2007.

Reichler, T. and Kim, J.: How well do coupled models simulate today's climate?, B. Am. Meteorol. Soc., 89, 303–311, 2008.

Reifen, C. and Toumi, R.: Climate projections: Past performance no guarantee of future skill?, Geophys. Res. Lett., 36, L13704, doi:10.1029/2009GL038082, 2009.

Shukla, J., DelSole, T., Fennessy, M., Kinter, J., and Paolino, D.: Climate model fidelity and projections of climate change, Geophys. Res. Lett., 33, L07702, doi:10.1029/2005GL025579, 2006.

Smith, I. and Chandler, E.: Refining rainfall projections for the Murray Darling Basin of south-east Australia – the effect of sampling model results based on performance, Clima. Change, 102, 377–393, 2010.

Stainforth, D. A., Allen, M. R., Tredger, E. R., and Smith, L. A.: Confidence, uncertainty and decision-support relevance in climate predictions, Philos. T. R. Soc. A, 365, 2145–2161, 2007.

Suppiah, R., Hennessy, K. L., Whetton, P. H., McInnes, K., Macadam, I., Bathols, J., Ricketts, J., and Page, C. M.: Australian climate change projections derived from simulations performed for IPCC 4th Assessment Reportm Aust. Met. Mag, 56, 131–152, 2007.

Taylor, K. E.: Summarizing multiple aspects of model performance in a single diagram, J. Geophys. Res., 106, 7183–7192, 2001.

van Oldenborgh, G. J., Philip, S. Y., and Collins, M: El Niño in a changing climate: a multi-model study, Ocean Sci., 1, 81–95, doi:10.5194/os-1-81-2005, 2005.

Watterson, I. G.: Calculation of probability density functions for temperature and precipitation change under global warming, J. Geophys. Res., 113, D12106, doi:10.1029/2007JD009254, 2008.

Whetton, P., McInnes, K. L., Jones, R. J., Hennessy, K. J., Suppiah, R., Page, C. M., and Durack, P. J.: Australian Climate Change Projections for Impact Assessment and Policy Application: A Review, CSIRO Marine and Atmospheric Research Paper 001, available at: www.cmar.csiro.au/e-print/open/whettonph_2005a.pdf, 2005.

Xu, C. Y.: Climate change and hydrologic models: A review of existing gaps and recent research developments, Water Resour. Manag., 13, 369–382, 1999.

Precipitation variability within an urban monitoring network via microcanonical cascade generators

P. Licznar[1], **C. De Michele**[2], **and W. Adamowski**[3]

[1]Faculty of Environmental Engineering, Wroclaw University of Technology, Wrocław, Poland
[2]Department of Civil and Environmental Engineering, Politecnico di Milano, Milan, Italy
[3]Institute of Environmental Engineering, John Paul II Catholic University of Lublin, Stalowa Wola, Poland

Correspondence to: P. Licznar (pawel.licznar@pwr.edu.pl)

Abstract. Understanding the variability of precipitation at small scales is fundamental in urban hydrology. Here we consider the case study of Warsaw, Poland, characterized by a precipitation-monitoring network of 25 gauges and microcanonical cascade models as the instrument of investigation.

We address the following issues partially investigated in literature: (1) the calibration of microcanonical cascade model generators in conditions of short time series (i.e., 2.5–5 years), (2) the identification of the probability distribution of breakdown coefficients (BDCs) through ranking criteria and (3) the variability among the gauges of the monitoring network of the empirical distribution of BDCs.

In particular, (1) we introduce an overlapping moving window algorithm to determine the histogram of BDCs and compare it with the classic non-overlapping moving window algorithm; (2) we compare the 2N–B distribution, a mixed distribution composed of two normal (N) and one beta (B), with the classic B distribution to represent the BDCs using the Akaike information criterion; and (3) we use the cluster analysis to identify patterns of BDC histograms among gauges and timescales.

The scarce representation of the BDCs at large timescales, due to the short period of observation (~ 2.5 years), is solved through the overlapping moving window algorithm. BDC histograms are described by a 2N–B distribution. A clear evolution of this distribution is observed, in all gauges, from 2N–B for small timescales, N–B for intermediate timescales and B distribution for large timescales.

The performance of the microcanonical cascades is evaluated for the considered gauges. Synthetic time series are analyzed with respect to the intermittency and the variability of

intensity and compared to observed series. BDC histograms for each timescale are compared with the 25 gauges in Warsaw and with other gauges located in Poland and Germany.

1 Introduction

Urban hydrology requires access to very precise information about the precipitation variability over small spatial and temporal scales. Widespread use of surface runoff models coupled to urban drainage networks increases the common request for rainfall data inputs at high temporal and spatial resolutions. As already estimated a decade ago by Berne et al. (2004), the necessary resolution of rainfall data as the input in hydrological models in Mediterranean regions was about 5 min in time and 3 km in space for urban catchments of ~ 1000 ha. For smaller urban catchments of ~ 100 ha, even higher resolutions of 3 min and 2 km were required. Results obtained with the application of operational semi-distributed urban hydrology models fully confirmed earlier observations from select study cases in England and France (Gires et al., 2012, 2013). These authors strongly recommend the use of radar data in urban hydrology, especially in the context of real-time control of urban drainage systems. In particular, they opted for X band radars (whose resolution is hectometric), as opposed to the more common C band radars, because they are affected by less uncertainty. Additionally, Gires et al. (2012) stated that small-scale rainfall variability under 1 km resolution cannot be neglected and should be accounted for in probabilistic ways in the real-time management of urban drainage systems. As a matter of fact, the implementation of radar techniques gained a rising popularity in major cities across the EU (for details refer to the Thames Tideway Tunnel (TTT), 2010).

Despite the obvious benefits of radar instruments, radar data are not always available for practical applications. Thus, current versions of even the most advanced computer rainfall-runoff urban drainage models do not consider radar data as rainfall input. Therefore the only possibility of accounting for spatial rainfall variability is to consider different point time series for each subcatchment (Gires et al., 2012). The vast majority of engineering practical calculations and modeling of drainage systems is still associated with point rainfall time series or their elaborations, such as intensity–duration–frequency (IDF) curves, depth–duration–frequency (DDF) relations or simplified design hyetographs. This explains the necessity of high temporal resolution of point rainfall measurements in urban catchments. It should be noted that time series at high temporal resolution (1–10 min) with a considerable record length (at least 20–30 years) are nowadays required, especially from the European perspective with respect to the probabilistic assessment of the urban drainage network functioning (Schmitt, 2000; BS EN 752-3, 1997) or the probabilistic assessment of retention volumes at hydraulic-overloaded storm-water systems (Deutsche Vereinigung für Wasserwirtschaft, Abwasser und Abfall e.V., 2006).

The strategy of using local precipitation time series as the basis of the probabilistic assessment of urban drainage systems has two important shortcomings. In the case of local precipitation data shortage, this strategy fails completely. Whereas in all other situations when some local precipitation data sets are accessible, questions and doubts about the representativeness and reliability of data arise. First of all, we consider the doubts regarding the temporal representativeness of data: short data sets could not describe (Willems, 2013) the multi-decadal oscillatory behavior of rainfall extremes in storm-water outflow modeling. Other doubts regarding the spatial representativeness of data include the recording of rainfall time series only in a limited number of gauges installed in selected subcatchments. This results in assigning the same time series to a group of neighboring subcatchments or, in critical but not rare cases, one time series for the whole urban drainage system, habitually collected by a gauge installed near the airport. Sometimes in situations of local precipitation shortage, time series from other locations are allowed by technical guidelines (Schmitt, 2000) only if there is compatibility in terms of annual precipitation totals and IDF values.

Finally, since most of the modeling activity is oriented to predict the future behavior (e.g., in the next 50 years) of drainage systems, the mere use of historical precipitation time series of the last 20–30 years could not be significant to represent the future scenarios. Alternatively, the generation of synthetic time series from precipitation models could represent probable precipitation scenarios to feed hydrodynamic urban drainage models and take into account the uncertainty associated with the discharge. However, it should be pointed out that the information content of historical precipitation records is not increased by precipitation models and synthetic data generation, which just provide an operational basis for the extraction of such information.

Thus, there is a strong motivation for the development of local precipitation models at high temporal resolutions. Many of them are based on the idea of precipitation disaggregation in time. Disaggregation refers to a technique generating consistent rainfall time series at some desired fine timescale (e.g., 5 min resolution) starting from the precipitation at a coarser scale (e.g., daily resolution). At the same time, as stressed by Lombardo et al. (2012), the downscaling techniques aim at producing fine-scale rain time series with statistics consistent with those of observed data. A general overview of rainfall disaggregation methods is given by Koutsoyiannis (2003). Among an ensemble of known techniques, random cascade models, especially microcanonical cascade models (MCMs), are quite often used. The popularity of the latter could be explained by their appeal to engineering applications, the assumption of mass conservation (i.e., rainfall depth conservation) across cascade levels and straight rules for the extraction of cascade generators from local precipitation time series (Cârsteanu and Foufoula-Georgiou, 1996). Olsson (1998), Menabde and Sivapalan (2000), Ahrens (2003) and Paulson and Baxter (2007) provide contributions demonstrating the potentiality of MCMs in rainfall downscaling. Molnar and Burlando (2005) and Hingray and Ben Haha (2005) highlight the application of MCMs in urban hydrology. Hingray and Ben Haha (2005) apply a continuous hydrological simulation to produce from synthetic rainfall series continuous discharge series used afterwards for the retention design. Recently, Licznar (2013) illustrates the possibility of substituting synthetic time series generated from MCMs for observed time series of the probabilistic design of storm-water retention facilities.

Two decades of random cascade applications to precipitation disaggregation has progressed the construction of generators. Quite soon the assumption of independence and identical distribution of the cascade weight generators at all timescales was questioned and found suitable only for a limited, rather narrow range of analyzed scales (Olsson, 1998; Harris et al., 1998). As an alternative, Marshak et al. (1994), Menabde et al. (1997) and Harris et al. (1998) promote the use of the so-called "bounded" random cascade; its weights distribution systematically evolves, decreasing the weights variance with the reduction of timescale. In addition, Rupp et al. (2009) suggest that microcanonical cascade weights should not only be timescale-dependent but also intensity-dependent. The common practice of assuming the beta distribution for MCM generators is questioned by Licznar (2011a, b), especially for sub-hourly timescales. Alternatively, MCM generators are assumed normal–beta (N–B) distributed, with

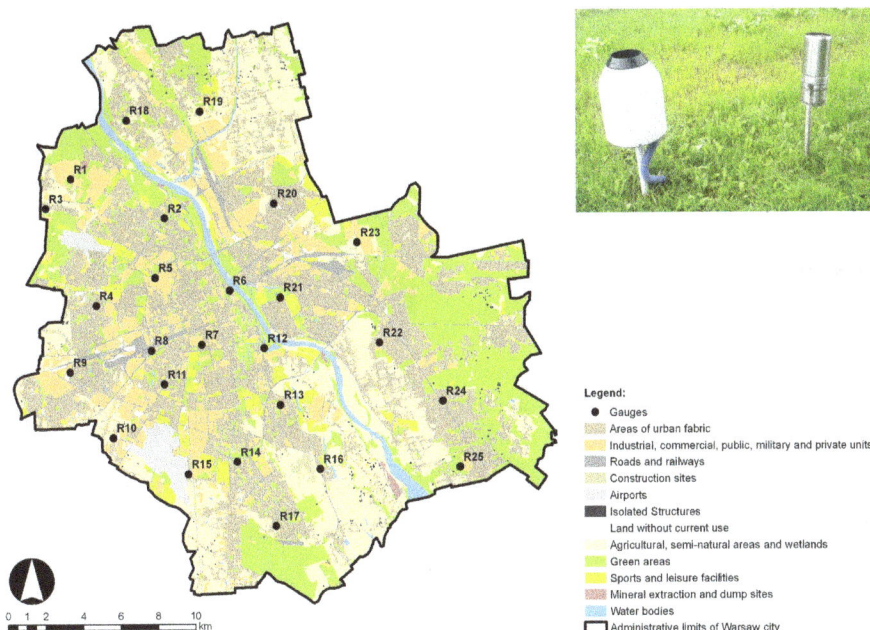

Figure 1. Map of 25 gauges composing the precipitation-monitoring network in Warsaw. Administrative limits of Warsaw city were marked in black. The land use classification was made with the Urban Atlas, which provides pan-European comparable land use and land cover data for large urban zones with more than 100 000 inhabitants (http://www.eea.europa.eu/data-and-maps/data/urban-atlas#tab-metadata). The average density of the network is one instrument for 20.7 km^2. MPS weighing-type TRwS 200E gauges were accompanied by standard Hellman gauges for the routine control of daily precipitation totals.

the atom at 0.5, or 3N–B distributed, composed of three N and one B distribution. For the sake of clarity, it should be stressed that B refers solely to the distribution of MCM generators and has nothing in common with the β model, the simplest cascade model often known as the monofractal model (for details refer to Over and Gupta, 1996).

Molnar and Burlando (2008) explore the variability of MCM generators on a large data set of 10 min time resolution, including 62 stations across Switzerland. These authors investigate seasonal and spatial variability in breakdown distributions to give indications concerning the parameters' estimation of MCM in ungauged locations. To our knowledge, there are only studies considering the large-scale variability (i.e., among different urban areas) of MCM generators, and there is a lack of knowledge concerning the small-scale variability (i.e., within an urban area).

It should be stressed that the fitting of cascade generators was relatively simple but extremely data-demanding. Observational precipitation time series in high resolution usually exceeding 20 years were unavoidable for the fit of the cascade parameters. This resulted in the prevailing practice of comparing the statistics of synthetic and observed time series. In the majority of studies, data originating from old-type manual gauges were subject to obvious uncertainty related to the precision of measurements as well as the resolution of records digitization. Simultaneously, the fitting of theoretical distributions to breakdown coefficients (BDCs) was, in

almost all cases, not supported by statistical tests confirming the correctness of achieved results or by the use of some information criteria to rank the theoretical distributions.

Keeping in mind the above discussed needs of urban hydrology, the current state of MCMs and severe limitations of this rainfall disaggregation technique, the goals of our study were the following:

1. Propose a methodology to calibrate microcanonical cascade generators in conditions of short time series;

2. Identify the probability distribution of BDCs through the use of information criterion;

3. Investigate the variability of empirical BDC distributions among a group of gauges;

4. Address the following questions of interest in urban hydrology: is it sufficient to use a single time series for the probabilistic assessment of the entire urban drainage system? Is it sufficient to fit just one MCM for the analysis of the whole city area? Could we continue the practice of supplying urban rainfall-runoff models by time series recorded outside city center by gauges located at the airport or over rural areas?

2 Data and methodology

2.1 Data

We used data belonging to a precipitation network of 25 gauges distributed across 517.24 km² of Warsaw, Poland (Fig. 1). The data set was the same used by Rupp et al. (2012) and consisted of a 1 min precipitation (both liquid and solid) time series recorded by electronic weighing-type gauges. All stations, TRwS 200E of MPS system Ltd (Fig. 2), were installed and operated by the Municipal Water Supply and Sewerage Company (MWSSC) in Warsaw. Prior to the network installation, studies about the location of the stations were done by the MWSSC to identify the configuration most representative of the precipitation variability within the urban area (Oke, 2006). Finding good places for the installation of gauges was possible due to the fact that the MWSSC in Warsaw operates a vast number of local water intakes and water- and sewage-pumping stations. Due to sanitary standards, all these installations had to occupy terrain with green areas to serve as buffers, e.g., for the spread of odors. In addition, all facilities were fenced and guarded for safety reasons. Therefore all instruments were placed on grass and the neighborhood met at least the requirements of class 2 or 3 as recommended by the WMO (2012). The majority of gauges (i.e., R1, R3, R5, R7, R8, R10, R12, R17, R18 and R19) were able to be installed on flat, horizontal surfaces, surrounded by an open area, thus meeting even requirements for class 1 instruments. In addition, gauge R15 was installed in perfect conditions on the ground at the Warsaw Fryderyk Chopin Airport.

Since the installation of the precipitation network in Warsaw was mainly motivated by the real-time control of the drainage system, all gauges (Fig. 1) were connected to a single data acquisition system. The accuracy of gauge measurements as claimed by the manufacturer was 0.1 %, and the data resolution was 0.001 mm for depth and 1 min for time. As previously mentioned by Rupp et al. (2012), field tests conducted prior to the operational use of the precipitation network have shown good agreement between simulated and recorded totals and have revealed a dampening/broadening of the input signal evident over the range of a few minutes. The last phenomenon – known as a "step response error" – was studied in detail in laboratory conditions for different gauge types by Lanza et al. (2005). They found that the step error of TRwS gauge was quite small in comparison to other gauges and equal to 3 min in laboratory conditions. Our short 15 min field test (as displayed on Fig. 2) suggested a dampening of the gauge-recorded signal for the first 3 min initial phase of the generated hyetograph and the slightly longer 5 min broadening at the final phase of hyetograph. Detailed discussion of the origins of gauge step response errors is beyond the scope of this paper and is in fact hard to realize since it is introduced by the gauge's inner microprocessor algorithm for data processing. This algorithm is known only by the gauge manufacturer and is not reported in the technical documentation.

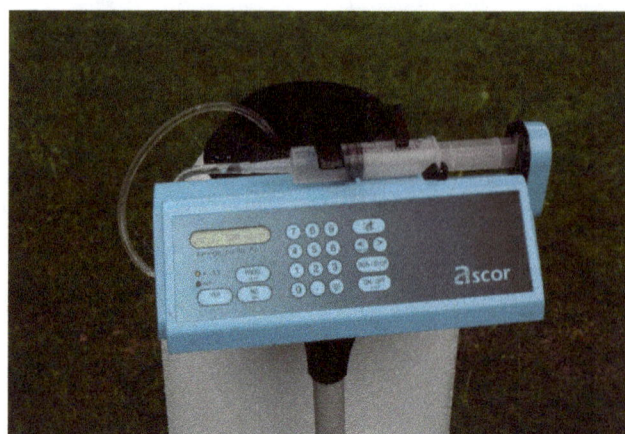

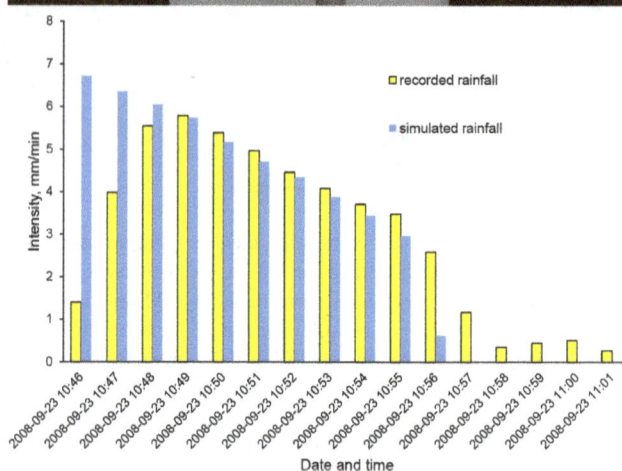

Figure 2. Weighing-type TRwS 200E gauge during some tests (upper panel). Rainfall is simulated by means of a precise medical pump. Sample of test results reporting simulated and recorded rainfall depths (lower panel).

In general, it could only be stated that in weighing-type electronic gauges, the weight of deposed precipitation was sampled by some electronic (often piezometric) sensor with some high temporal resolution at presumably kHz rate. Afterwards all samples were averaged over longer time windows, unknown to the user. This process was repeated for overlapping time windows, and the difference of the rainfall total of adjacent windows was calculated to obtain the temporal rainfall rate reported as instrument output at its recording time resolution. In addition, rainfall rates were always rounded regardless of the magnitude of real precipitation (resulting in additional rounding errors discussed afterwards). This procedure allowed for satisfactory smoothing of electronic sensor signal fluctuation due to wind effects and temperature changes. It allows for the introduction of some additional filters, cutting sudden signal jumps due to foreign-object deposition inside the open orifice of the gauge inner tank (e.g., falling leaves or acts of vandalism by throwing small stones or garbage).

Figure 3. Location of Polish and German precipitation gauges used during the comparison of Warsaw results with other studies.

In light of our personal experiences and test results of the WMO (Lanza et al., 2005), it could be stated that reliable precipitation recording at single minute scales by commercially available gauges is still the goal to be achieved and not a current reality. Having this in mind, as well as timescales of previous microcanonical cascade studies concerning urban hydrology realized on time series recorded by old-type gauges, we decided to work with the aggregated precipitation time series at 5 min resolution. The technique used to aggregate original 1 min data into 5 min time series is discussed afterwards; here we only mention that this operation was opposite to the rainfall total differentiation for adjacent time windows operated by the gauge microprocessor.

Despite the limited timespan of available data covering the period from the 38th week of 2008 to the 49th week of 2010, we believe that the Warsaw precipitation network might support good probing ground for the variability study in the microcanonical cascade parameters over small-scale urban areas. In fact, the Warsaw precipitation-monitoring network belongs to the biggest European urban gauge network. Its size can be compared only with similar networks of 25 gauges in Vienna (414.87 km^2) or with 24 gauges spread throughout Marseille (240.62 km^2) and Barcelona (100.4 km^2) (see TTT, 2010).

We compare the results of our study with those related to other Polish and German gauges. We limit our comparison to results previously published by Licznar et al. (2011a, b) for four gauges in Germany (gauges A, B, C and D represent local climates of different parts of western Germany) and one gauge in Wroclaw, Poland and yet unpublished results by Górski (2013) for a rain gauge in Kielce, Poland (Fig. 3). Our choice is motivated by the similarity of the used methodology

and the investigated range of timescales, as well as by the indispensable accessibility to precise recordings of the BDC histograms.

Finally, to investigate the existence of possible statistical bias induced by the calculation of BDCs on short precipitation records, we use additional data recorded by an old-type pluviograph gauge installed previously at the current location of gauge R7 on the ground of Lindley's Filters station. This pluviograph gauge was operated only in summer months from 1 May to 31 October. Data are in the form of 15 min rainfall time series read off the original paper strips with the resolution of 0.1 mm for depth, covering a period of 25 years (1983–2007).

2.2 Microcanonical cascade models

We use MCMs as in Licznar et al. (2011a, b). We consider the disaggregation of precipitation totals from 1280 min (quasi daily) into 5 min time series, assuming the branching number b equal to 2, and constructing cascades assembled from only nine levels ($n = 8, \ldots, 1, 0$) corresponding to timescales $\lambda = 2^n$ from $\lambda = 256$ to $\lambda = 1$ (Fig. 4). Precipitation-depth time series generated by such cascades are the products of the original precipitation total R_0 at timescale $\lambda = 256$, multiplied by the sequence of weights at the descending cascade levels:

$$R_{j,k} = R_0 \prod_{i=1}^{k} W_{f(i,j),i}, \tag{1}$$

where $j = 1, 2, \ldots 2^k - 1, 2^k$ marks the position in the time series at the kth cascade step. The sequence of randomly

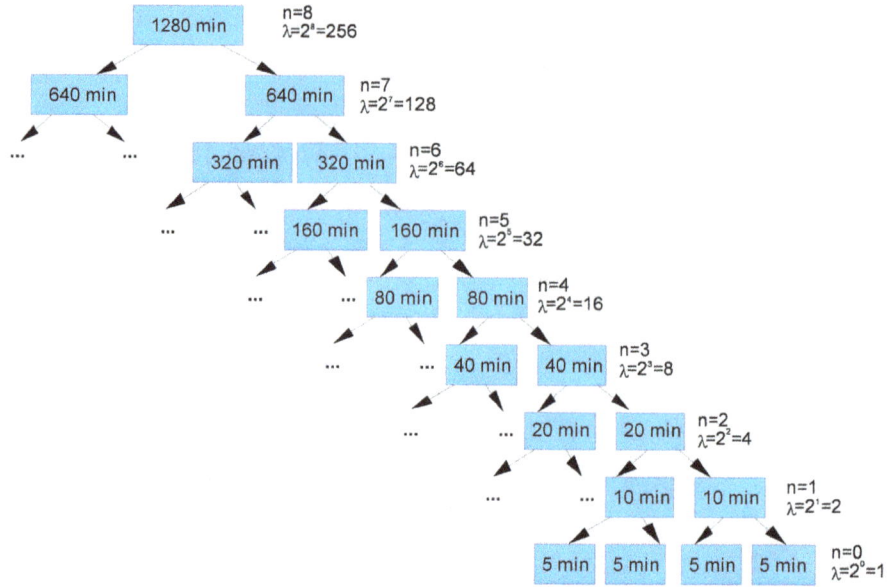

Figure 4. Schematic diagram of a developed microcanonical cascade model with branching number $b = 2$.

generated weights $W_{f(i,j),i}$ is steered at the following ith cascade step by the function $f(i,j)$, which rounds up $j/2^{k-i}$ to the closest integer. The weights in the microcanonical cascades are forced to sum to 1 so their pairs are always equal to W and $1-W$, where W is a two-sided truncated random variable from 0 to 1. The microcanonical assumption conserves the mass (precipitation depth in our case) at each branch and eliminates the risk of cascade degeneration. From an engineering perspective, this means that the downscaling process can be seen as opposite to precipitation summation realized by Hellman gauges, recording daily totals only, and a pragmatic solution for the generation of synthetic precipitation time series at 5 min resolution.

In our study we do not focus our attention on the disaggregation capabilities of microcanonical cascades already discussed in numerous papers. We concentrate on the small-scale variability of their generators W among gauges constituting the urban precipitation network. The obvious attractiveness of MCMs arises from the possibility of extracting the distribution of W from data on the basis of BDC studies (Cârsteanu and Foufoula-Georgiou, 1996). By definition, BDCs are generally calculated using non-overlapping adjacent pairs of precipitation time series:

$$\text{BDC}_{j,\tau} = \frac{R_{j,\tau}}{R_{j,\tau} + R_{j+1,\tau}} \qquad j = 1, 3, 5, \ldots, N_\tau - 1, \quad (2)$$

where $R_{j,\tau}$ is the precipitation amount for the time interval of length τ at position j in the time series, and N_τ is the length of time series at timescale τ. The calculation of BDCs with respect to Eq. (2) for Warsaw gauges is conducted only for non-zero pairs of R_j and R_{j+1}. Calculations are executed at aggregated intervals of length $2^n \tau_{\text{org}}$, where τ_{org} is the

original time step equal to 5 min and n is a cascade level, increasing from 0 to 8 with increasing cascade timescales λ from 1 to 256 (Fig. 4). Simultaneously for all analyzed timescales, BDC couples equal to 0/1 or 1/0 (when only one between R_j and R_{j+1} is 0) are separated from resulting data sets and their occurrence probabilities, $p_0(\text{LEFT})$ and $p_0(\text{RIGHT})$, respectively, are used to estimate intermittency probability p_0:

$$\Pr(\text{BDC}_n(j) = 0 \text{ or } \text{BDC}_n(j+1) = 0) \qquad (3)$$
$$= p_0(\text{LEFT}) + p_0(\text{RIGHT}) = p_0.$$

The probability p_0 is used within a MCM generator to take into account the intermittency characteristic of precipitation, forcing some portion of random weights W to be equal to 0.

The preliminary results have revealed an over-representation of BDC values equal to 1/2 or 1/3, 2/5 and 1/4 (2/3, 3/5 and 3/4, respectively), especially for small timescales, i.e., $\lambda = 1$ and $\lambda = 2$. Figure 5 (left panel) shows an example of BDC histogram for timescale $\lambda = 1$ with evident artificial spikes. Similar phenomenon was already reported by Rupp et al. (2009) and Licznar et al. (2011b) and explained as the result of the instrument or the recording precision of precipitation gauges. The magnitude of observed rounding errors for Warsaw gauges is, however, smaller than in the case of German gauges (Licznar et al., 2011b); the precipitation depths recorded with a better resolution of 0.001 mm still result in irregularity of BDC distribution, induced by sharp peaks at discrete BDC values, and hinder the identification of the theoretical distribution. In order to correct the rounding errors, a randomization procedure originally proposed by Licznar et al. (2011b) is applied. This type of procedure, also known as jittering,

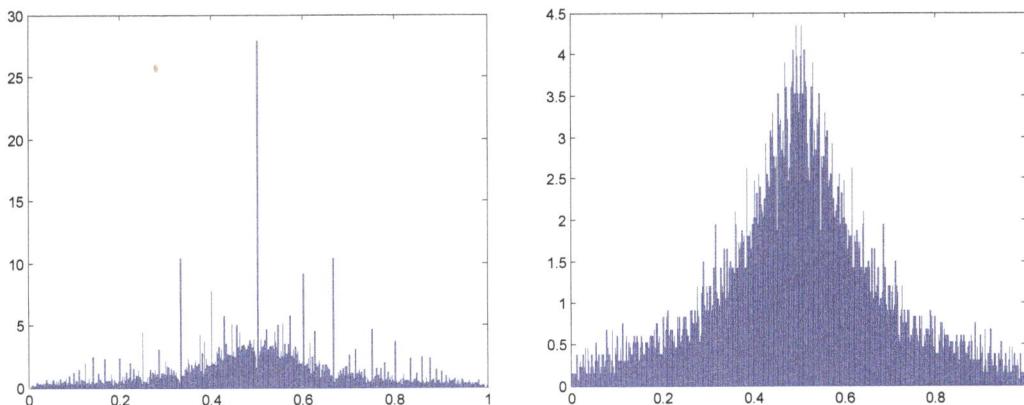

Figure 5. Comparison of BDC histograms for gauge R7 and timescale $\lambda = 1$, calculated according to the non-overlapping moving window algorithm and using original (left panel) and randomized (right panel) non-zero precipitation data. Horizontal axes show BDC range and vertical axes show the respective frequency values.

is fundamental to the analysis of data characterized by the presence of ties (De Michele et al., 2013). Thus, the original 1 min time series are slightly modified by adding some random corrections to the precipitation depths exceeding zero. Random correction values are sampled from the uniform distribution in the range $[-0.0005, 0.0005]$ mm, resulting in visible BDC histogram smoothing (Fig. 5, right panel). Note that the uniform distribution is used for the randomization of the rounding errors because, in the absence of information, it is the most intuitive distribution requiring less assumption (for more details, please see Licznar et al., 2011b).

Irregularities in BDC histograms are observed for timescales $\lambda > 8$. These are due to the decreasing sample size, calculated from a limited timespan of accessible data slightly exceeding 2 years. This issue is rather irrelevant in former studies (Molnar and Burlando 2005, 2008; Licznar et al., 2011a, b) realized on data series 10 or even 20 times longer. To solve this issue, we apply the overlapping moving window algorithm as an alternative to the classical non-overlapping moving window algorithm for the calculation of BDC values. Figure 6 shows the differences between the two algorithms for $\lambda = 1$. Switching from non-overlapping to overlapping moving window algorithm leads to an increase of the number of time segments for the calculation of BDC values. For time series of n data and a time window of size $m \leq n$, the number of non-overlapping windows is $\lfloor n/m \rfloor$, where the symbol $\lfloor \cdot \rfloor$ represents the integer part while the number of overlapping windows is $(n - m + 1)$. For large $n \gg m$, the overlapping moving window algorithm leads to almost m times the number of time segments available in the overlapping moving window algorithm. It should be stressed that the real strength of the overlapping moving window algorithm in analyzing distributions of BDC values can be observed for the largest timescales. The reason is that for small timescales most of the time segments are characterized by zero precipitation and thus not involved in the calculation

of BDCs, whereas for larger timescales, time segments are becoming larger and rarely characterized by zero precipitation. This phenomenon arises from the fractal properties of rainfall time series, and similar conclusions result from the "box-counting" analysis.

It is clear that the overlapping moving window algorithm is especially desired for limited observational data sets. However, its implementation for short time series may be characterized by a poor representativeness of BDC distributions due to multi-decadal oscillations of precipitation totals and extremes (Willems, 2013). To investigate the magnitude of the oscillations in the BDC distributions, we use historical time series from former old-type gauge R7 covering a 25-year period from 1983 to 2007 at 15 min resolution. For each year, there are only 6 months of data from May to October available. For this data set, we make the calculations of BDCs in seven time periods. First, we calculate BDCs for the 5-year periods 1983–1987, 1988–1992, 1993–1997, 1998–2002 and 2003–2007 using the overlapping moving window algorithm. We consider this temporal size (5 years × 6 months = 30 months) because it is comparable to the one available for electronic gauges. Afterwards, we repeat the same calculation with a 25-year size using both non-overlapping and overlapping moving window algorithms. As we work here with a coarser resolution (15 min instead of 5 min of electronic gauges), we perform the analysis with a smaller hierarchy of sub-daily timescales λ' from 1 to 32 and breakdown times from 15 to 30 min up to 480 to 960 min. For all calculations we perform the randomization of non-zero values. Since their reading precision was set to 0.1 mm, we introduce a random correction belonging to the uniform distribution in the range of $[-0.05, 0.05]$ mm.

To compare BDC histograms obtained for all analyzed timescales λ and λ', with theoretical functions, a probability distribution assembling two truncated (with truncation points at 0 and 1) N distributions (Robert, 1995) and one B

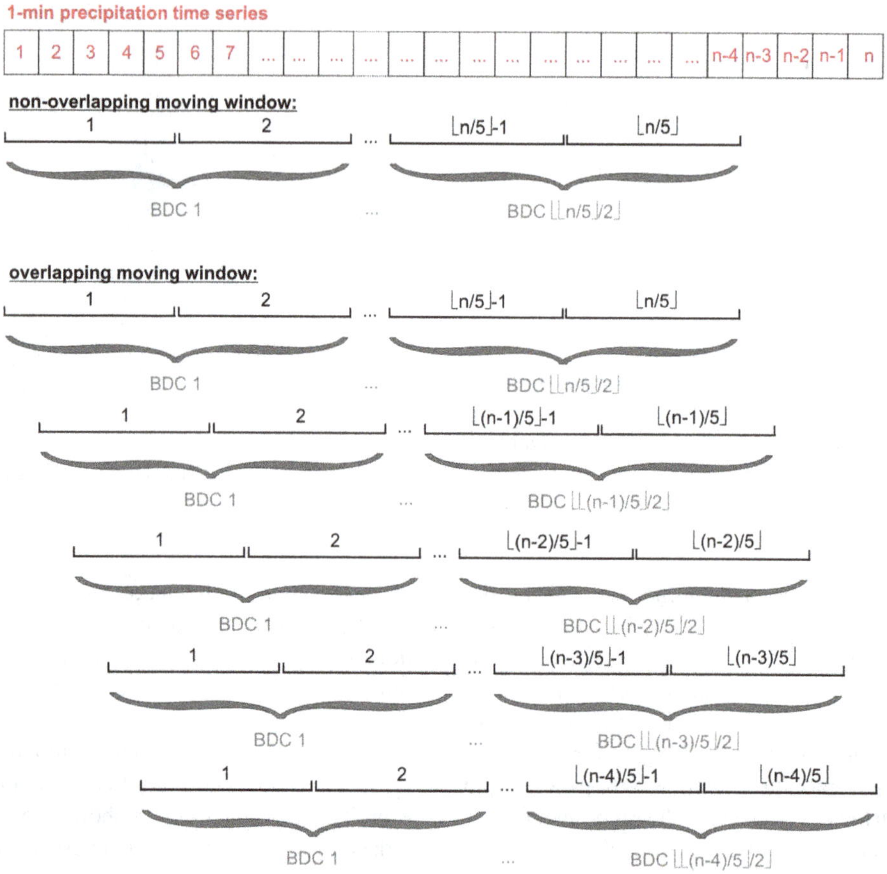

Figure 6. Example showing differences between non-overlapping and overlapping moving window algorithms for the calculation of BDCs in the case of 1 min precipitation time series and breakdown time of 5–10 min. Note that $\lfloor n \rfloor$ means the integer part of n, where n is the total length of 1 min precipitation time series.

symmetrical distribution is implemented. This distribution, indicated as 2N–B distribution, has the following density function:

$$p(w) = p_1 \left\{ \frac{1}{\sigma_1 \sqrt{2\pi}} e^{\frac{-(w-0,5)^2}{2\sigma_1^2}} \right\}$$
$$+ (1-p_1) \left\{ p_2 \left\{ \frac{1}{B(a)} w^{a-1}(1-w)^{a-1} \right\} \right.$$
$$\left. + (1-p_2) \left\{ \frac{1}{\sigma_2 \sqrt{2\pi}} e^{\frac{-(w-0,5)^2}{2\sigma_2^2}} \right\} \right\}, \tag{4}$$

where p_1 and p_2 are weights characterizing the contribution of the individual distributions within the 2N–B distribution, σ_1 and σ_2 are the scale parameters of truncated N distributions and $B(a)$ is the symmetrical B function parameterized by a.

The fitting of 2N–B distribution parameters is performed numerically by means of maximum likelihood estimation. It is very likely that the use of the model given in Eq. (4), governed by five parameters, can suffer from overparameterization in comparison to the most commonly used

B symmetrical distribution with only one parameter. Note that the application of goodness-of-fit tests (namely the Kolmogorov–Smirnov test or χ^2 test) at 1 or 5 % levels of significance gives negative results for both B and 2N–B distributions. This is because the large sample size of empirical BDCs leads to the rejection of the hypothesis, even in the case of very small differences between observed and theoretical distributions, as also pointed out in Licznar et al. (2011a). Here, we use the Akaike information criterion (AIC) as a measure of the relative quality of 2N–B and B models for given sets of empirical BDCs. AIC is the maximized value of the log-likelihood function (LL) penalized by the number of model parameters k:

$$\text{AIC} = 2k - 2\text{LL}. \tag{5}$$

The preferred distribution is the one with the minimum value of AIC.

2.3 Cluster analysis

To our knowledge, until now, the variability of MCM generators among a group of gauges was investigated comparing the value of the parameter of B distribution (Molnar and Burlando, 2008). Here, we prefer to compare directly the empirical distribution of BDCs instead of the parameters of the theoretical distribution, which are possibly biased by fitting errors. We encounter the same problems found in the implementation of statistical tests due to the large sample size. For this we use the cluster analysis to compare the shape of BDC histograms among the stations of the monitoring network in Warsaw and with other Polish and German gauges.

In particular, a hierarchical clustering is used. This is a data-mining tool applied to segment data into relatively homogeneous subgroups, or clusters, where the similarity of the records within the cluster is maximized (Larose, 2005). Prior to the application of the cluster analysis for each timescale and each site, the BDC histogram is sampled in 100 points selected at equal distance one from each other. These 100 values are the components of a vector representing the empirical BDC distribution. Note that a basic requirement of cluster analysis is the comparison of records of equal length. As all BDC distributions are left and right truncated in the interval (0, 1), sampling their histograms with a resolution of 0.01 produces vectors that describe well the shape of histograms. The clustering of these vectors (searching similar sites) is operated using the Euclidean distance. It is computed as

$$d_{Euclidean}(X, Y) = \sqrt{\sum_i (x_i - y_i)^2}, \qquad (6)$$

where x_i and y_i with $i = 1, \ldots, 100$ represent the ith component of X and Y vectors, respectively.

The Euclidean distance is a measure of similarity, not having, in general, a physical interpretation. Initially, in hierarchical clustering analysis, each vector is considered to be a tiny cluster of its own. Then, in following steps, the two closest clusters are aggregated into a new combined cluster. By replication of this operation, the number of clusters is reduced by one at each step and eventually sites are combined into a single huge cluster. During the agglomerative process, the distance between clusters is determined based on single-linkage criterion. In this case, the distance between clusters A and B is defined as the minimum distance between any element in cluster A and any element in cluster B. This single linkage is often termed the nearest-neighbor approach and tends to form long, slender clusters, clearly indicating similarities among clustered elements. As a final result of agglomerative clustering, a treelike cluster structure (named dendrogram) is created.

Dendrograms show similarities as well as dissimilarities between BDC distributions among the considered sites and they are prepared separately for all analyzed timescales.

In addition, the cluster analysis is also applied to the intermittency parameter, in this case comparing vectors of eight components: the p_0 value for the eight timescales $\lambda = 1, 2, 4, 8, 16, 32, 64, 128$.

3 Results and Discussion

Results are presented relative to gauge R7 for brevity. This station has been selected because of its localization in the strict city center, its installation in perfect meteorological conditions on the ground and the existence of former historical rainfall records. Results for the other gauges are qualitatively similar to those shown for R7.

3.1 Empirical BDC distributions

BDC histograms are calculated using the non-overlapping moving window algorithm and plotted in Fig. 7 for gauge R7 and a sequence of analyzed breakdown times. It is clearly visible that, despite the randomization procedure that removes pronounced peaks of histograms at certain specific BDC values, like 0.5 or 1/3, 2/5, 1/4 and 2/3, 3/5, 3/4, respectively (Fig. 5), the plots remain irregular, especially for timescales exceeding $\lambda = 8$, reducing the possibility of identifying the proper theoretical distribution. Visible irregularities of BDC histograms increase with increasing timescales, which is an obvious effect of decreasing data sets and thus decreasing populations of calculated BDC values do not allow the production of histograms of fine bins resolution (referring to the populations). Similarly, Fig. 8 reports the distributions of BDC calculated through the overlapping moving window algorithm. The comparison between Figs. 7 and 8 shows how the change of algorithm from a non-overlapping to an overlapping moving window brings evident smoothing of BDC histograms, occurring not only at larger timescales but also at small timescales. Note that the smoothness of BDC histograms in Fig. 8 is comparable with the quality of BDC histograms showed by Licznar et al. (2011b) for German gauges, derived using non-overlapping moving window algorithm for much longer precipitation time series ranging from 27 to 46 years of continuous records. The introduction of the overlapping moving window algorithm allows for the fitting of MCM parameters with the availability of extremely short time series (i.e., 2 years long) in the case of Warsaw gauges. The overall acceptance of overlapping moving window algorithm implementation, including for short rainfall time series, is discussed in Sect. 3.3.

3.2 Theoretical BDC distributions and their evolution along timescales

In Fig. 8 we also report the fitted theoretical distributions (2N–B distribution in solid red curves and B distribution in blue dashed lines) for each timescale considered. The visual comparison clearly indicates a better fit of 2N–B (or N–B in

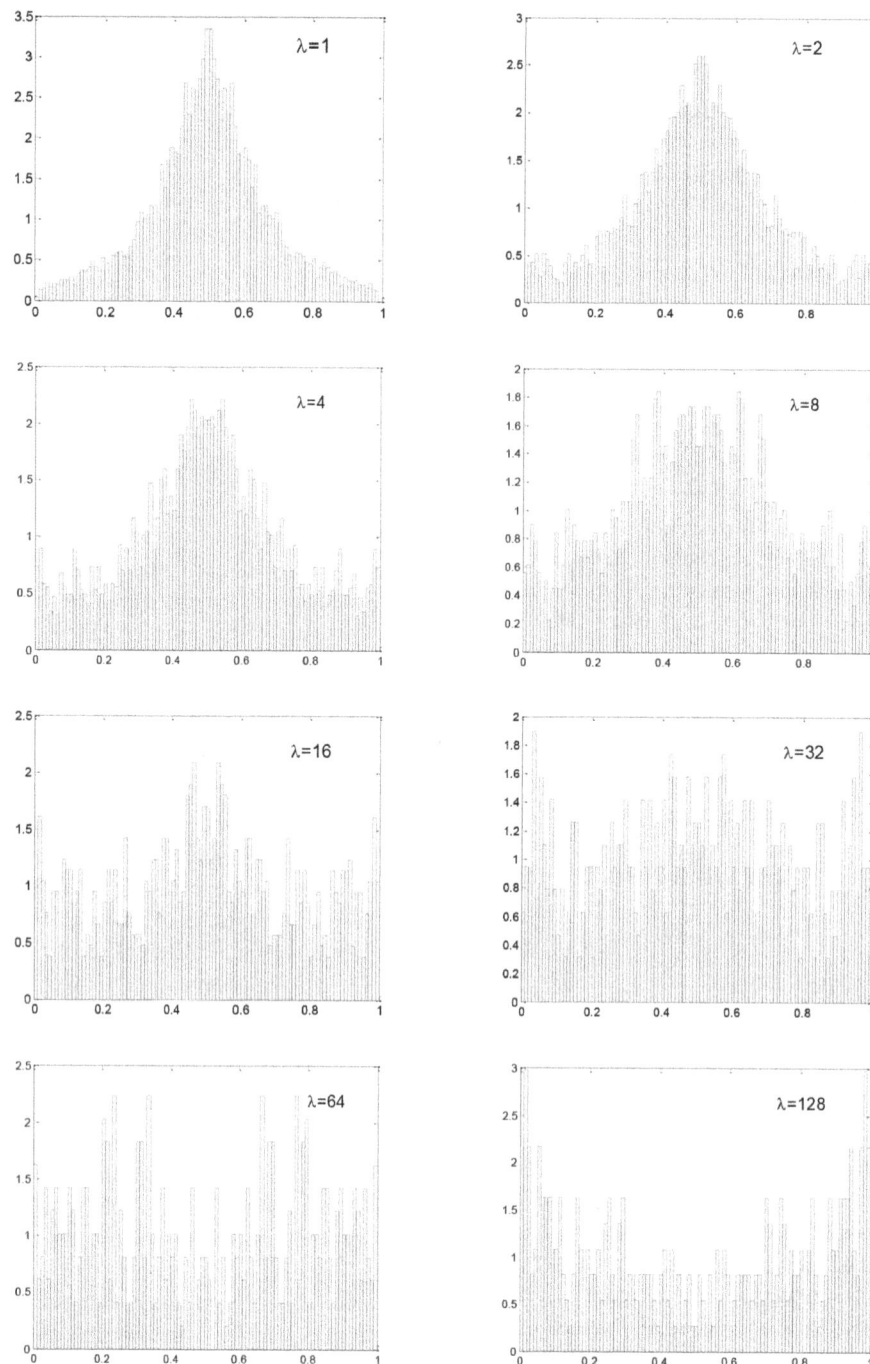

Figure 7. Histograms of BDC values for gauge R7 calculated according to the non-overlapping moving window algorithm and based on randomized precipitation time series. Horizontal axes show BDC range and vertical axes show the respective frequency values.

some cases) distribution for timescales smaller than $\lambda = 64$. In Fig. 8 it is possible to see how the distribution with the best fit changes from a B distribution at $\lambda = 128$ to a joined 2N–B for the smallest value of λ through a N–B distributions. This is in agreement with previous studies by Licznar et al. (2011a, b). This observation is supported by higher values of log-likelihood for 2N–B distribution (or the simplified N–B) in comparison to the B distribution (Table 2). These differences are in the range of thousands, and even after accounting for the number of model parameters, the AIC for 2N–B (or the simplified N–B) distributions are much smaller (or equal) than that of B distributions, confirming the visual result given in Fig. 8. Based on this, we prefer the 2N–B distribution with respect to the B distribution except for the case

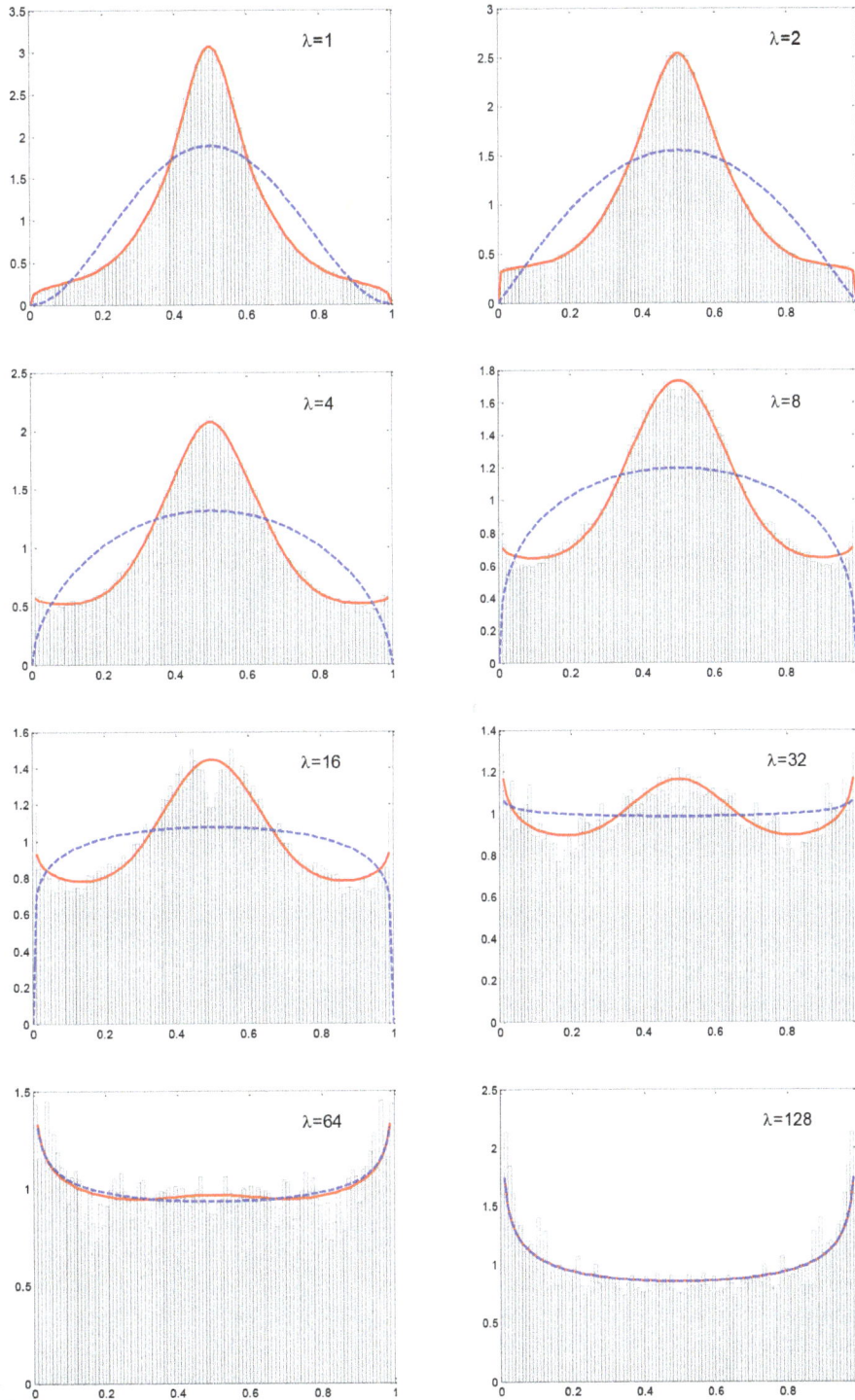

Figure 8. Histograms of BDC values calculated according to overlapping moving window algorithm and based on randomized gauge R7 precipitation times series. Horizontal axes show BDC range and vertical axes show the respective frequency values. The solid red curves represent the 2N–B probability density function, whereas the blue dashed curves represent the B probability density function.

Table 1. Values of p_1, p_2, a, σ_1 and σ_2 parameters at different timescales for gauge R7. The values of parameters are reported in bold, whereas their 95 % confidence limits are in italic.

Breakdown times	Timescale	p_1	p_2	a	σ_1	σ_2
5–10 min	$\lambda = 1$	**0.1541**	**0.3479**	**1.3350**	**0.0559**	**0.1341**
		0.1474	*0.3377*	*1.3097*	*0.0523*	*0.1300*
		0.1608	*0.3580*	*1.3604*	*0.0595*	*0.1383*
10–20 min	$\lambda = 2$	**0.0706**	**0.4036**	**1.0632**	**0.0559**	**0.1341**
		0.0644	*0.3950*	*1.0474*	*0.0523*	*0.1300*
		0.0768	*0.4121*	*1.0789*	*0.0595*	*0.1383*
20–40 min	$\lambda = 4$	**0.0212**	**0.5036**	**0.9437**	**0.0559**	**0.1341**
		0.0155	*0.4954*	*0.9325*	*0.0523*	*0.1300*
		0.0270	*0.5118*	*0.9548*	*0.0595*	*0.1383*
40–80 min	$\lambda = 8$	–	**0.6175**	**0.9484**	–	**0.1341**
		–	*0.6091*	*0.9390*	–	*0.1300*
		–	*0.6259*	*0.9579*	–	*0.1383*
80–160 min	$\lambda = 16$	–	**0.7548**	**0.9170**	–	**0.1341**
		–	*0.7494*	*0.9098*	–	*0.1300*
		–	*0.7601*	*0.9242*	–	*0.1383*
160–320 min	$\lambda = 32$	–	**0.8873**	**0.8929**	–	**0.1341**
		–	*0.8827*	*0.8873*	–	*0.1300*
		–	*0.8919*	*0.8985*	–	*0.1383*
320–640 min	$\lambda = 64$	–	**0.9797**	**0.8799**	–	**0.1341**
		–	*0.9758*	*0.8754*	–	*0.1300*
		–	*0.9835*	*0.8843*	–	*0.1383*
640–1280 min	$\lambda = 128$	–	**1.0000**	**0.7783**	–	**0.1341**
		–	*0.9973*	*0.7754*	–	*0.1300*
		–	*1.0027*	*0.7813*	–	*0.1383*

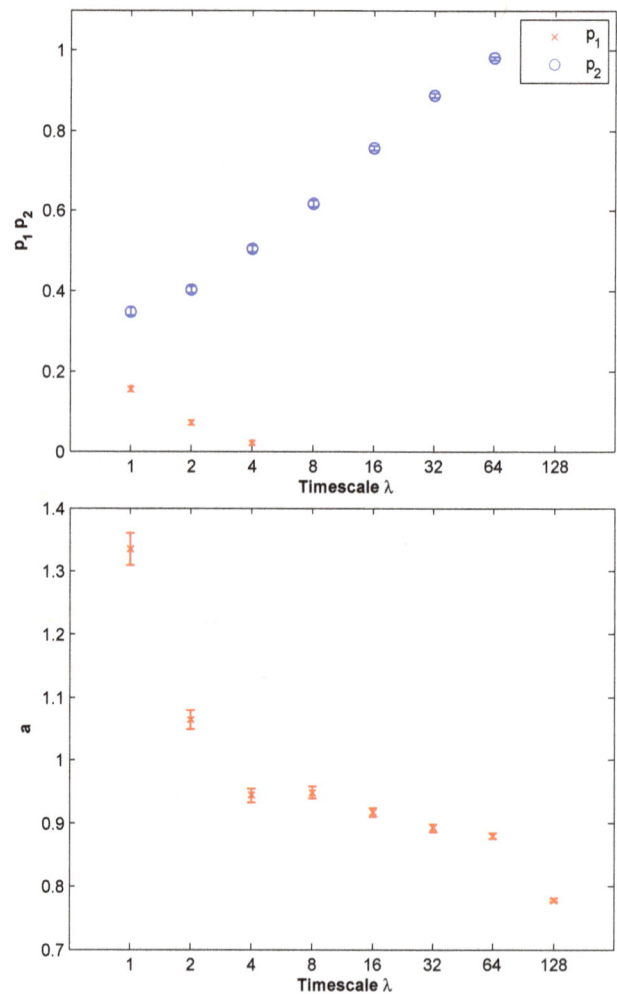

Figure 9. Value and 95 % confidence intervals of parameters of p_1, p_2 and a with λ for gauge R7. Horizontal axes are plotted at binary logarithm scale $\log_2$.

$\lambda = 128$. Analyzing the data reported in Table 2, it is worth noting the systematic increase of sample size n increasing the timescale.

From a practical point of view, a rapid increase in the number of BDCs, equal or close to 0.5, decreasing the timescale should be expected as a symptom of enclosing a limit of the precipitation temporal variability in a point by accessible instruments. The precipitation averaging over some small area of orifice and time intervals is inevitable for gauges; thus, for small timescales, most of the small-scale precipitation variability remains undetected and smoothed, leading to an over-representation of constant precipitation time intervals. From a theoretical point of view, it should be noted that bounded cascades allow the multiplicative weights (or precisely their distributions) to depend on the cascade level and converge to unity as the cascade proceeds. As a consequence, the simulated random process becomes smoother on smaller timescales (Lombardo et al., 2012), which in general mimics the dynamics of precipitation collected by gauges. In other words, as postulated by Marshak et al. (1994), Menabde et al. (1997) and Harris et al. (1998), the variance of weights reduces with every descending cascade level. As a simple extension of this rule, the increasing frequency of weights at the central part of their distribution plots has to be observed. The increase in the number of BDCs equal or close to 0.5 with decreasing timescale is well illustrated by empirical his-

tograms in well-known pioneering contributions to MCM applications for rainfall time series disaggregation published by Olsson (1998), Menabde and Sivapalan (2000) and Güntner et al. (2001). Quite recently, this behavior was also proved to be rainfall-intensity-dependent by Rupp et al. (2009).

For each analyzed timescale, we estimate the parameters of 2N–B probability distribution (or its simplifications N–B and B): p_1, p_2, a, σ_1 and σ_2. Table 1 gives the values for gauge R7 with their 95 % confidence limits. A good visual fit of empirical BDC distributions in Fig. 8 corresponds to quite narrow 95 % confidence limits of the fitted parameters (mostly invisible in Fig. 9 plots). The 95 % confidence limits do not exceed more than a few percent of the estimated values, with the sole exception of parameter p_1 for $\lambda = 4$, where the differences range up to 27 %. Additionally, the scale parameters of N distributions, σ_1 and σ_2, appear to be constant among analyzed timescales not only for gauge R7 but also for the other Warsaw gauges.

Table 2. Values of the Akaike information criterion (AIC) for the 2N–B distribution (model 1) – or its simplifications, N–B and B – and the B distribution (model 2) and the hierarchy of analyzed timescales λ at gauge R7. Calculations were based on estimates of the maximized value of the log-likelihood function (LL), known sample size (n) and number of model parameters (k).

Breakdown times	Timescale	n	Model 1 Distr.	k	LL	AIC(M1)	Model 2 Distr.	k	LL	AIC(M2)	Δ =AIC(M2) − AIC(M1)
5–10 min	$\lambda = 1$	132 940	2N–B	5	48 480	−96 950	B	1	36 307	−72 612	24 338
10–20 min	$\lambda = 2$	136 968	2N–B	5	32 272	−64 534	B	1	19 798	−39 593	24 941
20–40 min	$\lambda = 4$	144 778	2N–B	5	19 071	−38 132	B	1	8794	−17 585	20 547
40–80 min	$\lambda = 8$	159 272	N–B	3	11 119	−22 232	B	1	4464	−8927	13 305
80–160 min	$\lambda = 16$	185 014	N–B	3	4591.9	−9178	B	1	925	−1848	7330
160–320 min	$\lambda = 32$	230 716	N–B	3	1167.3	−2329	B	1	46	−91	2238
320–640 min	$\lambda = 64$	315 360	N–B	3	1543.70	−3081	B	1	1491	−2979	102
640–1280 min	$\lambda = 128$	501 092	B	1	12 614.40	−25 227	B	1	12 614	−25 227	0

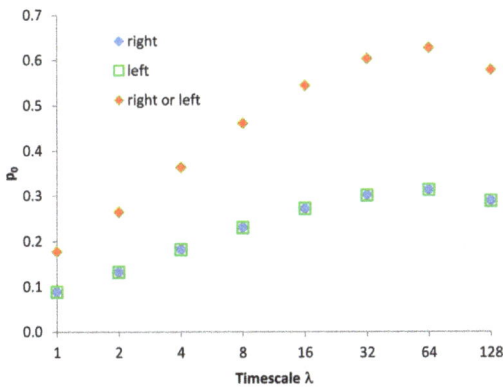

Figure 10. Variability of the intermittency parameter p_0 with λ for gauge R7. Horizontal axis is plotted at binary logarithm scale $\log_2$.

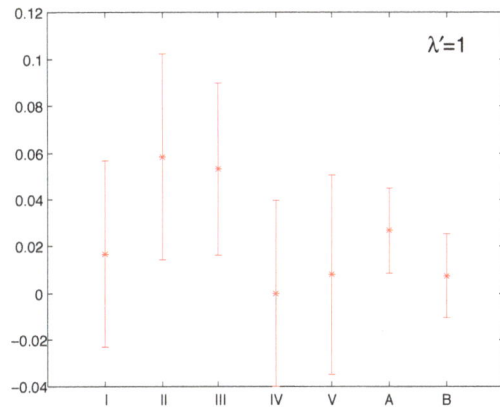

Figure 11. Value and 95 % confidence intervals of parameter p_1 at timescale $\lambda' = 1$ for gauge R7. Roman numerals I–V on horizontal axes indicate the 5-year ranges: 1983–1987, 1988–1992, 1993–1997, 1998–2002 and 2003–2007. Uppercase letters A and B indicate values calculated using the entire 25-year range of 1983–2007 and non-overlapping (A) and overlapping (B) moving window algorithm.

The variability of p_1, p_2 and a with λ is presented in Fig. 9 for gauge R7. A systematic decrease of p_1 down to 0 increasing the timescale is observed, denoting a decreasing importance of the first N within the 2N–B distribution. An opposite systematic increase of p_2 up to 1 increasing the timescale is observed, denoting a decreasing importance of the second N within the 2N–B distribution. The evolution of the B parameter a shows a fast reduction, with values below 1 noticed for the smallest scales, changing the B distribution shape from convex to concave. At larger timescales the reduction of a is hardly visible, with the sole exception of $\lambda = 128$. Figure 10 shows the variability of intermittency parameters p_0 with timescale λ. For all of them, the values of p_0(LEFT) match the values of p_0(RIGHT), which is in good agreement with previous studies by Molnar and Burlando (2005) and Licznar et al. (2011a, b). This could be interpreted as the proof of the fully random occurrence of intermittency in the precipitation time series. A systematical increase of p_0 with λ is observed, with the sole exception of a small drop at $\lambda = 128$. General increase of p_0 with timescale is a natural

outcome of the fractal properties of the geometric support of rainfall occurrence.

3.3 Performance of the overlapping moving window algorithm

The performance of the overlapping moving window algorithm was investigated in detail at gauge R7, where a 25-year time series at 15 min resolution was available. We calculated the parameters of 2N–B distribution for the hierarchy of sub-daily timescales λ' relative to the 5-year periods of 1983–1987, 1988–1992, 1993–1997, 1998–2002 and 2003–2007 (indicated afterwards with the Roman numerals I, II,..., V, respectively) and the whole 25-year data set (indicated in the next as case A) using the overlapping moving window algorithm. In addition, we calculate the parameters of 2N–B

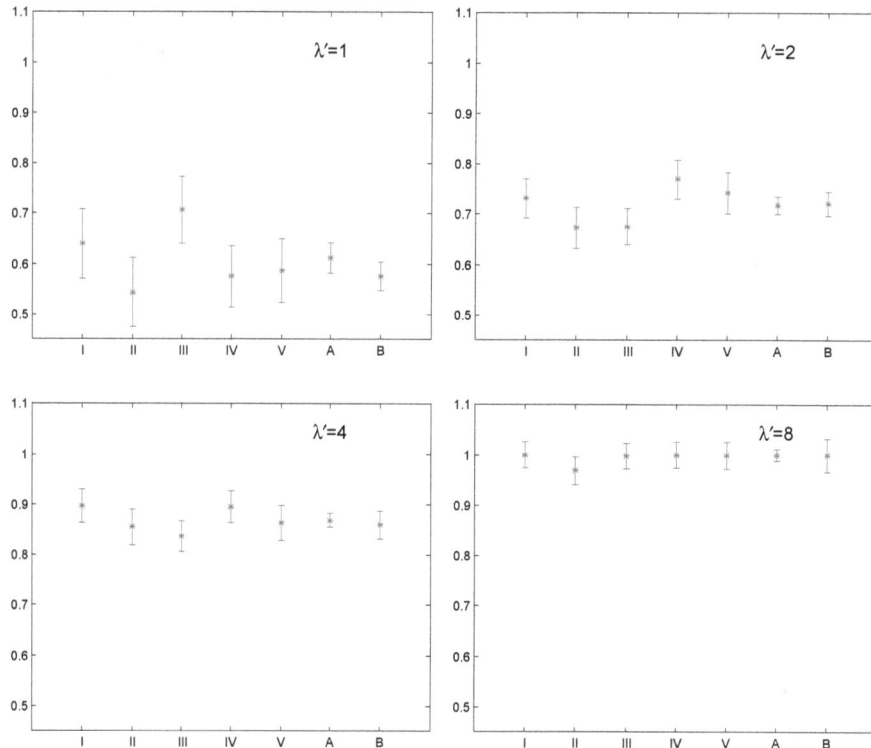

Figure 12. Value and 95 % confidence intervals of parameter p_2 at timescales $\lambda' = 1, 2, 4, 8$ for gauge R7. Roman numerals I–V on horizontal axes indicate respectively the 5-year ranges: 1983–1987, 1988–1992, 1993–1997, 1998–2002 and 2003–2007. Uppercase letters A and B indicate values calculated using the entire 25-year range of 1983–2007 and non-overlapping (A) and overlapping (B) moving window algorithm.

distribution also using the classical non-overlapping moving window algorithm over the whole 25-year data set (indicated in the next as case B). The results are shown in Figs. 11–13.

In general, the selected probability distribution is a B distribution for the largest timescales ($\lambda' = 16, 32$), a N–B distribution for $\lambda' = 2, 4, 8$ and a 2N–B distribution for $\lambda' = 1$ (the only exception is the period IV). The above listed timescales λ' are not compatible with timescales λ; however, transposing them on a coherent time axis leads to the conclusion that characteristic transitions from B to N–B and 2N–B distributions occur at approximately the same time ranges. The estimated parameters σ_1 and σ_2 appear to be constant among analyzed timescales and equal to 0.0646 and 0.1363, respectively. These values are very close to those reported in Table 1. Figure 11 shows the estimates of p_1 for $\lambda' = 1$, with a variability in the range 0–0.058 for the 5-year periods I–V. At the same time, the 95 % confidence limits of p_1 overlap each other partially and values estimated for cases A and B. Confidence limits for periods I–V are rather wide and are only reduced by 50 % for cases A and B. Note that here we work with 15 min time series and not 1 min time series as before.

A better agreement is observed for larger timescales, as illustrated in Figs. 12 and 13, with visibly narrow 95 % confidence limits; however, they still partial overlap one another.

For smaller timescales, larger oscillations of p_2 parameter can be observed over the periods I–V but, due to wider 95 % confidence limits, they overlap one another and those relative to cases A and B. The only exception is found for the period III at timescale $\lambda' = 1$.

For parameter a and $\lambda' = 1$, 95 % confidence limits for all calculations overlap except period V, which has slightly lower values. For $\lambda' = 2$ and $\lambda' = 4$, mutual overlay of 95 % confidence limits is noticed. Passing to $\lambda' = 8$ and $\lambda' = 16$, the overlapping among all pairs of periods from I to V is not always present; however, it is present with 95 % confidence limits drawn for case B. For $\lambda' = 32$, 95 % confidence limits for periods I–V and case A are extremely narrow.

Results reported above suggest good repeatability of BDC distributions calculated during all periods; this is graphically confirmed in Fig. 14, with the exception of period II and timescale $\lambda' = 1$. Probably this could be explained by the poor performance of the newly proposed overlapping moving window algorithm applied to low time resolution of the original time series. Our observations support the use of the overlapping moving window algorithm for BDC calculations in situations of short (about 2 years) precipitation time series access, while in previous microcanonical cascade studies (e.g., Molnar and Burlando, 2005, 2008), longer (e.g., about 20–30 years) time series are indispensable. In addi-

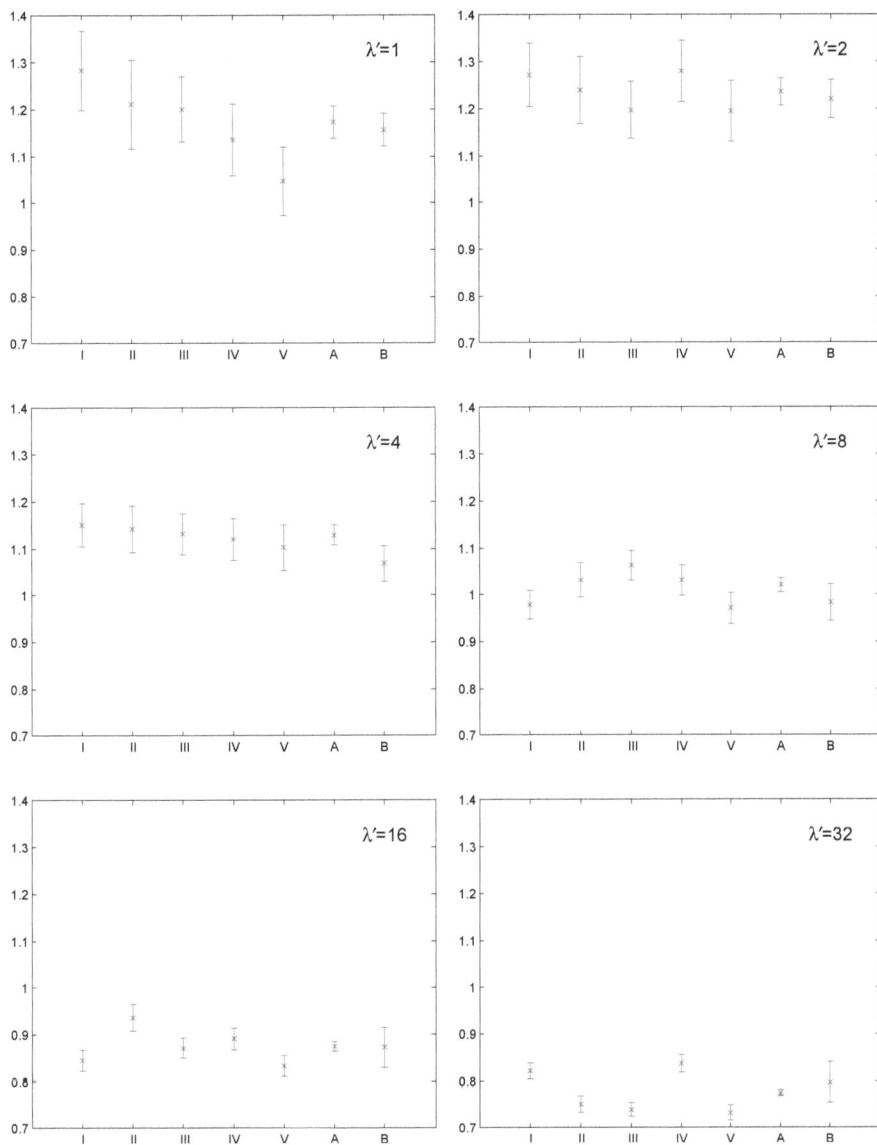

Figure 13. Value and 95 % confidence intervals of parameter a at timescales $\lambda' = 1, 2, 4, 8, 16, 32$ for gauge R7. Roman numerals I–IV on horizontal axes indicate the 5-year ranges: 1983–1987, 1988–1992, 1993–1997, 1998–2002 and 2003–2007. Uppercase letters A and B indicate values calculated using all 25-year range of 1983–2007 and non-overlapping (A) and overlapping (B) moving window algorithm.

tion, even in situations of longer precipitation time series access, BDC calculations by means of the proposed algorithm should be favored over the old non-overlapping moving windowtechnique because the new algorithm leads to narrowed 95 % confidence intervals of fitted BDC distributions parameters.

We do not claim here that the moving window technique combined with MCMs solves the problem of local precipitation time series shortage. It is obvious that rainfall statistics derived from short periods may be biased against long-term statistics (e.g., due to climate oscillations). Until now, to our best knowledge, there have been no attempts made to assess the possible bias of MCM generators due to precipitation os-

cillations driven by climate change. Hitherto contributions of MCM generators are mostly based on precipitation series that are not too long, presumably displaying only very weak, if any, oscillations and are always treated as a single data set.

Possible bias of MCM generators due to precipitation oscillations undoubtedly should be verified on other, much longer time series of better resolution, such as the 10 min time series collected in Uccle, Belgium (Willems, 2013). Simultaneously, only a detailed analysis based on long and complete precipitation time series covering at least a few decades could deliver us the answer to the question of whether the climate change effects could be retrieved via the temporal evaluation of microcanonical cascade generators.

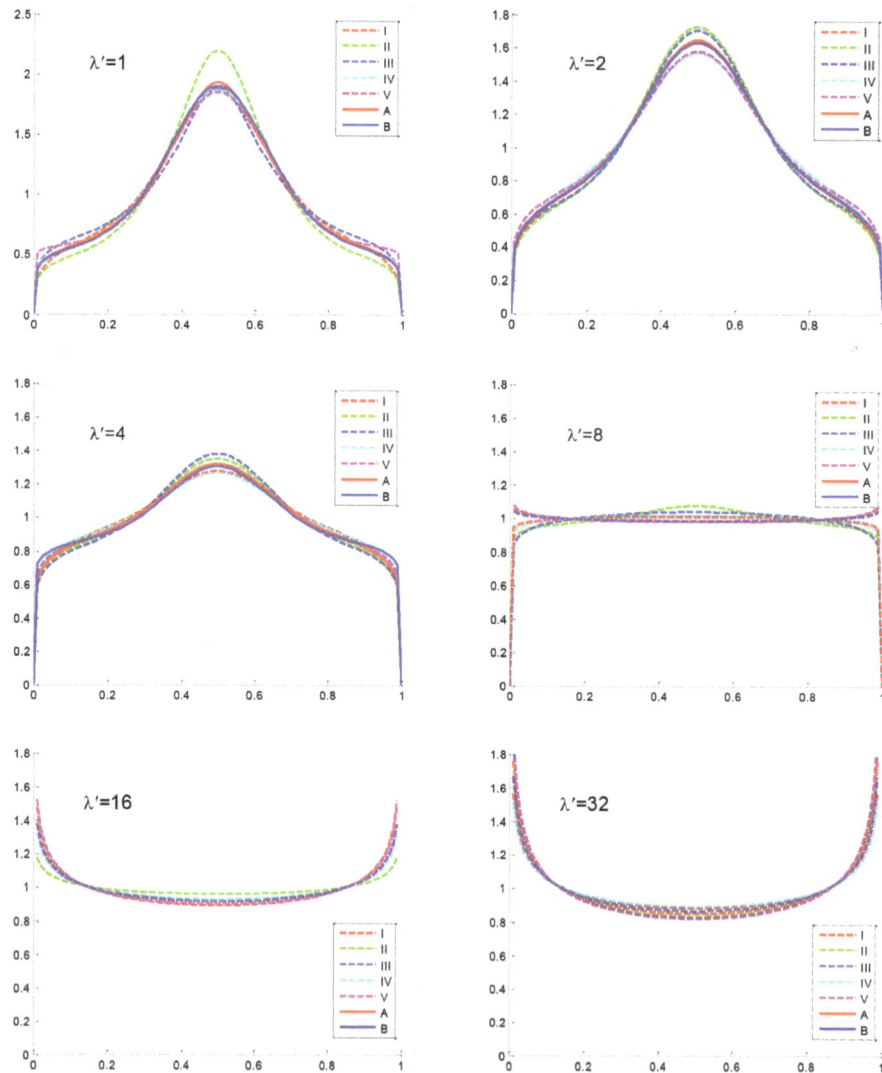

Figure 14. Variability of fitted theoretical BDC distributions histograms at timescales $\lambda' = 1, 2, 4, 8, 16, 32$ for gauge R7. Roman numerals I–V in legend indicate the 5-year ranges: 1983–1987, 1988–1992, 1993–1997, 1998–2002 and 2003–2007. Uppercase letters A and B indicate results calculated using all 25-year range of 1983–2007 and non-overlapping (A) and overlapping (B) moving window algorithm. In all plots, horizontal axes show BDC ranges and vertical axes show the frequency values.

From this perspective, the moving window technique could be of considerable usefulness in BDC distributions fitting for periods corresponding to 11-year solar spot cycles.

3.4 Performance of microcanonical cascade in disaggregation

As an additional check of the overall performance of the applied techniques (i.e., the randomization procedure, the overlapping moving window algorithm and the 2N-B probability distribution), we test the performance of microcanonical cascade in disaggregating the precipitation at the analyzed gauges. The MCM is used to generate 100 synthetic time series at 5 min resolution on the basis of the observed 1280 min precipitation totals (similar to Molnar and Bur-

lando, 2005; Licznar et al., 2011a, b). To evaluate the goodness of disaggregation, we compare the probability of zero precipitation at synthetic and observed time series for all analyzed timescales. Moreover, we calculate the survival probability function of non-zero synthetic precipitation amounts and compare it to the survival probability function of observed precipitation amounts. This analysis is limited to 5 min data, i.e., terminal results of the disaggregation most suitable for urban hydrology application. Special attention to the 5 min synthetic time series was also paid by other researchers (see e.g., Molnar and Burlando, 2005, 2008; Licznar et al., 2011a, b). An example of 56.3 mm event disaggregation is plotted in Fig. 15 for gauge R7. It should be stressed that the structure of the synthetic time series is composed

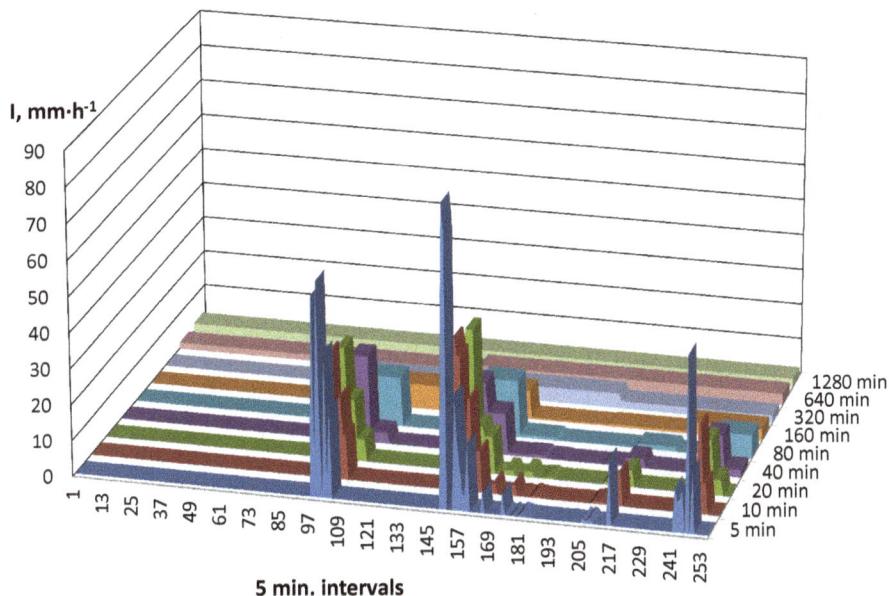

Figure 15. An example of precipitation disaggregation of a 56.3 mm event from 1280 to 5 min for gauge R7.

of uncorrelated segments like the one presented in Fig. 15. Thus, the synthetic time series is missing the correct autocorrelation structure of natural precipitation (for detail discussion see Lombardo et al., 2012). The expected value of the zero-precipitation probability $E(p_0)$ for observed and generated series is given in Fig. 16 for gauge R7. The synthetic values of $E(p_0)$ are calculated as the average over 100 MCM disaggregations. The differences in terms of $E(p_0)$ between observed and simulated are negligible (see Fig. 16). In addition, for comparison we also give the synthetic values of $E(p_0)$ for gauges R15 and R25.

Figure 17 shows the comparison between observed and simulated survival probability function of rainfall amount at 5 min for gauge R7. In Fig. 17, for gauge R7, we report the empirical survival probability function for a synthetic series out of 100 and the averaged function using all the generated series. In addition, for comparison, we give the averaged survival functions for gauges R15 and R25. At first glance, highest rainfall intensities drawn in Fig. 17 show strange behavior manifested by constant exceedance probability above a given precipitation threshold. This is especially pronounced for observed or synthetic series from a single MCM run. This is due to the very short rainfall time series used for the calculation of survival probability functions. According to multifractal theory, singularities in a small data set are very rare. Highest rainfall intensities as singularities are very rare in 2-year series. The behavior of both the synthetic functions for gauge R7 in Fig. 17 is very similar, with the sole exception of the extended and smoothed tail of the averaged function plot. Both the synthetic functions are placed above the observed function. This displacement reveals overprediction of 5 min precipitation depths, particularly at the range of intensities

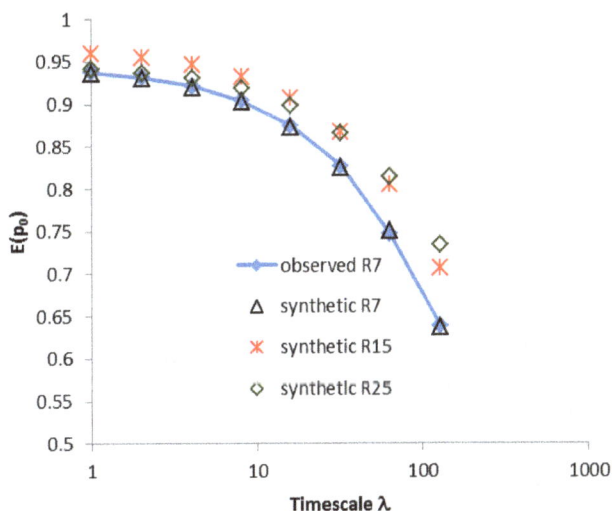

Figure 16. Comparison between observed for gauge R7 and synthetic series for gauges R7, R15 and R25 in terms of intermittency $E(p_0)$ for the considered timescales. The values for the generated data are calculated as the average of 100 disaggregation runs. The variability between runs was negligible and thus is not shown here.

from 0.3 to about 2.0 mm/5 min. It should be noted that the magnitude of dissimilarities between synthetic and observed survival functions for gauge R7 did not exceed the ones reported in other works, e.g., Molnar and Burlando (2005) and Licznar et al. (2011a, b). In comparison, the magnitude of dissimilarities between observed survival probability for gauge R7 and synthetic (average) survival probability function for other gauges R15 and R25 is much more pronounced.

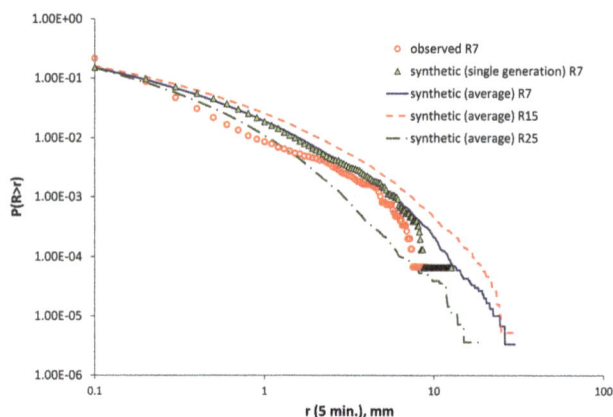

Figure 17. The survival probability function of 5 min precipitation amounts for the observed time series (circles) and the synthetic time series (triangles) generated by the disaggregation of 1280 precipitation amounts for gauge R7. The lines represent the average distributions calculated over the generation of 100 synthetic time series for gauge R7 and for comparison for gauges R15 and R25.

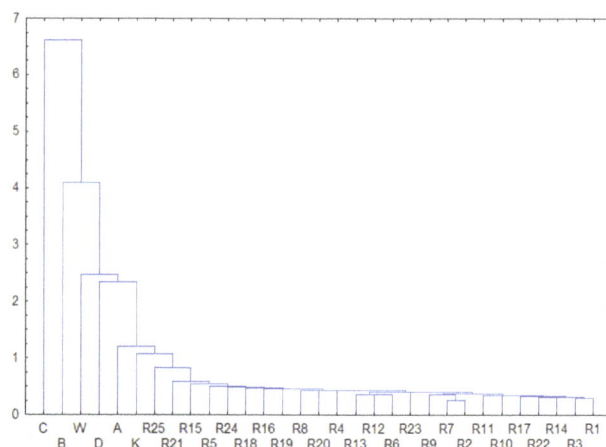

Figure 18. Dendrogram resulting from the cluster analysis of BDC histograms for $\lambda = 1$. The vertical scale shows binding distance, whereas names of gauges are given on the horizontal scale ("K" stands for Kielce gauge and "W" stands for Wroclaw).

3.5   Cluster analysis results and their interpretation

Dendrograms summarizing the results of the cluster analysis for BDC histograms are produced for each timescale and reported in Figs. 18 and 19 for $\lambda = 1$ and $\lambda = 128$, respectively. Results for the first four timescales, i.e., $\lambda = 1, 2, 4, 8$, are unsurprising and easily interpreted. All Warsaw gauges are grouped in a single cluster with similar shapes of BDC histograms; their interconnection on the dendrogram is placed at the level of binding distance equal to about 0.5. Only R25 seems to be characterized by a slightly different pattern of BDC histogram. However, gauge R25 has a behavior that is still much closer to other Warsaw gauges than to the other cities considered. For example, at $\lambda = 1$ gauge R25 is merged into the Warsaw gauges cluster at an Euclidean distance equal to 0.81, whereas the same occurs for the Kielce (the closest considered Polish city) gauge at the Euclidean distance equal to 1.07. For timescales $\lambda = 2, 4, 8$, gauge R25 merges the cluster of Warsaw gauges at quite similar Euclidean distances: 0.89, 0.83 and 0.81, respectively.

The dendrogram for $\lambda = 128$ is given in Fig. 19, being representative of timescales $\lambda = 16, 32, 64, 128$. From Fig. 19 it is possible to see the departure of gauge R15 from the cluster of other Warsaw gauges. The position of gauge R15 is isolated from other Warsaw gauges, and its Euclidean distance from the closest one is large and increases with greater timescale; it is equal to 1.8, 3.19, 3.88 and 8.03 for $\lambda = 16, 32, 64$ and 128, respectively. Simultaneously, the Euclidean distance from the cluster of Warsaw gauges to the nearest neighbor does not exceed 0.9, 1, 1.4 and 1.89 for $\lambda = 16, 32, 64$ and 128, respectively.

This last observation proves that the variability of BDC shapes among Warsaw gauges generally increases with a greater timescale. It may partly be explained by the already mentioned evolution of histogram shapes, the replacement of 2N–B distribution by less centered N–B and B-distribution characterized by a higher variance of BDC. In the specific case of gauge R15, the BDC histograms for the largest timescales are boldly concave (not shown for brevity) and their shapes are becoming similar to B symmetrical distributions parameterized by very small values of a: 0.76, 0.64, 0.54 and 0.45 for $\lambda = 16, 32, 64$ and 128, respectively.

In the last step we used the cluster analysis to investigate the variability among the gauges in terms of the intermittency parameter p_0, which is considered a vector with eight components as values corresponding to the considered timescales. Results are given in the form of a dendrogram in Fig. 20. With respect to p_0, all Warsaw gauges form one single chain-like cluster. Three gauges in the cluster, namely R14, R25 and R15, are characterized by the largest distances from the nearest neighbor with Euclidean distances equal to 0.079, 0.064 and 0.0614, respectively. The distance of gauges R15 and R25 from the other stations in the cluster is similar to observations made for Figs. 18 and 19. A possible, but not certain, explanation for gauge R14 could be its location close to gauge R15 in a weakly developed part of the city.

Unfortunately, we do not have access to other meteorological data to compare our results with other local climate conditions. To our knowledge, studies about microclimate or local turbulence have not been conducted for Warsaw. However, in our opinion the anomalous behavior of gauges R15 and R25 does not originate from random errors due to gauges installation. As previously mentioned, all gauges were installed in very good conditions, and R15 was an airport gauge. A plausible explanation of the anomalous behavior of gauges R25 and R15 could be found in its location. Gauge R25 was

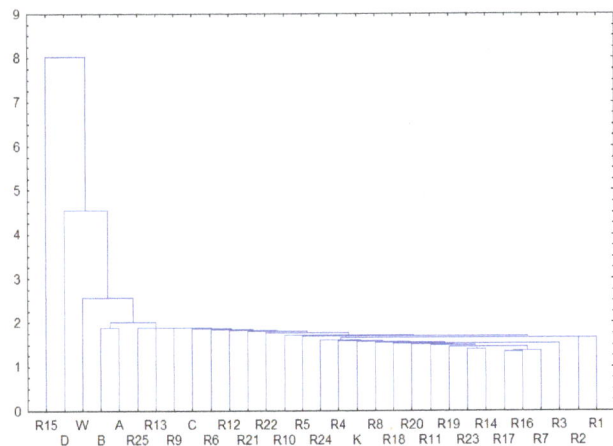

Figure 19. Dendrogram resulting from the cluster analysis of BDC histograms for the timescale $\lambda = 128$. The vertical scale shows binding distance, whereas names of gauges are given on the horizontal scale ("K" stands for Kielce gauge and "W" stands for Wroclaw).

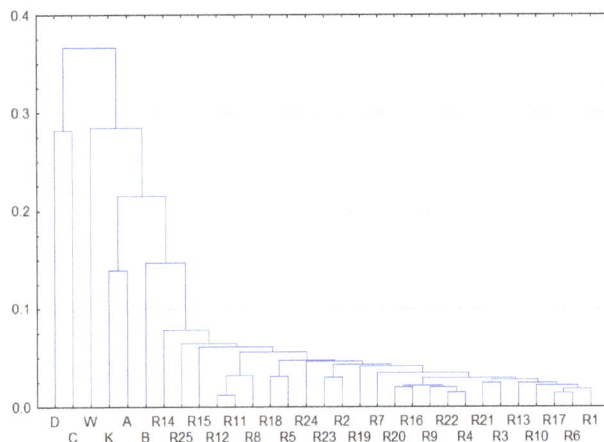

Figure 20. Dendrogram resulting from the cluster analysis of the intermittency parameter p_0. The vertical scale shows binding distance, whereas the names of gauges are given on the horizontal scale ("K" stands for Kielce gauge and "W" stands for Wroclaw).

located in a southeast suburban area in the close vicinity of a forested area and the Vistula river valley. This specific suburban area is also most frequently a place for the development of local convection processes (Prof. S. Malinowski, personal communication, 2013). The anomalous behavior of gauge R15 seems to arise from its specific location on the ground at the Warsaw airport. In the neighborhood of the instrument there are no high buildings and trees and the ground is covered only by short-cut grass. The local atmospheric turbulence conditions, additionally influenced by taking off and landing aircrafts, could have favored the different behavior of this station. In general, gauges R15 and R25 are the only instruments installed outside the areas of urban fabric (Fig. 1) in rather rural conditions of surrounding green areas. The suburban location of these gauges combined with the direct green surrounding reduces, or even minimalizes to zero, urban heat island effects. Peng et al. (2012) investigated the surface urban heat island intensity across 419 global big cities (including Warsaw). These authors showed that the distribution of daytime surface urban heat island intensity correlates negatively across cities with the difference of vegetation fractional cover and of vegetation activity between urban and suburban areas. Kłysik and Fortuniak (1999) found the occurrence of urban heat in the second biggest city in Poland, Łódź (about 120 km southwest), which is comparable to Warsaw flat topography. According to statistics calculated over many years at two stations – one in the center and one at the airport – over 80 % of nights were characterized by a surplus heat in town amounting generally to 2–4 °C and sporadically to 8 °C and more. Once more for Łódź, Fortuniak et al. (2006) investigated the data from two automatic stations: one urban and one rural. They found the relative humidity to be lower in the town, sometimes by more than 40 %, and

water vapor pressure differences to be either positive (up to 5 hPa) or negative (up to -4 hPa). Air-temperature differences between the urban and rural station exceeded 8 °C. It could be that similar processes occurred in Warsaw and affected local precipitation dynamics and thus gauges R7, R15 and R25. As a consequence, statistics of synthetic time series vary visibly in Figs. 16 and 17. However, the significance of these differences should be studied in more details in the future.

4 Conclusions

Keeping in mind the simplicity of the retrieval of microcanonical cascade generators from observational data, we proposed to use this technique for the local variability of very short precipitation time series within an urban monitoring network.

We considered a network of 25 gauges deployed in Warsaw over an area of 517.2 km². An attempt was made to define the generators of a MCM able to produce 5 min time series, as requested by urban hydrologists, through the disaggregation of quasi-daily precipitation totals. We showed that smooth distributions of BDC are possible for all analyzed timescales, even in the case of a limited length of time series, which in our case slightly exceeded 2 years only. This was made possible by the implementation of a randomization procedure and the use of an overlapping moving window algorithm for the calculation of BDCs.

The correctness of the overlapping moving window algorithm is checked using additional 15 min rainfall time series, 25 years long, at gauge R7. The algorithm is implemented for a hierarchy of sub-daily timescales and separate 5-year periods. The results of BDC calculations are compared to those obtained using all 25 years of data with both overlapping and non-overlapping moving window algorithms. Despite the coarse resolution of data and winter time gaps in the series, the results show a good agreement of BDC distributions calculated over the different periods, suggesting the correctness of the overlapping moving window algorithm, at least in central Poland.

To adequately describe the shapes of BDC histograms, we have implemented a special joined probability distribution, 2N–B, assembled from two N distributions and one B symmetrical distribution. A systematical evolution of BDC histograms from joined 2N–B through joined N–B up to B distributions was observed increasing the timescale. To test the use of more complicated models alternative to the classical B distribution, we suggested the AIC.

To check all the applied techniques (i.e., the randomization procedure, the overlapping moving window scheme and the 2N–B distribution), MCMs were used to disaggregate 1280 min precipitation into 5 min time series. The quality of the generated series was checked, comparing the statistical properties of these with those of observed series. In particular, we compared probabilities of zero precipitation and the survival probability functions of non-zero 5 min precipitation amounts for the considered timescales with agreement comparable to previous studies done in Switzerland, Germany and Poland.

As a main part of this study, we conducted an intercomparison of BDC histograms among the 25 Warsaw gauges and considered as a term of reference another six gauges located in Poland and Germany. The intercomparison was made, scale-by-scale, by means of cluster analysis. Resulting dendrograms for small timescales (i.e., $\lambda = 1, 2, 4, 8$) revealed rather little variability of BDC histograms among all Warsaw gauges in comparison to the variability exhibited with respect to the other external gauges. Only gauge R25 seemed to be characterized by a slightly different pattern. It might originate from the specific gauge location on the city boundary in the vicinity of forested areas and Vistula river valley.

Dendrograms obtained for large timescales (i.e., $\lambda = 16, 32, 64, 128$) also delivered a general picture of similarity among Warsaw gauges with the very clear exception of gauge R15. To our best knowledge, a possible explanation of this was its installation on the ground at the Warsaw airport, which was strongly man-modified and with turbulent local conditions. In addition, R25, R15 and R14 were also identified as gauges presenting slightly different behavior in terms of the intermittency parameter p_0.

As final remarks we can affirm that MCMs combined with cluster analysis could be used as a tool for the assessment of the spatial variability of local precipitation patterns among a group of gauges. This framework could be effectively implemented even in the case of very short observational series thanks to the proposed overlapping moving window algorithm. We believe that the use of this algorithm could increase the development and use of MCMs in urban hydrol-

ogy. At the same time, we are fully aware of the inherent MCM limitations in the quality of rainfall disaggregation and the necessity of additional verifications of the overlapping moving window algorithm for other gauges with longer and higher quality observational time series.

Returning to questions of interest in urban hydrology addressed at the end of the Introduction, we can formulate following answers:

1. Small precipitation variability within gauges located in city centers, as measured via microcanonical cascade generators, justifies the practice of a single time series use for the probabilistic assessment of the entire urban drainage system.

2. From current engineering needs in urban hydrology, it is enough to use only one fitted MCM for the precipitation time series disaggregation in Warsaw city. We suppose that this result could be valid even in larger urban areas, but the verification is necessary. We dissuade from the cascade generation fitted on precipitation time series collected at instruments located out of the city center in unrepresentative sites like, in our case, the ground at the airport.

3. We question the practice of using gauges from airports for urban hydrology.

Finally, we recommend further research to assess the influence of the local conditions on BDC histograms to find clearer explanations of observed anomalies. We also recognize the necessity of further tests in other cities and precipitation monitoring networks, especially in the case of cities with complicated orography and the presence of hydrological networks.

Acknowledgements. This project was financed by Polish National Science Center (NCN) funds allocated on the basis of decision no. 2011/03/B/ST10/06338. It was realized as a part of the scientific project "Spatiotemporal analysis and modeling of urban precipitation field." Precipitation data were provided by the Municipal Water Supply and Sewerage Company (MWSSC) in Warsaw, Poland. Authors also acknowledge the three anonymous reviewers for their suggestions and comments.

Edited by: R. Uijlenhoet

References

Ahrens, B.: Rainfall downscaling in an alpine watershed applying a multiresolution approach, J. Geophys. Res., 108, 8388, doi:10.1029/2001JD001485, 2003.

Berne, A., Delrieu, G., Creutin, J. D. and Obled, C.: Temporal and spatial resolution of rainfall measurements required for urban hydrology, J. Hydrol., 299, 166–179, doi:10.1016/j.jhydrol.2004.08.002, 2004.

BS EN 752-3: Drain and sewer systems outside buildings. Planning, British Standards Institution, ISBN 0580267822, 34 pp., 1997.

Cârsteanu, A. and Foufoula-Georgiou, E.: Assessing dependence among weights in a multiplicative cascade model of temporal rainfall, J. Geophys. Res., 101, 26363–26370, doi:10.1029/96JD01657, 1996.

De Michele, C., Salvadori, G., Vezzoli, R., and Pecora, S.: Multivariate assessment of droughts: Frequency analysis and dynamic return period, Water Resour. Res., 49, 6985–6994, doi:10.1002/wrcr.20551, 2013.

Deutsche Vereinigung für Wasserwirtschaft, Abwasser und Abfall e.V. (DWA): Arbeitsblatt DWA-A 117: Bemessung von Regenrückhalteräumen, Hennef, 2006.

Fortuniak, K., Klysik, K., and Wibig, J.: Urban–rural contrasts of meteorological parameters in Lodz, Theor. Appl. Climatol., 84, 91–101, doi:10.1007/s00704-005-0147-y, 2006.

Gires, A., Onof C., Maksimovic C., Schertzer D., Tchiguirinskaia I., and Simoes N.: Quantifying the impact of small scale unmeasured rainfall variability on urban hydrology through multifractal downscaling: a case study, J. Hydrol., 442–443, 117–128, doi:10.1016/j.jhydrol.2012.04.005, 2012.

Gires, A., Tchiguirinskaia, I., Schertzer, D., and Lovejoy, S.: Multifractal analysis of an urban hydrological model on a Seine-Saint-Denis study case, Urban Water J., 10, 195–208, doi:10.1080/1573062X.2012.716447, 2013.

Górski J.: Analysis of Precipitation Time Series for Needs of Urban Hydrology on Example of Kielce City. PhD thesis, Kielce University of Technology, Poland, 2013 (in Polish).

Güntner, A., Olsson, J., Calver, A., and Gannon, B.: Cascade-based disaggregation of continuous rainfall time series: the influence of climate, Hydrol. Earth Syst. Sci., 5, 145–164, doi:10.5194/hess-5-145-2001, 2001.

Harris, D., Menabde, M., Seed, A., and Austin, G.: Breakdown coefficients and scaling properties of rain fields, Nonlin. Processes Geophys., 5, 93–104, doi:10.5194/npg-5-93-1998, 1998.

Hingray, B. and Ben Haha, M.: Statistical performances of various deterministic and stochastic models for rainfall series disaggregation, Atmos. Res., 77, 152–175, 2005.

Kłysik, K. and Fortuniak, K.: Temporal and spatial characteristics of the urban heat island of Łódź, Poland, Atmos. Environ., 33, 3885–3895, 1999.

Koutsoyiannis, D.: Rainfall disaggregation methods: Theory and applications, in: Proc. Workshop on Statistical and Mathematical Methods for Hydrological Analysis, edited by: Piccolo, D. and Ubertini, L., Universitá di Roma "La Sapienza", Rome, http://itia.ntua.gr/en/docinfo/570, 2003.

Lanza, L., Leroy, M., Alexandropoulos, C., Stagi, L., and Wauben, W.: WMO laboratory intercomparison of rainfall intensity gauges, Final report, IOM Report No. 84, WMO/TD No. 1304, 2005.

Larose, D. T.: Discovering knowledge in data: an introduction to data mining, John Wiley & Sons, Inc., Hoboken, New Jersey, USA, 222 pp., 2005.

Licznar, P.: Stormwater reservoir dimensioning based on synthetic rainfall time series, Ochrona Srodowiska, 35, 27–32, 2013.

Licznar, P., Łomotowski, J., and Rupp, D. E.: Random cascade driven rainfall disaggregation for urban hydrology: An evaluation of six models and a new generator, Atmos. Res., 99, 563–578, doi:10.1016/j.atmosres.2010.12.014, 2011a.

Licznar, P., Schmitt, T. G., and Rupp, D. E.: Distributions of microcanonical cascade weights of rainfall at small timescales, Acta Geophysica, 59, 1013–1043, doi:10.2478/s11600-011-0014-4, 2011b.

Lombardo, F., Volpi, E., and Koutsoyiannis, D.: Rainfall downscaling in time: theoretical and empirical comparison between multifractal and Hurst-Kolmogorov discrete random cascades, Hydrolog. Sci. J., 57, 1052–1066, doi:10.1080/02626667.2012.695872, 2012.

Marshak, A., Davis, A., Cahalan, R., and Wiscombe, W.: Bounded cascade models as nonstationary multifractals, Phys. Rev. E, 49, 55–69, 1994.

Menabde, M. and Sivapalan, M.: Modeling of rainfall time series and extremes using bounded random cascades and Levy-stable distributions, Water Resour. Res., 36, 3293–3300, doi:10.1029/2000WR900197, 2000.

Menabde, M., Harris, D., Seed, A., Austin, G., and Stow, D.: Multiscaling properties of rainfall and bounded random cascades, Water Resour. Res., 33, 2823–2830, doi:10.1029/97WR02006, 1997.

Molnar, P. and Burlando, P.: Preservation of rainfall properties in stochastic disaggregation by a simple random cascade model, Atmos. Res., 77, 137–151, doi:10.1016/j.atmosres.2004.10.024, 2005.

Molnar, P. and Burlando, P.: Variability in the scale properties of high-resolution precipitation data in the Alpine climate of Switzerland, Wat. Resour. Res., 44, W10404, doi:10.1029/2007WR006142, 2008.

Oke, T.: Initial guidance to obtain representative meteorological observations at urban sites, Instruments and Observing Methods, Report no. 81, World Meteorological Organization, WMO/TD-No. 1250, 2006.

Olsson, J.: Evaluation of a scaling cascade model for temporal rain-fall disaggregation, Hydrol. Earth Syst. Sci., 2, 19–30, doi:10.5194/hess-2-19-1998, 1998.

Over, T. M. and Gupta, V. K.: A space-time theory of mesoscale rainfall using random cascades, J. Geophys. Res., 101, 26319–26331, 1996.

Paulson, K. S. and Baxter, P. D.: Downscaling of rain gauge time series by multiplicative beta cascade, J. Geophys. Res., 112, D09105, doi:10.1029/2006JD007333, 2007.

Peng, S., Piao, S., Ciais, P., Friedlingstein, P., Ottle, C., Bréon, F.-M., Nan, H., Zhou, L., and Myneni, R. B.: Surface urban heat island across 419 global big cities, Environ. Sci. Technol., 46, 696–703, doi:10.1021/es2030438, 2012.

Robert, C. P.: Simulation of truncated normal variables, Stat. Comput., 5, 121–125, 1995.

Rupp, D. E., Keim, R. F., Ossiander, M., Brugnach, M., and Selker, J. S.: Time scale and intensity dependency in multiplicative cascades for temporal rainfall disaggregation, Water Resour. Res., 45, W07409, doi:10.1029/2008WR007321, 2009.

Rupp, D. E., Licznar, P., Adamowski, W., and Leśniewski, M.: Multiplicative cascade models for fine spatial downscaling of rainfall: parameterization with rain gauge data, Hydrol. Earth Syst. Sci., 16, 671–684, doi:10.5194/hess-16-671-2012, 2012.

Schmitt, T. G.: ATV-DVWK Kommentar, ATV-A 118 Hydraulische Berechnung von Entwässerungssystemen, DWA, Hennef, 2000.

Thames Tideway Tunnel (TTT): Needs Report, Appendix B, Report on Approaches to UWWTD Compliance in Relation to CSO's in

major cities across the EU, available at: http://aim.prepared-fp7.
eu/viewer/doc.aspx?id=28 (last access: 6 January 2015), 2010.

Willems, P.: Multidecadal oscillatory behaviour of rainfall extremes, Clim. Chang., 120, 931–944, doi:10.1007/s10584-013-
0837-x, 2013.

World Meteorological Organization (WMO): WMO-No. 8, Guide
to Meteorological Instruments and Methods of Observation, 7
edition, World Meteorological Organization, Geneva, Switzerland , 2012.

4

What made the June 2013 flood in Germany an exceptional event? A hydro-meteorological evaluation

K. Schröter[1,3]**, M. Kunz**[2,3]**, F. Elmer**[1,3]**, B. Mühr**[2,3]**, and B. Merz**[1,3]

[1]Helmholtz Centre Potsdam, GFZ German Research Centre for Geosciences, Section Hydrology, Potsdam, Germany
[2]Karlsruhe Institute of Technology, Institute for Meteorology and Climate Research, Karlsruhe, Germany
[3]CEDIM – Center for Disaster Management and Risk Reduction Technology, Potsdam, Germany

Correspondence to: K. Schröter (kai.schroeter@gfz-potsdam.de)

Abstract. The summer flood of 2013 set a new record for large-scale floods in Germany for at least the last 60 years. In this paper we analyse the key hydro-meteorological factors using extreme value statistics as well as aggregated severity indices. For the long-term classification of the recent flood we draw comparisons to a set of past large-scale flood events in Germany, notably the high-impact summer floods from August 2002 and July 1954. Our analysis shows that the combination of extreme initial wetness at the national scale – caused by a pronounced precipitation anomaly in the month of May 2013 – and strong, but not extraordinary event precipitation were the key drivers for this exceptional flood event. This provides additional insights into the importance of catchment wetness for high return period floods on a large scale. The database compiled and the methodological developments provide a consistent framework for the rapid evaluation of future floods.

1 Introduction

In June 2013, wide parts of central Europe were hit by large-scale flooding. Particularly southern and eastern Germany were affected, but also other countries such as Austria, Switzerland, the Czech Republic, Poland, Hungary, Slovakia, Croatia and Serbia. Almost all main river systems in Germany showed high water levels: the Elbe between Coswig and Lenzen, the Saale downstream of Halle, and the Danube at Passau experienced new record water levels. Severe flooding occurred especially along the Danube and Elbe rivers, as well as along the Elbe tributaries Mulde and Saale. In the Weser and Rhine catchments exceptional flood magnitudes were, however, observed only locally in some smaller tributaries. The area affected most in the Rhine catchment was the Neckar with its tributaries Eyach and Starzel. In the Weser catchment the Werra sub-catchment was affected most – in particular the discharges in the Hasel and Schmalkalde tributaries were on an exceptional flood level (BfG, 2013). As a consequence of major dike breaches at the Danube in Fischerdorf near Deggendorf, at the confluence of the Saale and Elbe rivers at Rosenburg, and at the Elbe near Fischbeck, large areas were inundated with strong impacts on society in terms of direct damage and interruption of transportation systems (see Fig. A1 in the Appendix for geographic locations).

Estimates on overall losses caused by the flooding in central Europe are in the range of EUR 11.4 (Munich Re, 2013) to 13.5 billion (Swiss Re, 2013), whereof EUR 10 billion occurred in Germany alone. Official estimates of economic loss for Germany amount to EUR 6.6 billion (Deutscher Bundestag, 2013) with an additional EUR 2 billion of insured losses (GDV, 2013). These numbers are about 60 % of the total loss of EUR 14.1 billion (normalized to 2013 values) in Germany caused by the extreme summer flood in August 2002 (Kron, 2004; Thieken et al., 2005) which remains the most expensive natural hazard experienced in Germany so far.

The June 2013 flood was an extreme event with regard to magnitude and spatial extent as well as its impact on society and the economy (Blöschl et al., 2013; Merz et al., 2014). The Forensic Disaster Analysis (FDA) Task Force of the Centre for Disaster Management and Risk Reduction Technology (CEDIM) closely monitored the evolution of the flood in June 2013 including the impacts on people, transportation and economy in near real time. In this way CEDIM made science-based facts available for the identification of major event drivers and for disaster mitigation. The first phase of this activity was done by compiling scattered information available from diverse sources including in situ sensors and remote sensing data, the internet, media and social sensors as well as by applying CEDIM's own rapid assessment tools. Two reports were issued: the first report focused on the meteorological and hydrological conditions including comparisons to major floods from the past (CEDIM, 2013a), while the second one focused on impact and management issues (CEDIM, 2013b).

The subsequent phase of this FDA activity focused on the research question: what made the flood in June 2013 an exceptional event from a hydro-meteorological point of view? This question is analysed in this paper. We expect this analysis to improve the understanding of key drivers of large-scale floods and thus contribute to the derivation of well-founded and plausible extreme scenarios.

In this context, the statement of BfG (2013) and Blöschl et al. (2013) that high initial soil moisture played an important role for the generation of this extreme flood are an interesting starting point. Klemes (1993) reasoned that high hydrological extremes are more due to unusual combinations of different hydro-meteorological factors than to unusual magnitudes of the factors themselves. On the one hand, catchment wetness state is an important factor for the generation of floods (Merz and Blöschl, 2003). As such it is a useful indicator in flood early warning schemes (e.g. Van Steenbergen and Willems, 2013; Alfieri et al., 2014; Reager et al., 2014) and is also incorporated in procedures for extreme flood estimation (e.g. Paquet et al., 2013). On the other hand the contribution of catchment wetness to extreme floods has been shown to be of decreasing importance with increasing return periods of rainfall (e.g. Ettrick et al., 1987; Merz and Plate, 1997). However, the interaction of various hydro-meteorological factors, primarily rainfall and soil moisture, has been studied mainly for small-scale catchments (e.g. Troch et al., 1994; Perry and Niemann, 2007). Only few studies examined the interplay of various hydro-meteorological factors for large-scale floods. One example is the work of Nied et al. (2013) who investigated the role of antecedent soil moisture for floods in the Elbe catchment (ca. $150\,000\,\mathrm{km}^2$) and emphasized the increased occurrence probability of large-scale floods related to large-scale high soil moisture.

In this study, we examine key meteorological and hydrological characteristics of the June 2013 flood and compare them to two other large-scale high-impact events, the August 2002 and July 1954 floods in Germany. The factors considered are antecedent and event precipitation, initial streamflow conditions in the river network and flood peak discharges. We evaluate these factors in a long-term context in terms of recurrence intervals using extreme value statistics based on a 50-year reference period. For this period the set

of large-scale floods in Germany identified by Uhlemann et al. (2010) are updated and now comprises 74 flood events. Hence, the analysis is deliberately limited to the national borders of Germany in order to be able to compare the 2013 flood with the event set of Uhlemann et al. (2010). For a coherent comparison of the events we use available long-term data sets of precipitation and discharge observations. Besides the statistical analysis we derive different indices to rank the spatial extent and magnitude of the hydro-meteorological factors.

The spatial extent and hydrological severity of large-scale floods in Germany has been analysed by Uhlemann et al. (2010) in terms of flood peak discharges using a specifically developed flood severity index. In our study we enhance this framework to include antecedent and event precipitation as well as initial streamflow as additional hydro-meteorological factors. We introduce severity indices for these factors to evaluate their relative importance among the event set. Precipitation and flood peak discharges are key figures which are commonly used to characterize cause and effect of floods. The antecedent precipitation index is a well-established parameter to approximate catchment wetness (Teng et al., 1993; Ahmed, 1995). Even though there are reasonable objections against API as it disregards soil and land use characteristics which influence soil hydrological processes, it provides sufficient information to compare the potential wetness between different large-scale floods. Initial streamflow is usually not considered in hydrological analyses of flood events but is a very relevant factor for dynamic flood routing processes (Chow, 1959) as it controls the load of a river section. The inclusion of this factor within a statistical analysis of large-scale flood events is, to the knowledge of the authors, done for the first time.

The paper is organized as follows. Section 2 describes the data and methods used to conduct the hydro-meteorological analysis of the June 2013 flood and the set of large-scale flood events. Section 3 describes the meteorological situation associated with the flood in June 2013 and presents the results from the analysis of antecedent and event precipitation, initial river flow conditions and flood peak discharges. Detailed comparisons with the extreme summer floods of August 2002 and July 1954 are made. The section concludes with a sensitivity analysis of the procedure. In Sect. 4 we discuss the key findings and provide recommendations for future work. A map of geographical locations mentioned in the paper can be found in the Appendix as well as some additional information regarding sensitivities.

Table 1. Data sources, resolution and analysis methods for hydro-meteorological parameters.

Hydro-meteorological factors		Data source	Spatial resolution	Temporal resolution	Analysis/classification
Precipitation		REGNIE DWD[1]	1 km^2	Daily	Maximum 3 day totals R3d. extreme value statistics based on annual series
				Event-based	Precipitation index for all large-scale floods
Initial catchment state	Antecedent precipitation index API	REGNIE DWD[1]	1 km^2	Daily	API quantification 30 days ahead of R3d; extreme value statistics based on partial series conditional on past flood events
				Event-based	Wetness index for all past flood events
	Ratio of initial river flow to mean annual flood	Discharge gauges BfG[2]/WSV[3] and hydrometric services of federal states	Point information; 162 gauges and related sub-basins	Daily mean	Extreme value statistics based on partial series conditional on past flood events
				Event-based	Initial hydraulic load index for all past flood events
Peak flood discharge		Discharge gauges BfG[2]/WSV[3] and hydrometric services of federal states	Point information; 162 gauges and related sub-basins	Daily mean	Extreme value statistics based on annual maximum series
				Event-based	Flood severity index for all past flood events

[1] German Weather Service; [2] German Federal Institute of Hydrology; [3] Water and Shipment Administration.

2   Data and methods

2.1   Data

2.1.1   Database of large-scale floods

For the analysis of the meteorological and hydrological conditions prior to and during large-scale flood events in Germany and their relation to the climatological context, a consistent database of precipitation and discharge data was compiled. For this, we considered a set of large-scale floods which had been first determined in a consistent way by Uhlemann et al. (2010) for the period from 1952 to 2002. In this study, we used an updated event set from 1960 to 2009. These flood events are identified from daily mean discharge records at 162 gauges in Germany by screening these time series for the occurrence of peak discharges above a 10-year flood and significant flood peaks at other gauges within a defined time window that accounts for the time shift between hydraulically coherent peak flows. According to Uhlemann et al. (2010), large-scale floods are characterized by a spatial extent of mean annual flooding which affects at least 10 % of the river network considered in Germany. Applying this criterion, 74 large-scale floods are identified in the reference period 1960–2009. For each flood we derive consistent samples for hydro-meteorological factors including antecedent and event precipitation, initial streamflow conditions and peak discharges. A compilation of hydro-meteorological factors and related data sources, their spatial and temporal resolution, and the methods applied is presented in Table 1.

2.1.2   Meteorological data sets

For the triggering of large-scale floods the amount and spatial variability of precipitation are more important than the small-scale temporal variability. For this reason, we used 24 h precipitation sums of REGNIE (regionalized precipitation totals) both for the reference period 1960–2009 and for the single events 2013 (April–June) and 1954 (June–July). The data set, compiled and provided by the German Weather Service (Deutscher Wetterdienst, DWD), is interpolated from climatological stations to an equidistant grid of 1×1 km^2. The interpolation routine considers several geographical factors such as altitude, exposition or slope by distinguishing between background monthly climatological fields and daily anomalies (see Rauthe et al., 2013 for further details). In cases of convective or orographic precipitation, where a very high density of stations is required, it can be expected that REGNIE underestimates the actual spatial variability of precipitation. However, since large-scale flood events are mainly driven by advective precipitation, this effect is of minor importance in the present study. Additionally, weather charts and sounding data are used to describe the characteristics of the atmosphere on the days with maximum rainfall.

2.1.3 Hydrological data sets

We use time series of daily mean discharges from 162 gauging stations operated by the water and shipment administration (WSV), the German Federal Institute of Hydrology (BfG) or by hydrometric services of the federal states. The same selection of gauges has been used by Uhlemann et al. (2010) to compile the set of large-scale flood events in Germany. These gauges have provided continuous records since 1952 and have a drainage area larger than $500\,km^2$. Basin areas vary from $521\,km^2$ to $159\,300\,km^2$ with a median of $3650\,km^2$ including a high percentage of nested catchments. For the flood in June 2013 raw data of daily mean discharges were available for 121 gauges mainly covering the central, southern and eastern parts of Germany which have been affected most by flooding.

Based on the procedure proposed by Uhlemann et al. (2010), the point observations of discharge peaks at the 162 gauges are regionalized to represent the flood situation in a particular river stretch and its associated catchment area. The regionalization scheme uses the location of the gauges and the hierarchical Strahler order (Strahler, 1957) which accounts for the branching complexity of the river network. A gauge is assumed as representative for an upstream river reach until the next gauge and/or the Strahler order of the river stretch decreases by two orders. In the downstream direction, a gauge is representative until the Strahler order of the river changes by one order or a confluence enters the river which has the same Strahler order or one order smaller. The total length of the river network considered amounts to $13\,400\,km$.

2.2 Methods

For the statistical analysis of the hydro-meteorological factors and their consistent comparison within the set of large-scale flood events, a clear event definition including its onset and duration is required. The start of an event determines the point in time for which we evaluate the different hydro-meteorological factors instantaneously (e.g. initial streamflow) forward (event precipitation, peak discharges) and backward in time (antecedent precipitation). Due to temporal dynamics of the precipitation fields across Germany, flood triggering precipitation affects different catchment areas at different days. Therefore, we do not consider a fixed event start date for the whole of Germany, but one that may vary in space and time, that is, from one grid point to another or from one sub-catchment to another, respectively.

2.2.1 Definition of event start dates

We considered two different definitions of the event start date. The first one is related to the onset of the large-scale floods compiled in the event set by Uhlemann et al. (2010). It considers the flood response in the spatial series of mean daily discharges recorded at 162 gauges in Germany taking significant hydraulically coherent peak flows into account. The second is based on the maximum precipitation that triggers the floods. For this we quantify the highest 3-day precipitation totals (R3d) at each REGNIE grid point within a centred 21-day time window that spans from 10 days ahead to 10 days after the event start of a large-scale flood. The duration of the chosen time window considers the time lag which links flood-triggering precipitation with discharge response (e.g. Duckstein et al., 1993) and the travel times of flood waves along the river-course (e.g. Uhlemann et al., 2010). Considering the R3d totals excludes local-scale convective precipitation, which is relevant for local or flash floods but not for large-scale floods (Merz and Blöschl, 2003).

2.2.2 Event precipitation

The first day of the R3d period defines the meteorological event start for a given grid point. Depending on the space-time characteristics of the precipitation fields, these days will be more or less correlated for adjoined grid points. We have performed this analysis for maximum precipitation totals of 3 to 7 days duration and found that this variation does not imply considerable changes in the meteorological event start date. As shown in Fig. A2 in the Appendix the spatial pattern of the 7-day totals do not differ largely from the R3d patterns for the flood events investigated. Therefore we use R3d as a reasonable figure for the meteorological start date of event precipitation.

For the statistical evaluation of event precipitation, annual maximum 3-day precipitation totals are determined for the reference period from 1960 to 2009 and for the two events of 1954 and 2013. Using extreme value statistics, return periods are determined for the event-triggering R3d totals independently for each grid point.

2.2.3 Antecedent precipitation

The meteorological event starts (first day of maximum R3d) are used to calculate antecedent precipitation backward in time. We use the antecedent precipitation index (API) according to Köhler and Linsley (1951) as a proxy for the wetness conditions in a catchment in the period before the event precipitation. The relation between surface soil moisture content and different versions of the API was shown, for instance by Blanchard et al. (1981) or Teng et al. (1993). We quantify API over a 30-day period prior to the meteorological event start dates at each grid point for each event of the large-scale flood set. API is given by the sum of daily precipitation weighted with respect to the time span (here: $m = 30$ days) of rainfall occurrence before the reference day:

$$\text{API}(x, y) = \sum_{i=1}^{30} k^i R_i(x, y)(m - i), \qquad (1)$$

where $R_i(x, y)$ is the 24 h total at a specific grid point (x, y) and i represents the day prior to the 3-day maximum, which ensures that event precipitation and antecedent precipitation are clearly separated. Usually a value between 0.8 and 0.98 is used for the depletion constant k (Viessman and Lewis, 2002). The potentiation of k with the number of days i assigns continuously decreasing weights to rainfall that occurred earlier. This relation approximates the decrease of soil moisture due to evapotranspiration and percolation to deeper soil layers. In our study we selected a mean value of $k = 0.9$. For the statistical analysis of API and thus the calculation of return periods we use partial series which are derived using the meteorological event start dates identified for the 74 large-scale flood events in the period 1960–2009.

2.2.4 Precipitation and wetness indices

To further evaluate the importance of the hydro-meteorological factors R3d and API and to rank their spatial extent and magnitude for the floods in June 2013, August 2002 and July 1954 among the set of large-scale floods we introduce precipitation and wetness severity indices as aggregated measures:

$$S_X^k = \frac{1}{\Gamma} \sum_{i,j} \left\{ \frac{X_{i,j}^k}{X_{i,j}^{5\,\mathrm{yr\,RP}}} \right\} \left| X_{i,j}^k \geq X_{i,j}^{5\,\mathrm{yr\,RP}} \right., \tag{2}$$

where X is either R3d or API and 5 yr RP denotes the values for a 5-year return period. In this formulation, values of R3d and API, respectively, are considered at REGNIE grid points i, j that exceed the 5-year return values. For each event k the sum of the ratios of R3d and API to the 5-year return period are normalized with the mean area size Γ represented by the total number of REGNIE grid points in Germany.

2.2.5 Initial hydraulic load

To transfer the meteorological event start dates, possibly varying from grid cell to grid cell, to the discharge time series given at gauge locations, we need to spatially integrate and hence to average the event start dates for individual grid points within hydrological sub-basins. We use the sub-catchments of the 162 river gauges as spatial units. The resulting "areal mean" dates per sub-catchment are used as the event start date for the hydrological analyses.

The streamflow situation at the beginning of the flood event provides information on the initial hydraulic load of the river cross-section. An already increased discharge level may considerably strain the discharge capacity of a river section, and thus the superposition of the subsequent flood wave may increase the load on flood protection schemes and may aggravate inundations. For the statistical analysis of the initial streamflow conditions, we normalize the discharge values by calculating the ratio of the daily mean discharge on the event start date (Q_i) and the mean annual flood (MHQ = mean of

annual maximum discharges) for each of the $n = 162$ gauges. For each gauge a partial series is created by evaluating the ratio of Q_i and MHQ for the areal mean event start dates in the corresponding sub-catchment which are derived using the meteorological event start dates identified for the 74 large-scale flood events in the period 1960–2009.

Further, we introduce an initial load severity index representing the spatially weighted sum of the initial hydraulic load level in the river network for each event k:

$$S_{Q_i}^k = \sum_n \left\{ \lambda_n \cdot \left(\frac{Q_i}{\mathrm{MHQ}} \right)_n \right\} \left| \left(\frac{Q_i}{\mathrm{MHQ}} \right)_n \geq \left(\frac{Q_i}{\mathrm{MHQ}} \right)_n^{5\,\mathrm{yr\,RP}} \right., \tag{3}$$

where 5-yr RP denotes the flow ratio with a 5-year return period and the weights λ_n correspond to the ratio of the river stretch length (l_n) associated with a certain gauge and the total length of the river network: $\lambda_n = \frac{l_n}{\sum_n l_n}$.

2.2.6 Peak discharge

Peak discharge (Q_p) is a key figure to characterize the magnitude of a flood at a specific location. Q_p is the integrated outcome of hydrological and hydraulic processes upstream of that location and provides important information for numerous water resources management issues in particular flood estimation and flood design. For the statistical evaluation of the observed flood peaks at each of the 162 gauges we use the annual maximum series (AMS) of daily mean discharges. We evaluate the spatial flood extent and magnitude using an aggregated measure of event severity. For this purpose we calculate the length of the river network L for which during event k the peak discharge Q_p exceeds the 5-year return period:

$$L^k = \sum_n \{\lambda_n \cdot 100\} \left| Q_{p_n}^k \geq Q_{p_n}^{5\,\mathrm{yr\,RP}} \right., \tag{4}$$

where 5-yr RP denotes the discharge with a 5 year return period and the weights λn are defined as explained above. The flood severity index represents a weighted sum of peak discharges Q_p normalized by a 5-year flood using λn as weights:

$$S_{Q_p}^k = \sum_n \left\{ \lambda_n \cdot \frac{Q_{p_n}^k}{Q_{p_n}^{5\,\mathrm{yr\,RP}}} \right\} \left| Q_{p_n}^k \geq Q_{p_n}^{5\,\mathrm{yr\,RP}} \right., \tag{5}$$

2.2.7 Extreme value statistics

To calculate exceedance probabilities and return periods (Tn) for the various hydro-meteorological factors, i.e. R3d, API, Q_i/MHQ and Q_p, observed for the June 2013, August 2002 and July 1954 floods, we applied the classical generalized extreme value distribution (Embrechts et al., 1997). Most appropriate and widely used in the case of precipitation is the Fisher–Tippett type I extreme value distribution, also known

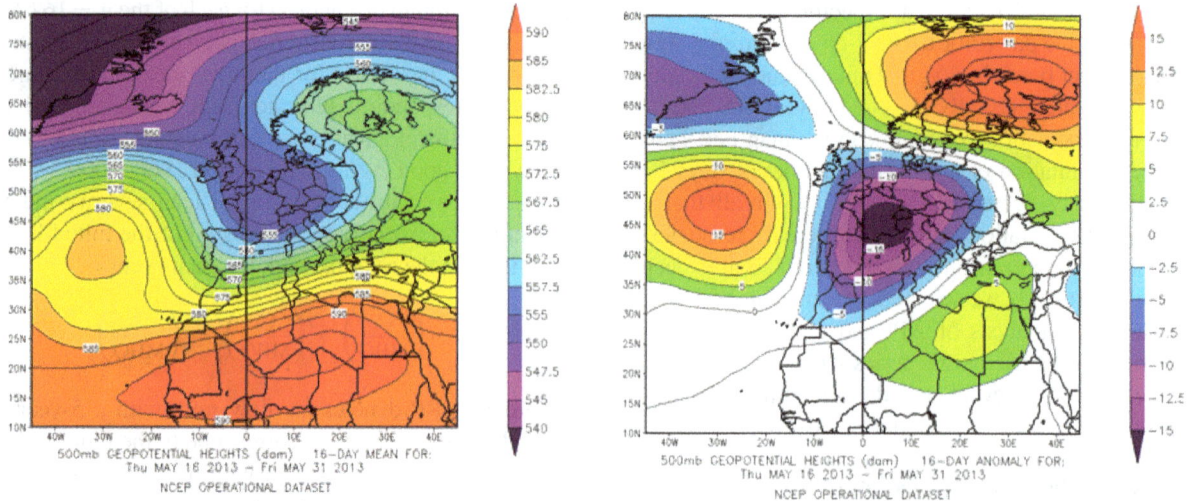

Figure 1. 500 hPa geopotential height, 16 day mean for 16–31 May 2013 (left) and anomaly in respect to the climatology based on 1979–1995 (right). Credit: Data/image provided by the NOAA/OAR/ESRL PSD, Boulder, CO, USA, from their Web site at http://www.esrl.noaa.gov/psd/ (last access: 4 April 2014).

as Gumbel distribution, with a cumulative distribution function (CDF) of

$$F(x) = \exp\left[-\exp\left(-\frac{x - \beta}{\alpha}\right)\right], \tag{6}$$

where α is the scale parameter affecting the extension in the x-direction and β is the mode that determines the location of the maximum. This distribution is also suitable for the Q_i/MHQ samples. For the statistical analysis of Q_p we fit a generalized extreme value distribution to the AMS of daily mean discharges. The CDF of the generalized extreme value distribution has a function of

$$F(x) = \exp\left\{-\left[1 + \frac{\gamma(x - \zeta)}{\delta}\right]^{-1/\gamma}\right\}, \tag{7}$$

where δ is the scale parameter affecting the extension in the x-direction, ζ is a location parameter and γ is a shape parameter.

3 Results

3.1 Meteorological conditions

Large-scale central European floods are mainly caused by the interaction of upper-level pressure systems, associated surface lows and the continuous advection of moist and warm air over long distances. In 2013, the second half of the month of May was exceptionally wet across most of central Europe due to the unusual persistence of an extended upper-air low-pressure system (trough; Fig. 1, left) that triggered several surface lows. The persistence of the quasi-stationary trough is reflected by a strong negative geopotential anomaly

compared to the long-term mean (1979–1995) over France, Switzerland and northwestern Italy (Fig. 1, right). This trough was flanked by two upper-air high-pressure systems over northeastern Europe and the North Atlantic Ocean, which caused a blocking situation. Therefore, Atlantic air masses from the west were prevented from entering central Europe. On the other side, warm and humid air masses were repeatedly advected from southeastern Europe northwards and eventually curved into Germany and Austria.

The intense and widespread rain that finally triggered the 2013 flood occurred at the end of May/ beginning of June. Responsible for the heavy rainfall was a cut-off low that moved slowly with its centre from France (29 May) over northern Italy (30 May; Fig. 2a) to Eastern Europe (1 June; Fig. 2b). In the latter region, three consecutive surface lows were triggered by short-wave troughs that travelled around the cut-off low (CEDIM, 2013a). On the northeastern flank of the upper low and near the secondary surface lows, warm and moist air masses were advected into central Europe. Grams et al. (2014) identified evapotranspiration from continental landmasses of central and Eastern Europe as the main moisture source. Due to the significant horizontal pressure gradient in the lower troposphere that prevailed from the end of May to the first days of June, there was a constant and strong northerly flow of moist and warm air which caused substantial rain enhancement on the northern side of the west-to-east oriented mountain ranges, e.g. the Alps, Ore Mountains and Swabian Jura.

In summary, the combination of large-scale lifting at the downstream side of the troughs, orographically induced lifting over the mountains, and embedded convection in the mainly stratiform clouds due to unstable air masses resulted in prolonged and widespread heavy rainfall.

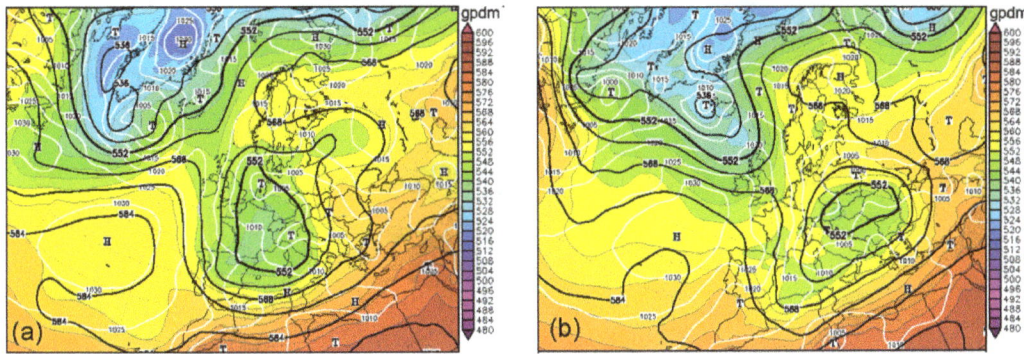

Figure 2. Weather charts for 30 May (**a**) and 1 June 2013 (**b**) 00:00 UTC with analysis of 500 hPa geopotential height (black lines), surface pressure (white lines) and 1000/500 hPa relative topography (colours) from the Global Forecast System (GFS). Image credit: wetter3.de (last access: 7 May 2014).

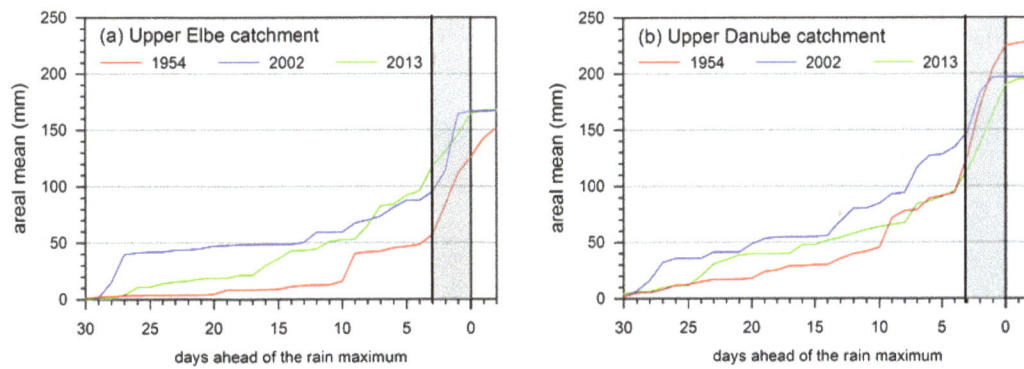

Figure 3. Time series of cumulated areal mean precipitation for the upper Elbe catchment in Germany up to the inflow of the Saale River (**a**; area: 63 171 km^2) and for the upper Danube catchment in Germany (**b**; area: 51 379 km^2). The x-axis marks the days prior to the 3-day maximum precipitation totals. Event precipitation is highlighted in grey.

3.2 Precipitation

Highest precipitation totals within the 30-day period prior to the flood event start dates can be observed between 3 and 4 days ahead of the flood event start date (indicated by the zero in Fig. 3), as shown by the time series of cumulated areal precipitation averaged over the upper Elbe (Fig. 3a) and Danube (Fig. 3b) catchments. Note that these characteristics are almost the same for the other two floods considered, 2002 and 1954, respectively. Especially for the Elbe catchment in May 2013, rain totals were high up to 17 days prior the event start, and higher compared to the other events (if the large totals 28 days ahead of the 2002 flooding are neglected). For the whole month of May 2013, the precipitation averaged over Germany was 178 % of the long-term average for the period 1881–2012 (DWD, 2013). To better explain differences and similarities of the three flood events considered, we analysed both maximum 3-day precipitation totals (R3d) as event precipitation and precipitation in the month before the flooding in terms of API. In both cases, the quantities are calculated independently at each grid point of the REGNIE gridded precipitation data (see Sect. 2.2).

3.2.1 Event precipitation

Maximum 3-day totals (R3d) in 2013 show high values in excess of 60 mm over southern and eastern Germany (Fig. 4, left). The highest rain maximum with R3d = 346 mm was observed at the DWD weather station of Aschau-Stein (31 May–3 June 2013, 06:00 UTC), which is situated in the Bavarian Alps at an elevation of 680 m a.s.l. This station also recorded the maximum 24 h rain sum of 170.5 mm on 1 June 2013 (from 1 June 06:00 UTC until 2 June 2013 06:00 UTC). On that day, peak rainfall was recorded at many other stations in the federal states of Bavaria, Saxony and Baden-Württemberg. Overall, the R3d maxima were registered almost homogeneously between 30 May and 1 June 2013 (Julian day 152, Fig. 5 left). At the upper reaches of Danube and Elbe (German part) the maxima occurred 1 day later. Over the very eastern parts, especially near Dresden and Passau, the temporal difference was even 2 days. This consecutive shift of the main precipitation fields in the west-to-east direction, i.e. following the flow direction of the Danube, caused an additional amplification of the high-water peaks.

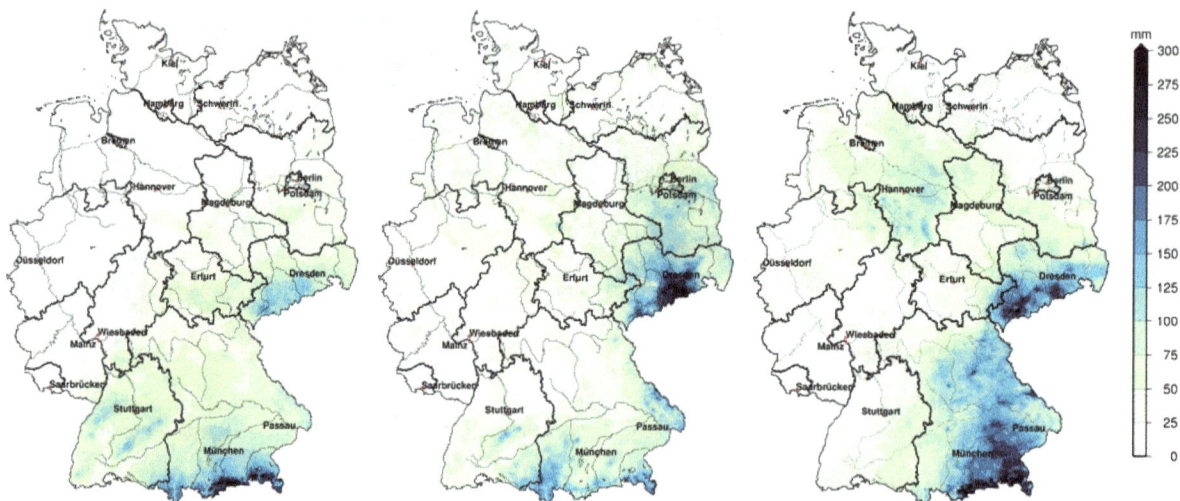

Figure 4. The 3-day maximum precipitation according to REGNIE data sets for June 2013 (left), August 2002 (middle) and July 1954 (right).

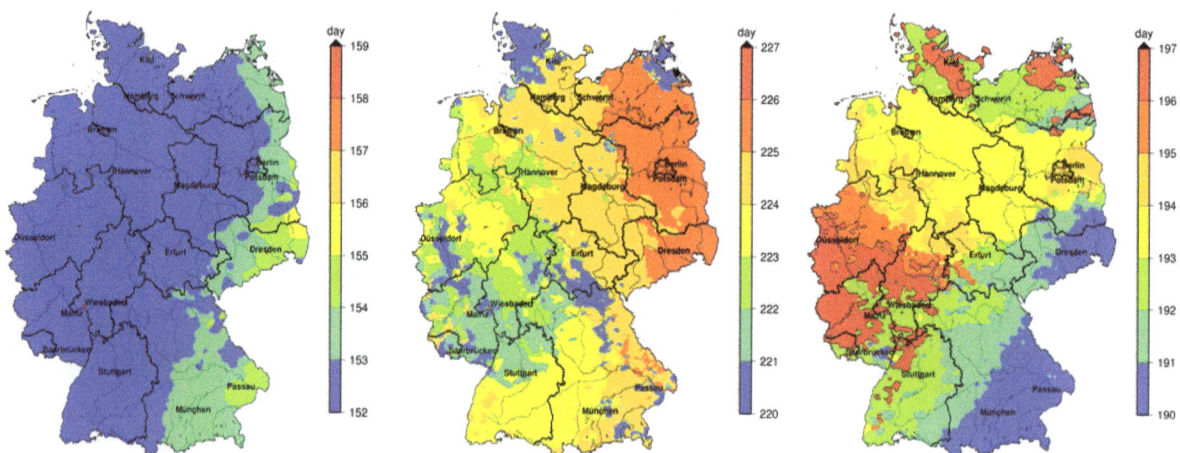

Figure 5. Day of the year at each REGNIE grid point where the event related maximum R3d total according to Fig. 4 occurred (end of the 3-day total) for June 2013 (left), August 2002 (middle) and July 1954 (right). The day 152 corresponds to 1 June, 220 to 8 August, and 190 to 9 July. The indicated days refer to the end of R3d.

Even if the flood-related rainfall in 2013 was mainly driven by meso-scale processes such as uplift related to the troughs and advection of moist air masses, the R3d map suggests that additional orographically induced lifting over the mountains increased the rain totals substantially. Highest rain sums occurred along the crests of the Ore Mountains (near Dresden), the Black Forest and Swabian Jura (west and east of Stuttgart, respectively), the Alpine foothills (south of Munich) and the Bavarian Alps. Overall, the rain enhancement over the low-mountain ranges estimated from the ratio between areal rainfall over the mountains and adjacent lowlands was between 200 and 310 %. This substantial local-scale increase in precipitation can be plausibly explained by the characteristics of the air mass on the large scale. First of all, the lifting condensation level (LCL), which represents the level of the cloud base in the case of synoptic-scale or orographic lifting, was very low on the first 3 days of June as observed at the sounding stations at Munich, Stuttgart, Meiningen and Kümmersbruck. The pressure levels were only around 920 hPa, i.e. near the surface (e.g. at Kümmersbruck the LCL was on average 924.7 hPa / 765 m a.s.l). A low LCL ensures that a large amount of atmospheric moisture, which decreases almost exponentially with elevation, basically can be converted into rain. Furthermore, precipitable water (pw) – as the vertical integral of the specific water vapour content – was large, with values of up to 26 mm. The sounding at Stuttgart, for example, measured a pw value of 25.9 mm (1 June 2013, 12:00 UTC), which is even above the 90 % percentile ($pw_{90} = 23.7$ mm) obtained from all heavy precipitation events between 1971 and 2000 at the same station according to the study of Kunz (2011). Together with high horizontal wind speeds between 20 and

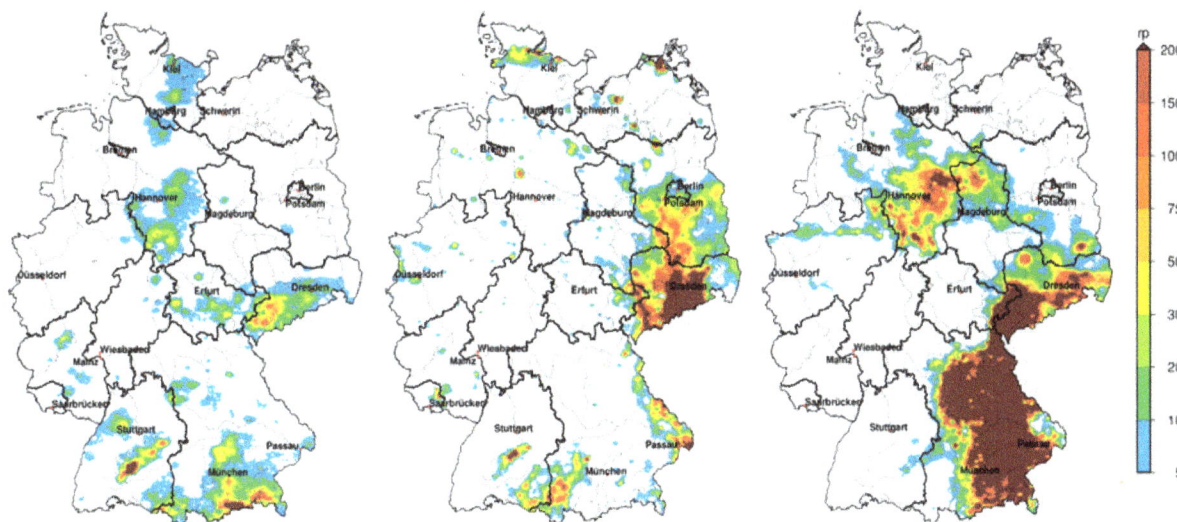

Figure 6. Return periods of 3-day maximum precipitation for each REGNIE grid point derived from data of the period from 1960 to 2009 for the corresponding rain totals displayed in Fig. 4: June 2013 (left), August 2002 (middle), and July 1954 (right).

$75 \, \text{km h}^{-1}$ (850 hPa; around 1500 m a.s.l.) this led to a substantial increase of the incoming water vapour flux (F_{wv}). This quantity can be considered as an upper limit of the conversion of moisture into precipitation (Smith and Barstad, 2004; Kunz, 2011) Thus, the high F_{wv} values observed during the first days of June 2013 plausibly explain the substantial orographic rainfall enhancement over the mountains.

To relate the June 2013 precipitation event to the climatological context, we quantify statistical return periods based on REGNIE data for the period from 1960 to 2009. In Fig. 6 (left), the return periods are displayed only in the range between 5 and 200 years. The estimated values of the return periods have been truncated to 200 years as statistical uncertainty substantially increases for larger return periods due to the short observation period of 50 years. Over the southwestern parts of the Ore Mountains, the Swabian Jura and the very southern border of Bavaria, the return periods are in the range between 5 and 20 years. Only a limited number of grid points show peak values in excess of 100 or even 200 years, for example the aforementioned station of Aschau-Stein. Thus, one can conclude that the rainfall was unusually but not extraordinarily high, and hence cannot fully explain the dimension of the 2013 flood.

The most important rainfall characteristics that were decisive for the 2013 flood can be summarized as: (i) high – but not extraordinary – 3-day totals over parts of the Danube and Elbe catchments; (ii) substantial rainfall increase over the mountains that was decisive for the onset of the flooding; and (iii) areal precipitation occurring almost simultaneously with a slight temporal shift of 2 days between the western and eastern parts of Germany.

These meteorological conditions differ largely from those prevailing during the floods in 2002 and 1954. Areal 3-day rain totals averaged over the upper Elbe catchment (Germany

only, upstream of the confluence of Elbe and Saale) were 49.3 mm compared to 75.9 mm in 2002 and 68.8 mm in 1954. Over the upper Danube catchment (Germany only), the mean areal rain was 75.7 mm compared to 62.5 and 111.2 mm in 2002 and 1954, respectively.

The most striking feature in 2002 was the extreme precipitation over the Ore Mountains reaching values of 312 mm in the 24 h before 13 August 2002, 06:00 UTC, at the station of Zinnwald-Georgenfeld (Ulbrich et al., 2003). The R3d totals (Fig. 4, middle) show a larger area at the eastern parts of the Ore Mountains with values in excess of 300 mm. However, additional high rain totals were only observed at the southern border of Bavaria as well as over the Swabian Jura. This distribution is mainly caused by northerly flow in conjunction with a so-called Vb weather situation (Ulbrich et al., 2003). Comparable to the 2013 event, flood-triggering precipitation occurred with a shift of 2 days between the southern and eastern parts of Germany that correspond to the Danube and Elbe catchments, respectively (Fig. 5, middle). Note that the regions with larger temporal differences in the occurrence of R3d maxima are not associated with high amounts of precipitation (see Fig. 4). Application of extreme value statistics to R3d totals yields return periods of more than 200 years for the maxima. Return periods around 100 years are estimated for the lowlands north of the Ore Mountains (Fig. 6, middle). Precipitation in that region also contributed to the large increase in runoff of the Elbe.

In 1954, most parts of Bavaria experienced 3-day accumulated rainfalls in excess of 150 mm (Fig. 4, right). This was even the case for the lowlands in the north of Bavaria. Near the Alps as well as over the western parts of the Ore Mountains, R3d reached values of 300 mm or even more. These extreme totals recorded within a time shift of only 1 day (Fig. 5, right) correspond to statistical return periods of more

Figure 7. Antecedent precipitation index (API) over 30 days for the floods in June 2013 (left), August 2002 (middle) and July 1954 (right). See text for further details.

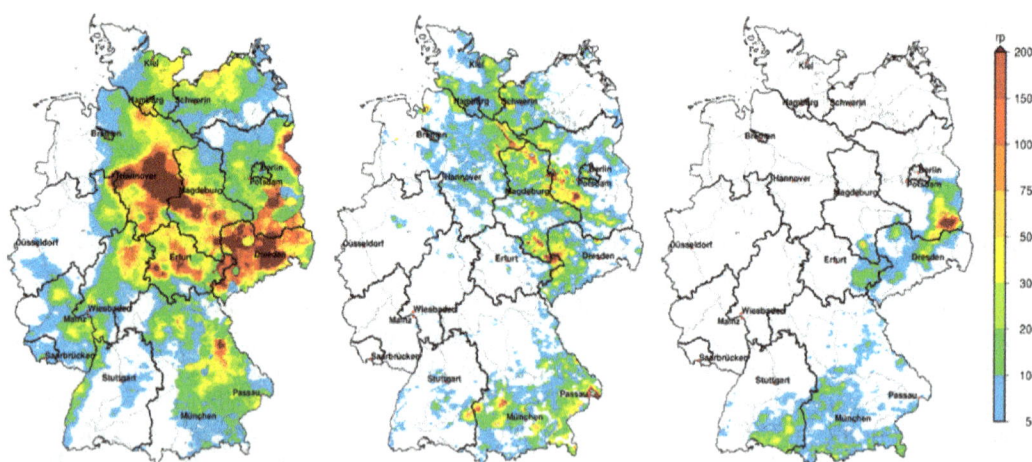

Figure 8. Return periods of the API displayed in Fig. 7 derived from 30-day API of large-scale floods in the period from 1960 to 2009: June 2013 (left), August 2002 (middle) and July 1954 (right).

than 200 years covering more than half of Bavaria (Fig. 6, right). Thus, considering only the observed precipitation directly prior to the onset of the flooding, 1954 was certainly the most extreme event that occurred within the last 60 years.

The same conclusions can be drawn when considering 7-day instead of 3-day maxima (see Fig. A2 in the Appendix). Of course, the rain totals increase for the longer accumulation period, for example over the Ore Mountains in 2013 or in Bavaria for 2002. The estimated return periods, especially in the Elbe and Danube catchments, are less affected by these changes – with the exception of an area in the north of Munich, where return periods in excess of 100 years can be identified for June 2013. Note that the high return periods for 7-day precipitation totals in June 2013 which are visible in Northwest Bavaria are related to the Rhine catchment (see Fig. A1).

3.3   Initial catchment state

3.3.1   Antecedent precipitation

In the next step, we assess initial catchment wetness by means of the antecedent precipitation index (API). This proxy is based on the starting date of R3d (day of the year shown in Fig. 5 minus 3 days) and computed independently at each grid point of REGNIE. API reached high values between 100 mm and in excess of 150 mm over large parts of Germany, especially – and most importantly – over the catchments of Elbe and Danube (Fig. 7, left). At a large number of grid points, especially in the upper Elbe catchment, the return periods are between 100 and 200 years, at some points even in excess of the latter (Fig. 8, left). Note that the maximum that occurred between Hanover and Magdeburg was related to considerable flooding at the Aller, Oker and Leine rivers in the Weser catchment for which no discharge data were available. The high rain totals in the month of May, especially

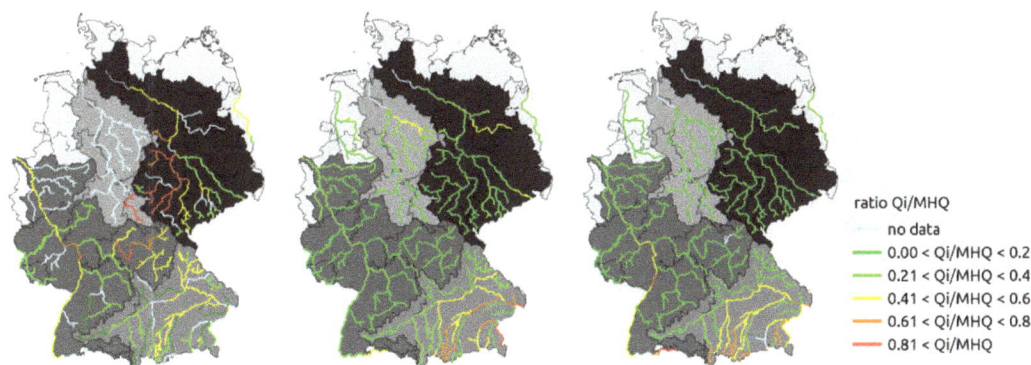

Figure 9. Initial flow ratio at meteorological event start Q_i normalized for MHQ (calculated from AMS 1950–2009) for June 2013 (left), August 2002 (middle) and July 1954 (right).

those at the end of May (recall the increasing weighting of rain totals in API with decreasing temporal distance to R3d), resulted in very wet catchments and filling of storage capacities and thus very favourable conditions for high runoff coefficients.

Regarding the initial moisture conditions, it is found that API was significantly lower prior to the floods in 1954 and 2002, respectively (Fig. 7). In both cases, high values of API up to 150 mm can be observed only over parts of the Bavarian Alps related to orographic precipitation induced by northerly flow directions. Whereas in 2013 the maxima of API correspond well with those of R3d, this is not the case for the two other events. Especially over the Ore Mountains and north of them, where highest rainfall was observed, API was below 50 mm in both cases, yielding return periods below 20 years at most of the grid points (Fig. 8). The same applies to the API in the Danube catchment in 1954. Both in 2002 and 1954 high API values indicate that the initial wetness was comparatively high, but in general not in those regions where the event precipitation was highest (compare Fig. 4 and Fig. 7). Apart from areal precipitation as described above, this is the major difference to the 2013 event.

3.3.2 Initial hydraulic load

As a consequence of the large amounts of rainfall accumulated during the month of May, reflected by the extended areas of high API, also the initial hydraulic load in the river network was already clearly increased at the beginning of the event precipitation in 2013. In general, the pattern of increased initial hydraulic load in the rivers shown in Fig. 9 (left) resembles the spatial distribution of high API values (Fig. 7, left). This mostly applies to the central and southeastern parts of Germany. Most prominent in this regard were the Saale River and its tributaries Wipper and Bode in the western part of the Elbe catchment with an initial flow ratio above 0.8 of MHQ. The Rhine, upper Main, Danube, with tributaries Naab and Isar and the Werra River were also affected. Note that for many gauges in the Weser and lower

Rhine catchments no discharge data are available for the June 2013 flood (see Fig. A1 in the Appendix for geographic locations).

In comparison, for the August 2002 and July 1954 floods the initial hydraulic load of the river network was clearly lower with few exceptions (Fig. 9). In August 2002, basically the Danube and its tributaries Inn, Isar, Lech and Regen showed a noticeable increase of initial river discharge (ca. 0.5 of MHQ). These catchments showed also high API values. Similarly, at the beginning of the July 1954 flood increased river discharges of about 0.4 to 0.8 of MHQ for the Danube and its southern tributaries are visible. Also the middle and upper parts of the Rhine show increased initial hydraulic loads in this range. The lower coincidence of regions of increased initial hydraulic load with regions of increased API for the July 1954 flood (compare Fig. 7 and 9) suggests that the increased initial hydraulic load particularly along the Rhine was induced by different mechanisms than high amounts of antecedent precipitation, presumably due to snow-melt in the alpine headwaters of the Rhine.

From the statistical extreme value analysis applied to the Q_i/MHQ samples at each gauge we obtain an estimate for the return period of the specific initial river flow situation for the June 2013, August 2002 and July 1954 floods. The results presented in Fig. 10 show that for the June 2013 flood the initial flow ratios observed in central Germany, in particular at the upper Main (Rhine catchment), Werra (Weser catchment), Wipper, Saale, Weiße Elster, Mulde (Elbe catchment) and Naab and Vils (Danube catchment), exhibit return periods in the range of 10–50 years, in some river stretches even above 100 years. For the events in August 2002 and July 1954 comparable extremes are only observed for few river stretches in the Danube catchment including the Regen, upper Isar, Ilz, Inn and Salzach rivers in 2002 and the upper Iller, Lech and Isar rivers in 1954.

The initial hydraulic load of the river network (13 400 km) was clearly increased in June 2013 given the comparison to other large-scale flood events from the last 50 years. Hence,

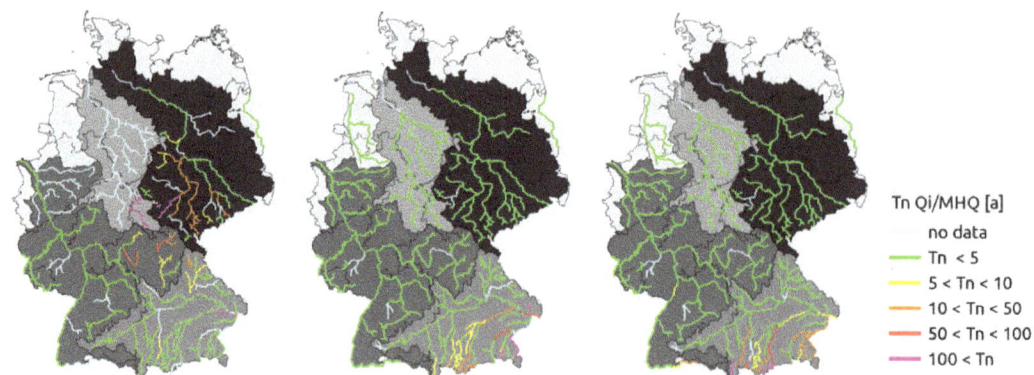

Figure 10. Return periods of initial flow ratio at meteorological event start (Q_i normalized for MHQ) derived from Q_i / MHQ ratios of large-scale floods in the period from 1969 to 2009: June 2013 (left), August 2002 (middle) and July 1954 (right).

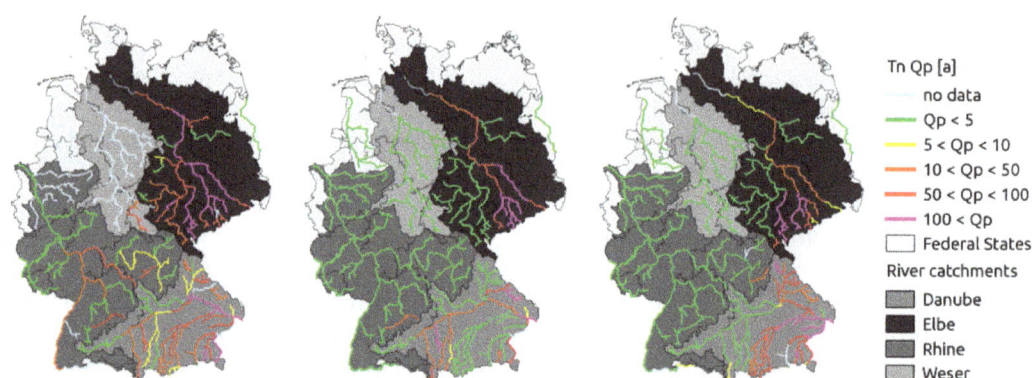

Figure 11. Regionalized return periods (Tn) of flood peak discharges for June 2013 (left), August 2002 (middle) and July 1954 (right). Gauge data were made available by the Water and Shipping Management of the Fed. Rep. (WSV) prepared by the Federal Institute for Hydrology (BfG) and environmental state offices of the federal states.

the aggravating effect of increased initial hydraulic load was stronger in June 2013 than in August 2002 and July 1954. However, extraordinarily high initial flow ratios occurred only in some river stretches, namely the Saale River and its tributaries.

3.4 Peak flood discharges

In June 2013, 45 % of the total river network considered in Germany showed peak discharges above a 5-year flood. As can be seen in Fig. 11 (left), all major catchments showed flooding, namely the Weser, Rhine, Elbe and Danube catchments. Particularly the Elbe and Danube rivers and many of their tributaries were affected by extraordinarily high flood levels. In the Elbe catchment, flood peak discharges exceeded a return period of 100 years along the whole Elbe stretch between Dresden and Wittenberge (Brandenburg), the Mulde, and the tributaries of the Saale River, Weiße Elster and Ilm. In the Danube catchment, the section of the Danube downstream of Regensburg as well as the Inn and Salzach rivers experienced peak discharges with return periods above 100 years. In addition, the Isar, Naab and Iller rivers showed flood

peaks above 50-year return periods. Further, in the Rhine catchment, the Neckar and parts of the Main as well as the Werra River in the Weser catchment experienced peak discharges above the 50 year return period. New record water levels were registered at the Elbe between Coswig and Lenzen (along a total length of 250 km), at the Saale downstream of Halle, and at the Danube in Passau. Severe flooding occurred especially along the Danube and Elbe rivers, as well as along the Elbe tributaries Mulde and Saale, in most cases as a consequence of dike breaches. It is remarkable that large parts of catchments affected by flooding did not receive exceptional amounts of rain (see Fig. 4). In particular, this applies to the upstream parts of the Saale, Werra and Main catchments. However, these regions show high amounts of antecedent precipitation and substantial initial hydraulic load.

The August 2002 and July 1954 floods show peak discharges in the order of 100-year return periods at the Elbe between Dresden and Wittenberg (Saxony-Anhalt), in parts of the Mulde, Regen and Mindel and of 50 years at the Freiberger and Zwickauer Mulde and the Elbe downstream

Table 2. Severity indices for June 2013, August 2002 and July 1954 floods.

Index	June 2013	August 2002	July 1954
Precipitation index (S_{R3d})	16.9	30.1	55.2
Wetness index (S_{API})	114.1	47.3	21.1
Initial hydraulic load index (S_{Q_i})	12.7	6.0	6.1
Flood severity index (S_{Q_p})	74.6	35.4	49.8

of Wittenberg (Saxony-Anhalt) to Wittenberge (Brandenburg) (see Fig. 11, middle and right panels). In July 1954 return periods of 100 years occurred at the Weiße Elster and Mulde in the Elbe catchment and the Isar, Rott and Inn in the Danube catchment. Flood peaks with a return period of 50 years were observed at the Danube-downstream Regensburg, the Naab, Inn and Salzach as well as the upper Isar rivers. However, as can be seen in Fig. 11 (middle and right), the river stretches with high-magnitude flood peaks are clearly less extended in August 2002 and July 1954: the index L describing the spatial flood extent amounts to 19 % in August 2002, 27 % in July 1954 and 45 % in June 2013 (see Fig. A1 in the Appendix for geographic locations).

The major differences of the flood in June 2013 in comparison to August 2002 and July 1954 are that the Elbe, the Mulde and the Saale rivers were affected simultaneously by extraordinary flooding which by superposition of flood waves resulted in unprecedented flood levels particularly in the middle part of the Elbe. Further, nearly all tributaries of the Danube showed flood responses and jointly contributed to the record flood along the Danube downstream of Regensburg. Also the Rhine and Weser catchments were considerably affected even though the magnitude of the peak discharges was not as extreme as in the Elbe and Danube catchments.

3.5 Index-based classification

We evaluate the importance of the individual hydro-meteorological factors within the different flood events using the severity indices introduced in Sect. 2.3. The precipitation-, wetness-, initial hydraulic load- and flood severity indices enable us to compare the 74 past large-scale flood events with regard to the spatial extent and magnitude of each hydro-meteorological factor. This allows for the identification of singularities in terms of extreme situations associated with individual events. The index values for the June 2013, August 2002 and July 1954 events are listed in Table 2.

Among these events, the June 2013 flood is characterized by the highest wetness, initial hydraulic load and flood severity indices which are more than twice the values of the August 2002 flood and with regard to wetness more than five times the value of the July 1954 flood. In contrast, the precipitation index of July 1954 exceeds the value of June 2013

by a factor of 3 and is nearly twice as high as for the August 2002 event. These proportions emphasize the prominent role of extreme antecedent precipitation and increased initial hydraulic load in the river network as key factors for the formation of the extreme flood in June 2013.

Figure 12 shows a scatterplot of the precipitation and wetness indices of the 74 past large-scale floods in Germany. The June 2013 flood is the most extreme in terms of the wetness index, whereas the July 1954 flood is by far the most extreme in terms of the precipitation index. To explore the relationship between precipitation and wetness indices as flood drivers and the flood severity index as dependent variable, we apply a locally weighted scatterplot smooth (LOWESS) model (Cleveland, 1979). For this locally weighted linear least-squares regression, the tri-cube weight function and a span of 50 % are used. The span specifies the percentage of data points that are considered for estimating the response value at a certain location. The performance of the LOWESS model to explain the variation of flood severity is expressed in terms of root mean square error (RMSE) which can be interpreted as the standard deviation of unexplained variance.

The inclined orientation of the response surface indicates that both precipitation and wetness are equally relevant factors to explain resulting flood severity. According to this model, flood severity index values above around 0.5 (normalized values) increase approximately proportionate with precipitation and wetness severity. However, both the concave shape of the response surface, visible for precipitation and wetness index values below 0.5 (normalized values), and the moderate performance of the LOWESS model to explain variability of flood severity (RMSE = 13.2) suggest that additional factors and characteristics influence this relationship. The spatial variability and the corresponding degree of areal overlaps of the factors as well as other hydrological processes, for instance snow melt or seasonal variations in base flow, play a role in this regard.

3.6 Sensitivity Analysis

To check the robustness of our evaluation of the flood in June 2013, it is important to revisit the specifications of parameters of the methodology. Besides, depending on the focus of the analysis the use of different return periods as reference levels for the assessment of severity may be of interest. We examine the implication of varying duration of event precipitation and antecedent precipitation index period as well as different values for the depletion constant for the calculation of API, as well as different return periods as reference levels for the calculation of severity indices following a one-at-a-time sensitivity analysis design (Saltelli et al., 2000). The scenarios examined are listed in Table 3. To assess the implications of these variations on the evaluation of the flood events, we are interested in the changes in the ranking of the flood events with regard to different severity indices. For this purpose, we compare the reference set-up which has been

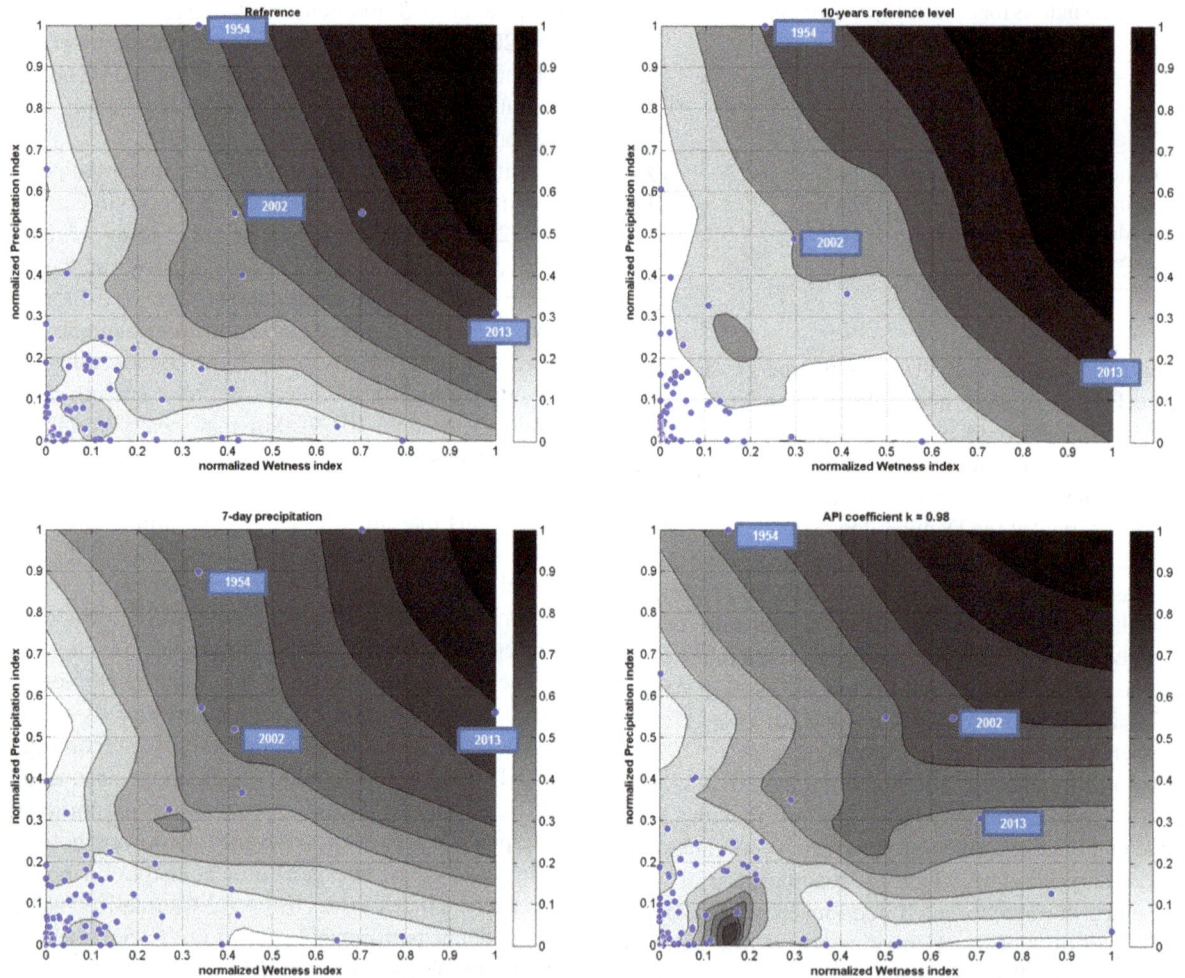

Figure 12. Locally weighted scatterplot smooth (LOWESS) model for the relationship between precipitation and wetness indices as predictors for the flood severity index (grey colour code) of past large-scale flood events in Germany. Top left: Reference (5-year return period as reference level for severity indices, R3d, API 30 days, $k = 0.9$); top right: 10-year return period as reference level for severity indices; bottom left: R7d; bottom right: API k0.98. Note that all severity indices have been normalized to the respective maximum values and that the upper right corners do not contain observed data.

Table 3. Variation scenarios examined within sensitivity analysis.

Scenario	Code	Reference	Variation
Duration event precipitation	R7d	3 days	7 days
Duration antecedent precipitation	API15	30 days	15 days
Depletion constant API	API k0.8	$k = 0.9$	$k = 0.8$
	API k0.98	$k = 0.9$	$k = 0.98$
Return period reference level flood severity	S10a	5 years	10 years
	S25a	5 years	25 years
Return period reference level precipitation severity	P10a	5 years	10 years
	P25a	5 years	25 years
Return period reference level wetness severity	W10a	5 years	10 years
	W25a	5 years	25 years
Return period reference level initial hydraulic load severity	I10a	5 years	10 years
	I25a	5 years	25 years

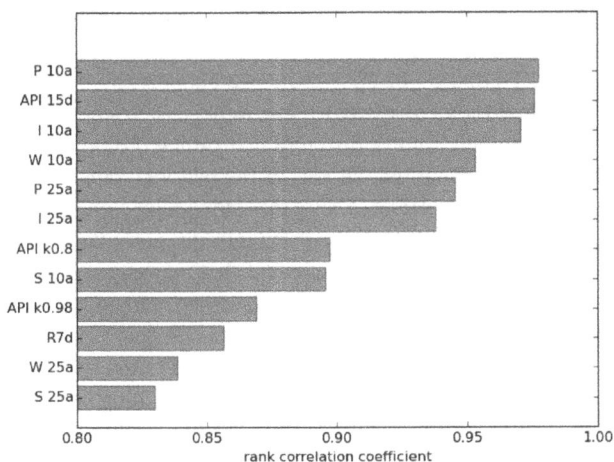

Figure 13. Spearman's rank correlation coefficients between the reference scenario and the variations examined within the sensitivity analysis; for the scenario definition see Table 3.

used to introduce the methodology to the outcomes from the different variations in terms of Spearman's rank correlation coefficient (ρ). Lower rank correlations mean larger differences in the outcomes and indicate a larger sensitivity to these variations.

The implications of these variations are moderate. The correlation coefficients between the reference scenario and these variations are above 0.83 (see Fig. 13). The most sensitive variations are related to changing the return periods used as reference level for the calculation of flood severity and wetness severity indices to 25 years (S25a and W25a). This is followed by duration of event precipitation (R7d) and increasing the depletion constant, i.e. the weight of earlier precipitation within the calculation of API (API k0.98). We track the implications on the outcomes of the LOWESS model for these variations (see Fig. 12 bottom left and right panels for R7d and API k0.98). Further, we examine the changes in LOWESS model outcome for the variation of return periods used as reference level for the calculation of severity indices, i.e. 10 and 25 years. Increasing the return period used as reference level for the calculation of severity indices implies a reduced range of precipitation or discharge observations, and hence, an increased focus on local extremes. For the 25-year level this leads to a pronounced clustering of precipitation and wetness index values below 5 (not shown). Exceptions are the floods in July 1954, August 2002 and June 2013. Using a 10-year return period as reference level the scattering of data points is also low resulting in a less well-defined model for precipitation indices below 0.3 (normalized values) and wetness indices below 0.2 (normalized values, Fig. 12 top right panel). The interpolated surface indicates a stronger inclination towards the wetness index which suggests that flood severity increases disproportionately with catchment wetness. Varying the duration of event precipitation to 7 days (R7d) shifts the attention to events which are

more related to west cyclonic circulation patterns, and thus is rather associated with winter floods (Beurton and Thieken, 2009) but also with the autumn flood in October 1998 (Uhlemann et al., 2014). Accordingly, the October 1998 flood yields the highest precipitation severity index in Fig. 12 (bottom left panel). The increase of the depletion coefficient k within API corresponds to an almost equal weighting of the precipitation over the antecedent precipitation period. As a result three floods achieve higher wetness indices than the flood in June 2013 even though the flood severity of these events is clearly lower. According to the resulting LOWESS model interpolation (see Fig. 12 bottom left panel) the importance of catchment wetness for flood severity is reduced.

Overall, across the variation scenarios examined the relationship between precipitation and wetness indices as flood drivers and the flood severity index as dependent variable is largely comparable. The floods of July 1954, August 2002 and June 2013 remain among the most severe events and mainly determine the shape of the LOWESS model response surface in the region of high severity indices. Hence, the main finding of the index-based classification which points out that both precipitation and wetness are equally relevant factors to explain flood severity remains valid.

4   Conclusions

This study provides new insights into the characteristics of hydro-meteorological factors that caused the flood in June 2013 and presents a statistical evaluation of the associated return periods. The data-based approach further comprises aggregated index values which consider both the spatial extent and magnitudes of the different hydro-meteorological factors and allows for the comparison to past and future large-scale flood events. The results of this analysis proved robust against variations in parameters within the calculation procedure. The large-scale flood database and the methodological framework developed enable the rapid assessment of future floods based on precipitation and discharge observations.

The results illustrate that the sequence of prevalent circulation patterns in May 2013 introduced an important boundary condition for the extraordinary precipitation anomaly observed. For this flood, diverse hydro-meteorological factors showed exceptional characteristics.

First, the development of event precipitation and in particular the substantial orographic rainfall enhancement was driven by a very low lifting condensation level in combination with high amounts of precipitable water in the atmosphere. This was continuously sustained by the strong influx of high water vapour resulting from a strong and persistent flow of air from the north to northeast.

Second, during the weeks before the onset of the flood, enormous amounts of antecedent precipitation occurred over large parts of Germany. As the areas of high antecedent and event precipitation were amply overlapping, the wet initial conditions strongly intensified the runoff response to event precipitation. Hence, particularly the large areal superposition and interplay of event precipitation and wet initial catchment conditions proved to be key drivers for the exceptional hydrological severity of the flood in June 2013. In the Saale catchment the increased initial hydraulic load in the river network was an additional aggravating factor. In the Danube, the movement of the event precipitation field from west to east, i.e. following the streamflow direction, amplified the superposition of the flood waves from the tributaries.

Third, the spatial extent of high-magnitude flood peaks marks a new record for large-scale floods in Germany for at least the last 60 years and set new record water levels along extensive river sections in Germany.

In comparison, the flood in August 2002 was triggered in Germany by extremely intense precipitation which was relatively localized in the Ore Mountains. Initial wetness showed considerably high values in some parts of Germany but these areas did not coincide largely with event precipitation. The flooding in July 1954 was for the main part caused by exceptional amounts of event precipitation affecting large parts of Bavaria. In comparison to August 2002 and June 2013, initial wetness was a less important factor in Germany. However, at the northern ridge of the Alps initial wetness contributed to flood generation in the Salzach and Inn rivers (Blöschl et al., 2013).

Our results show that the influence of catchment wetness is a considerable factor for high-return period, large-scale floods in Germany. In this regard we support the hypothesis that hydrological extremes are rather a consequence of unusual combinations of different hydro-meteorological factors than of unusual magnitudes of the factors themselves as stated by Klemes (1993). Using the knowledge gained about the characteristics, the range of magnitudes and interactions of the various hydro-meteorological factors associated with large-scale floods from the past 60 years, we can advance the derivation of plausible extreme scenarios. In this regard, the database compiled for large-scale floods in Germany may be analysed concerning the possibilities of coinciding extremes of individual hydro-meteorological factors, as for instance the combination of initial wetness observed in June 2013 and event precipitation as in July 1954. Of course, the development of such scenarios requires an in-depth analysis of synoptic meteorological situations and the corresponding transition of related weather conditions. The hydrological evaluation of such extreme scenarios could provide new insights for large-scale flood hazard assessment, planning scenarios for national disaster response, spatial risk as well as cumulated flood losses. These insights may find further use in advanced approaches for flood frequency analysis and design flood estimation (e.g. Merz and Blöschl, 2008; Paquet et al., 2013).

Appendix A

Figure A1. Outline map of geographic locations referred to in the text.

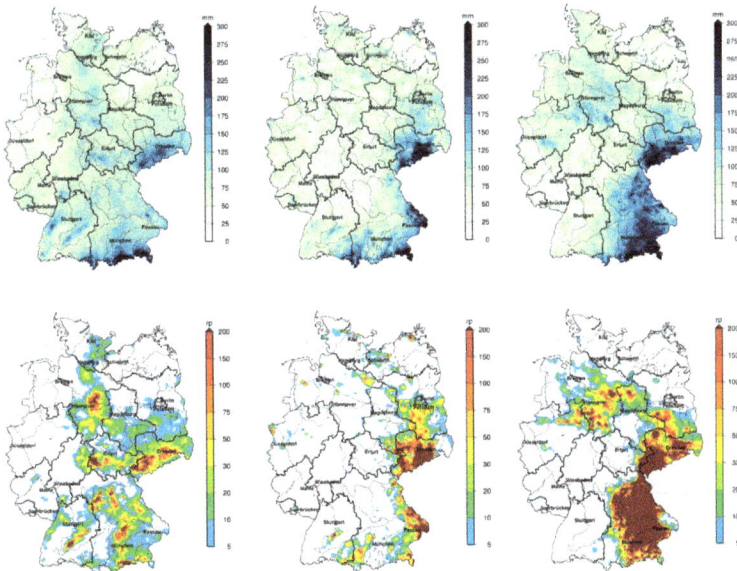

Figure A2. As Figs. 4 and 6, but for 7-day maximum precipitation (top line: 7-day maximum precipitation; bottom line: return periods – June 2013, left; August 2002, middle; July 1954, right).

Acknowledgements. We thank Michel Lang, Christian Reszler, Massimiliano Zappa and two anonymous referees for their thoughtful comments and helpful suggestions to improve the paper.

The study was undertaken and financed under the framework of CEDIM – Center for Disaster Management and Risk Reduction Technology. We gratefully thank the German Weather Service (DWD) for providing REGNIE data, Bavarian State Office of Environment (LfU), Baden-Württemberg Office of Environment, Measurements and Environmental Protection (LUBW), Brandenburg Office of Environment, Health and Consumer Protection (LUGV), Saxony State Office of Environment, Agriculture and Geology (SMUL), Saxony-Anhalt Office of Flood Protection and Water Management (LHW), Thüringen State Office of Environment and Geology (TLUG), Hessian Agency for the Environment and Geology (HLUG), Rhineland Palatinate Office of Environment, Water Management and the Factory Inspectorate (LUWG), Saarland Ministry for Environment and Consumer Protection (MUV), Office for Nature, Environment and Consumer Protection North Rhine-Westphalia (LANUV NRW), Lower Saxony Office for Water Management, Coast Protection and Nature Protection (NLWKN), Water and Shipping Management of the Fed. Rep. (WSV), prepared by the Federal Institute for Hydrology (BfG) who provided discharge data.

The service charges for this open access publication have been covered by a Research Centre of the Helmholtz Association.

Edited by: N. Ursino

References

Ahmed, N. U.: Estimating soil moisture from $6 \cdot 6$ GHz dual polarization, and/or satellite derived vegetation index, Int. J. Remote Sens., 16, 687–708, doi:10.1080/01431169508954434, 1995.

Alfieri, L., Pappenberger, F., and Wetterhall, F.: The extreme runoff index for flood early warning in Europe, Nat. Hazards Earth Syst. Sci., 14, 1505–1515, doi:10.5194/nhess-14-1505-2014, 2014.

Beurton, S. and Thieken, A.: Seasonality of floods in Germany, Hydrol. Sci. J., 54, 62–76, doi:10.1623/hysj.54.1.62, 2009.

BfG: Länderübergreifende Analyse des Juni Hochwassers 2013, Bundesanstalt für Gewässerkunde, Deutscher Wetterdienst, Koblenz, 2013.

Blanchard, B. J., McFarland, M. J., Schmugge, T. J., and Rhoades, E.: Estimation of Soil Moisture with Api Algorithms and Microwave Emission1, JAWRA J. Am. Water Resour. Assoc., 17, 767–774, doi:10.1111/j.1752-1688.1981.tb01296.x, 1981.

Blöschl, G., Nester, T., Komma, J., Parajka, J., and Perdigão, R. A. P.: The June 2013 flood in the Upper Danube Basin, and comparisons with the 2002, 1954 and 1899 floods, Hydrol. Earth Syst. Sci., 17, 5197–5212, doi:10.5194/hess-17-5197-2013, 2013.

CEDIM: June 2013 Flood in Central Europe – Focus Germany Report 1 – Update 2: Preconditions, Meteorology, Hydrology, Center for Disaster Management and Risk Reduction Technology, Potsdam, available at: http://www.cedim.de/download/ FDA_Juni_Hochwasser_Bericht1-ENG.pdf, last access: 17 October, 2013a.

CEDIM: June 2013 Flood in Central Europe – Focus Germany Report 2 – Update 1: Impact and Management, Center for Disaster Management and Risk Reduction Technology, Potsdam, available at: http://www.cedim.de/download/ FDA-Juni-Hochwasser-Bericht2-ENG.pdf, last access: 17 October, 2013b.

Chow, V. T.: Open-channel hydraulics, McGraw-Hill., New York, NY, USA, 710 pp., 1959.

Cleveland, W. S.: Robust Locally Weighted Regression and Smoothing Scatterplots, J. Am. Stat. Assoc., 74, 829–836, doi:10.2307/2286407, 1979.

Deutscher Bundestag: Bericht zur Flutkatastrophe 2013: Katastrophenhilfe, Entschädigung, Wiederaufbau, 2013.

Duckstein, L., Bárdossy, A., and Bogárdi, I.: Linkage between the occurrence of daily atmospheric circulation patterns and floods: an Arizona case study, J. Hydrol., 143, 413–428, doi:10.1016/0022-1694(93)90202-K, 1993.

DWD: Deutschlandwetter im Frühling 2013, Wetter Klima – Dtsch. Wetterd. – Presse, available at: http://www.dwd.de/bvbw/ generator/DWDWWW/Content/Presse/Pressemitteilungen/ 2013/20130529__DeutschlandwetterimFruehling, templateId=raw,property=publicationFile.pdf/20130529_ DeutschlandwetterimFruehling.pdf, last access: 6 June 2013.

Embrechts, P., Klüppelberg, C., and Mikosch, T.: Modelling Extremal Events – for Insurance and Finance, Springer, Berlin, Heidelberg, available at: http://www.springer.com/mathematics/ quantitative+finance/book/978-3-540-60931-5 (last access: 20 June 2014), 1997.

Ettrick, T. M., Mawdlsey, J. A., and Metcalfe, A. V.: The influence of antecedent catchment conditions on seasonal flood risk, Water Resour. Res., 23, 481–488, doi:10.1029/WR023i003p00481, 1987.

GDV: Erste Schadenbilanz: Hochwasser 2013 verursacht 180.000 versicherte Schäden in Höhe von fast 2 Milliarden Euro, available at: http://www.presseportal.de/pm/39279/2505807/, last access: 2 July 2013.

Grams, C. M., Binder, H., Pfahl, S., Piaget, N., and Wernli, H.: Atmospheric processes triggering the central European floods in June 2013, Nat. Hazards Earth Syst. Sci., 14, 1691–1702, doi:10.5194/nhess-14-1691-2014, 2014.

Klemes, V.: Probability of extreme hydrometeorological events – a different approach, in: Extreme Hydrological Events: Precipitation, Floods and Droughts, IAHS, Yokohama, 213, 167–176, 1993.

Köhler, M. A. and Linsley, R. K., Jr.: Predicting runoff from storm rainfall, US Weather Bureau, Washington, D.C., p. 9, 1951.

Kron, W.: Zunehmende Überschwemmungsschäden: Eine Gefahr für die Versicherungswirtschaft?, DCM, Meckenheim, Würzburg, 47–63, 2004.

Kunz, M.: Characteristics of Large-Scale Orographic Precipitation in a Linear Perspective, J. Hydrometeorol., 12, 27–44, doi:10.1175/2010JHM1231.1, 2011.

Merz, R. and Blöschl, G.: A process typology of regional floods, Water Resour. Res., 39, 1340, doi:10.1029/2002WR001952, 2003.

Merz, B. and Plate, E. J.: An analysis of the effects of spatial variability of soil and soil moisture on runoff, Water Resour. Res., 33, 2909–2922, doi:10.1029/97WR02204, 1997.

Merz, B., Elmer, F., Kunz, M., Mühr, B., Schröter, K. and Uhlemann-Elmer, S.: The extreme flood in June 2013 in Germany, Houille Blanche, 1, 5–10, doi:10.1051/lhb/2014001, 2014.

Merz, R. and Blöschl, G.: Flood frequency hydrology: 1. Temporal, spatial, and causal expansion of information, Water Resour. Res., 44, W08432, doi:10.1029/2007WR006744, 2008.

Munich Re: Natural Catastrophes 2013 Analyses, assessments, positions, Munich Re, Munich, 65 pp., 2013.

Nied, M., Hundecha, Y., and Merz, B.: Flood-initiating catchment conditions: a spatio-temporal analysis of large-scale soil moisture patterns in the Elbe River basin, Hydrol. Earth Syst. Sci., 17, 1401–1414, doi:10.5194/hess-17-1401-2013, 2013.

Paquet, E., Garavaglia, F., Garçon, R., and Gailhard, J.: The SCHADEX method: A semi-continuous rainfall–runoff simulation for extreme flood estimation, J. Hydrol., 495, 23–37, doi:10.1016/j.jhydrol.2013.04.045, 2013.

Perry, M. A. and Niemann, J. D.: Analysis and estimation of soil moisture at the catchment scale using EOFs, J. Hydrol., 334, 388–404, doi:10.1016/j.jhydrol.2006.10.014, 2007.

Rauthe, M., Steiner, H., Riediger, U., Mazurkiewicz, A., and Gratzki, A.: A Central European precipitation climatology – Part I: Generation and validation of a high-resolution gridded daily data set (HYRAS), Meteorol. Z., 22, 235–256, doi:10.1127/0941-2948/2013/0436, 2013.

Reager, J. T., Thomas, B. F., and Famiglietti, J. S.: River basin flood potential inferred using GRACE gravity observations at several months lead time, Nat. Geosci., 7, 588–592, doi:10.1038/ngeo2203, 2014.

Saltelli, A., Chan, K., and Scott, E. M.: Sensitivity Analysis, John Wiley & Sons Ltd, Chichester, UK, 475 pp., 2000.

Smith, R. B. and Barstad, I.: A Linear Theory of Orographic Precipitation, J. Atmospheric Sci., 61, 1377–1391, doi:10.1175/1520-0469(2004)061< 1377:ALTOOP> 2.0.CO;2, 2004.

Van Steenbergen, N. and Willems, P.: Increasing river flood preparedness by real-time warning based on wetness state conditions, J. Hydrol., 489, 227–237, doi:10.1016/j.jhydrol.2013.03.015, 2013.

Strahler, A. N.: Quantitative analysis of watershed geomorphology, Trans. Am. Geophys. Union, 38, 913–920, doi:10.1029/TR038i006p00913, 1957.

Swiss Re: Sigma preliminary estimates: natural catastrophes and man-made disasters in 2013 cost insurers worldwide USD 44 billion, Prelim. Catastr. Estim. 2013 Swiss Re – Lead. Glob. Reinsur, available at: http://www.swissre.com/media/news_releases/nr_20131218_sigma_natcat_2013.html, last access: 18 December 2013.

Teng, W. L., Wang, J. R., and Doraiswamy, P. C.: Relationship between satellite microwave radiometric data, antecedent precipitation index, and regional soil moisture, Int. J. Remote Sens., 14, 2483–2500, doi:10.1080/01431169308904287, 1993.

Thieken, A. H., Müller, M., Kreibich, H., and Merz, B.: Flood damage and influencing factors: New insights from the August 2002 flood in Germany, Water Resour. Res., 41, 1–16, doi:10.1029/2005WR004177, 2005.

Troch, P. A., Smith, J. A., Wood, E. F., and de Troch, F. P.: Hydrologic controls of large floods in a small basin: central Appalachian case study, J. Hydrol., 156, 285–309, doi:10.1016/0022-1694(94)90082-5, 1994.

Uhlemann, S., Thieken, A. H., and Merz, B.: A consistent set of trans-basin floods in Germany between 1952–2002, Hydrol. Earth Syst. Sci., 14, 1277–1295, doi:10.5194/hess-14-1277-2010, 2010.

Uhlemann, S., Thieken, A. H., and Merz, B.: A quality assessment framework for natural hazard event documentation: application to trans-basin flood reports in Germany, Nat. Hazards Earth Syst. Sci., 14, 189–208, doi:10.5194/nhess-14-189-2014, 2014.

Ulbrich, U., Brücher, T., Fink, A. H., Leckebusch, G. C., Krüger, A., and Pinto, J. G.: The central European floods of August 2002: Part 1 – Rainfall periods and flood development, Weather, 58, 371–377, doi:10.1256/wea.61.03A, 2003.

Viessman, W. and Lewis, G. L.: Introduction to Hydrology, 5th edition., Prentice Hall, New York, USA, 612 pp., 2002.

5

Calibration approaches for distributed hydrologic models in poorly gaged basins: implication for streamflow projections under climate change

S. Wi[1], Y. C. E. Yang[1], S. Steinschneider[1], A. Khalil[2], and C. M. Brown[1]

[1]Department of Civil and Environmental Engineering, University of Massachusetts Amherst, USA
[2]The World Bank, Washington, DC, USA

Correspondence to: S. Wi (sungwookwi@gmail.com)

Abstract. This study tests the performance and uncertainty of calibration strategies for a spatially distributed hydrologic model in order to improve model simulation accuracy and understand prediction uncertainty at interior ungaged sites of a sparsely gaged watershed. The study is conducted using a distributed version of the HYMOD hydrologic model (HYMOD_DS) applied to the Kabul River basin. Several calibration experiments are conducted to understand the benefits and costs associated with different calibration choices, including (1) whether multisite gaged data should be used simultaneously or in a stepwise manner during model fitting, (2) the effects of increasing parameter complexity, and (3) the potential to estimate interior watershed flows using only gaged data at the basin outlet. The implications of the different calibration strategies are considered in the context of hydrologic projections under climate change. To address the research questions, high-performance computing is utilized to manage the computational burden that results from high-dimensional optimization problems. Several interesting results emerge from the study. The simultaneous use of multisite data is shown to improve the calibration over a stepwise approach, and both multisite approaches far exceed a calibration based on only the basin outlet. The basin outlet calibration can lead to projections of mid-21st century streamflow that deviate substantially from projections under multisite calibration strategies, supporting the use of caution when using distributed models in data-scarce regions for climate change impact assessments. Surprisingly, increased parameter complexity does not substantially increase the uncertainty in streamflow projections, even though parameter equifinality does emerge. The results suggest that increased (excessive) parameter complexity does not always lead to increased predictive uncertainty if structural uncertainties are present. The largest uncertainty in future streamflow results from variations in projected climate between climate models, which substantially outweighs the calibration uncertainty.

1 Introduction

In an effort to advance hydrologic modeling and forecasting capabilities, the development and implementation of physically based, spatially distributed hydrologic models has proliferated in the hydrologic literature, supported by readily available geographic information system (GIS) data and rapidly increasing computational power. Distributed hydrologic models can account for spatially variable physiographic properties and meteorological forcing (Beven, 2012), improving simulations compared to conceptual, lumped models for basins where spatial rainfall variability effects are significant (Ajami et al., 2004; Koren et al., 2004; Reed et al., 2004; Khakbaz et al., 2012; Smith et al., 2012) and for nested basins (Bandaragoda et al., 2004; Brath et al., 2004; Koren et al., 2004; Safari et al., 2012; Smith et al., 2012). The benefits of distributed modeling have been recognized by the U.S. National Oceanic and Atmospheric Administration's National Weather Service (NOAA/NWS) and demonstrated in the Distributed Model Intercomparison Project (DMIP) (Reed et al., 2004; Smith et al., 2004, 2012, 2013). Importantly, distributed hydrologic models can evaluate hydrological response at interior ungaged sites, a benefit not afforded by lumped models. The use of distributed hydrologic modeling for interior point streamflow estimation is particularly relevant for poorly gaged river basins in developing countries, where reliable predictions at interior sites are often required to inform water infrastructure investments. As international development agencies begin to integrate climate change considerations into their decision-making processes (e.g., Yu et al., 2013), these investments need to be robust under both current climate conditions and possible future climate regimes.

Despite their roots in physical realism, distributed hydrologic models can suffer from substantial uncertainty. A major source of uncertainty originates from the proper identification of parameter values that vary across the watershed, especially when observed streamflow data is only available at one or a few points (Exbrayat et al., 2014). Parameters can be discretized across the watershed in several ways (Flugel, 1995; Efstratiadis et al., 2008; Khakbaz et al., 2012): uniquely for each grid cell or hydrologic response unit (fully distributed), based on sub-basins whose boundaries do not necessarily ensure homogenous characteristics (semi-distributed) or, in the simplest case, a single parameter set for all model grid cells (lumped). With limited data, the parameter identification problem, particularly for the fully distributed case, can be impractical or infeasible (Beven, 2001). The parameterization challenge has spurred substantial advances in understanding appropriate calibration techniques for distributed hydrologic models. Many studies have attempted to reduce the dimensionality of the calibration problem to alleviate the issue of equifinality (Beven and Freer, 2001), which is the phenomenon whereby multiple parameter sets produce indistinguishable model performance. This work has found favorable results when the parametric complexity of the distributed model is aligned with the data available for calibration (Leavesley et al., 2003; Ajami et al., 2004; Eckhardt et al., 2005; Frances et al., 2007; Zhu and Lettenmaier, 2007; Cole and Moore, 2008; Pokhrel and Gupta, 2010; Khakbaz et al., 2012). There has also been extensive research exploring the use of multiple objectives and different operational procedures to understand parameter estimation tradeoffs and identifiability for distributed model calibration, with great success (Madsen, 2003; Efstratiadis and Koutsoyiannis, 2010; Li et al., 2010; Kumar et al., 2013).

Despite these advances, important questions still persist. It still remains difficult to compare the uncertainty that emerges from different operational calibration procedures for multisite applications (i.e., whether gages in series should be used sequentially or simultaneously for calibration) and under different levels of parametric complexity. Due to the computational burden required to calibrate distributed models, this uncertainty is problematic to explore. Furthermore, in poorly gaged basins, it is challenging to quantify the lost accuracy and increased uncertainty for interior flow estimation when a distributed model is calibrated only at an out-

let gage (which is often all that is available in developing-country river basins). In the case of significant spatial variability in the basin properties that influence runoff generation (e.g., permeability, vegetation, and slope), accurate runoff predictions are unlikely at interior locations based only on the lumped information obtained at the basin outlet (Anderson et al., 2001; Cao et al., 2006; Breuer et al., 2009; Lerat et al., 2012; Smith et al., 2012; Wang et al., 2012). The extent of this error and uncertainty is not well understood for heterogeneous basins due to the computational expense required to explore this issue. Finally, rarely have the implications of these calibration issues been explicitly examined for possible future climate conditions, which is required in climate change impact studies. This question has been explored for lumped, conceptual models (Wilby, 2005; Steinschneider et al., 2012), but has been difficult to evaluate for computationally expensive distributed models.

This study addresses the above research challenges by focusing on the following four questions: (1) how does calibration procedure for using multisite data affect the accuracy and uncertainty of distributed models used for streamflow predictions at ungaged sites; (2) what effects does increased parameter complexity have on distributed model calibration and prediction; (3) how much degradation in model accuracy and uncertainty can be expected for interior flow estimation based on a calibration procedure using only the basin outlet; and (4) how do different calibration formulations for a distributed model alter projections of streamflow at ungaged sites under climate change conditions? These questions are considered in an application of a distributed version of the daily HYMOD hydrologic model to the Kabul River basin in Afghanistan and Pakistan. To address these research questions, high-performance computing is utilized to manage the computational burden that often hinders such explorations (Laloy and Vrugt, 2012; Zhang et al., 2013).

2 Study area

The Kabul River basin ($67\,370\,km^2$) is a plateau surrounded by mountains located in the eastern central part of Afghanistan (Fig. 1). It is the most important river basin of Afghanistan, containing 35 % of the country's population. While it encompasses just 12 % of the area of Afghanistan, the basin's average annual streamflow (about 24 billion cubic meters) is about 26 % of the country's total streamflow volume (World Bank, 2010).

Water resources from the basin are shared by Afghanistan and Pakistan and serve as a water supply source for more than 20 million people. The shared use of transboundary water between these two countries is central in establishing regional water resources development for this area (Ahmad, 2010). It is crucial to develop tools that can support engineering plans for existing and potential water infrastructure to take full advantage of the water resources in the basin. The government

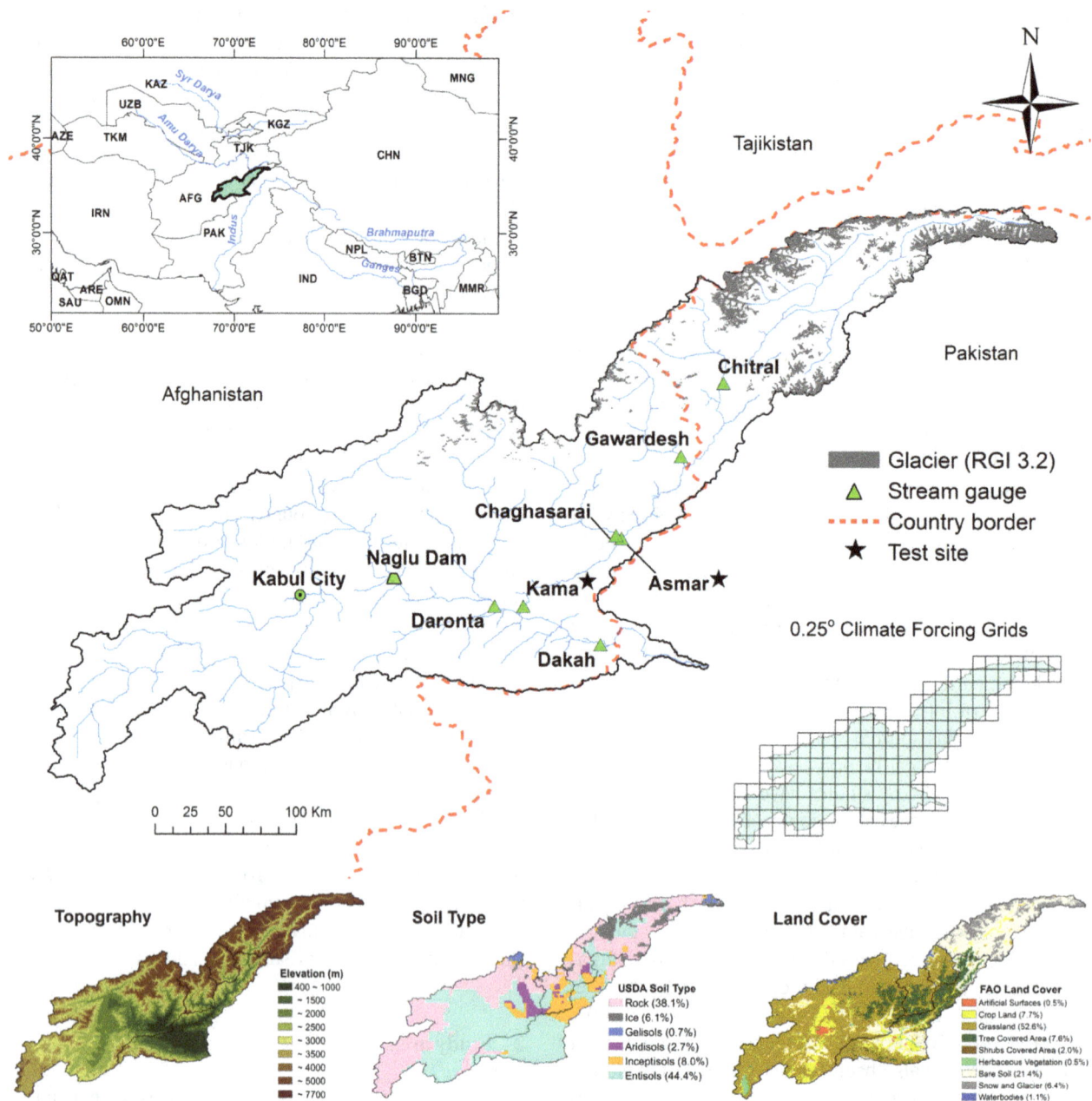

Figure 1. Kabul River basin.

of Afghanistan has developed comprehensive plans for new hydropower projects on the Kabul River owing to its advantageous topography for the development of water storage and hydropower (IUCN, 2010), and recently reached an agreement with the Pakistan government to work on a 1500 MW hydropower project on the Kunar River (one of major tributary in the Kabul River basin) as part of the joint management of common rivers between the two countries (DAWN, 2013). The streamflow regime of the Kabul River can be classified as glacial with maximum streamflow in June or July and minimum streamflow during the winter season. Approximately

70 % of annual precipitation (475 mm) falls during the winter season (November–April). While the dominant source of streamflow in winter is baseflow and winter rainfall, glaciers and snow cover are the most important long-term forms of water storage and, hence, the main source of runoff during the ablation period for the basin (Shakir et al., 2010). In total, 2.9 % (1954 km^2) of the basin is glacierized based on the Randolph Glacier Inventory version 3.2 (Pfeffer et al., 2014). The meltwater from glaciers and snow produce the majority (75 %) of the total streamflow (Hewitt et al., 1989). Table 1 provides the climates and geophysical properties of

Table 1. Streamflow gaging stations in the Kabul River basin.

Data source	Station name	River	Data period		Physiographic property			Basin climate		
			Start	End	Drainage area (km^2)	Glacier area (%)	Mean elev. (m)	Mean annual Prcp. (mm)	Mean annual mean Temp. (°C)	Mean annual flow (mm)
USGS/ GRDC	Dakah	Kabul	2/1968	7/1980	67370	2.9	2883	418	7.7	282
USGS/ GRDC	Pul-i-Kama	Kunar	1/1967	9/1979	26005	7.3	3446	446	5.6	573
USGS	Asmar	Kunar	3/1960	9/1971	19960	9.4	3716	483	4.1	651
GRDC	Chitral	Kunar	1/1978	12/1981	11396	14.4	4126	518	2.1	698
USGS	Gawardesh	Landaisin	5/1975	6/1978	3130	2.1	3707	555	4.5	521
USGS/ GRDC	Chaghasarai	Pech	2/1960	2/1979	3855	0.4	3141	482	7.4	535
USGS/ GRDC	Daronta	Kabul	10/1959	9/1964	34375	0.3	2722	350	8.0	165

each sub-watershed delineated by the stations located inside the Kabul Basin (Fig. 1). Two different climate patterns are distinguishable across the sub-basins. The sub-basins on the Kunar River tributary (Kama, Asmar, Chitral, Gawardesh, and Chaghasarai) receive moderate annual precipitation and are highly affected by snow and glacier covers. All of these sub-basins have high ratios of mean annual flow to mean annual precipitation, with the ratios for the Kama, Asmar, Chitral, and Chaghasarai sub-basins larger than 1. Conversely, the Daronta sub-basin contains only minimal glacial cover, and is relatively dry. Daronta is also much less productive, with annual streamflow far below the other sub-basins with an average of only 165 mm yr^{-1}.

Issues of shared water resources between Afghanistan and Pakistan in the Kabul River basin are becoming complex due to the impacts of climatic variability and change (IUCN, 2010). The vulnerability of glacial streamflow regimes to changes in temperature and precipitation (Stahl et al., 2008; Immerzeel et al., 2012; Radic et al., 2014) highlights the need to assess the impact of climate change on future water availability in this area.

3　Data and models

3.1　Data

Gridded daily precipitation and temperature products with a spatial resolution of 0.25 °C were gathered between calendar years 1961 and 2007 from the Asian Precipitation Highly Resolved Observational Data Integration Towards Evaluation (APHRODITE) data set (Yatagai et al., 2012). There has been some concern regarding underestimation of precipitation in APHRODITE for some regions of Asia (Palazzi et al., 2013); our preliminarily data analysis (intercomparison of precipitation products between five different databases) confirmed this for the Kabul River basin (shown in Fig. S1 in

the Supplement). Thus, the APHRODITE precipitation was bias-corrected by the precipitation product from the University of Delaware global terrestrial precipitation (UD) data set (Legates and Willmott, 1990). Daily series of bias-corrected APHRODITE precipitation were coupled with APHRODITE temperature for 160 0.25 °C grid cells to produce a climate forcing data set for the distributed domain of the Kabul River basin model.

This study used the set of global climate change simulations from the World Climate Research Programme's Coupled Model Intercomparison Project Phase 5 (CMIP5) multimodel ensemble (Talyor et al., 2012). Monthly climate outputs of GCMs (general circulation models) were downscaled to a daily temporal resolution and 0.25 °C spatial resolution based on the bias-correction spatial disaggregation (BCSD) statistical downscaling method introduced by Wood et al. (2004).

Monthly streamflow observations for seven locations in the Kabul River basin (Fig. 1) were gathered between calendar years 1960 and 1981 from two data sources: the Global Runoff Data Centre (GRDC) database and the United States Geological Survey (USGS) database (Table 1). Streamflow data were not collected in Afghanistan after September 1980 until recently because stream gaging was discontinued soon after the Soviet invasion of Afghanistan in 1979 (Olson and Williams-Sether, 2010). Though measurements were taken at a daily time step, data are only made available for public use at monthly aggregated levels, calculated using the mean of the daily values. The available monthly streamflow observations at each station were used for calibrating and validating the distributed hydrologic model (Fig. 2). Kama and Asmar stations are treated as ungaged sites because they align with the potential dam project on the Kunar River tributary. The two gage stations are left out of the processes of multisite calibrations in order to evaluate the model's ability to predict streamflow at interior ungaged sites. Furthermore, half of the

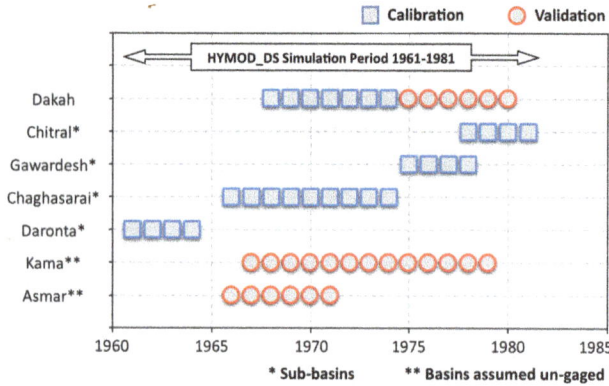

Figure 2. Streamflow data usage for the model calibration and validation.

record at the Dakah station, located at the basin outlet, is also used for validation purposes.

The Randolph Glacier Inventory version 3.2 (RGI 3.2) data set (Pfeffer et al., 2014) was used to extract glacial coverage in the Kabul River basin, which totaled 5.7 % of the basin area (Fig. S2). In the hydrological modeling process, the model needs to be informed by reliable estimates on volume of water retained in glaciers, especially for future simulations under warming conditions. We followed the method proposed in Grinsted (2013), which uses multivariate scaling relationships to estimate glacier and ice cap volume based on elevation range and area. Specifically, the scaling law including area and elevation range factors was applied to estimate glacier/ice cap volume when the glacier depth exceeded 10 m. Otherwise, glacier/ice cap volume was estimated with the area–volume scaling law. The elevation range spanned by each individual glacier is estimated using the global digital elevation model (DEM) from the shuttle radar topography mission (SRTMv4) in 250 m resolution (Jarvis et al., 2008). Density of ice ($0.9167\,\mathrm{g\,cm^{-3}}$) is applied to calculate glacier/ice cap volume in meters of water equivalent.

The database for land covers and soil types of the Kabul River basin (Fig. 1) are provided by the Food and Agriculture Organization of the United Nations (Latham et al., 2014) and United States Department of Agriculture – Natural Resources Conservation Service Soils (USDA-NRCS, 2005), respectively.

3.2 Distributed Hydrologic Model (HYMOD_DS)

In this study the lumped conceptual hydrological model HYMOD (Boyle, 2001) is coupled with a river routing model to be suitable for modeling a distributed watershed system. We name it HYMOD_DS denoting the distributed version of HYMOD. Snow and glacier modules have been introduced to enhance the modeling process for glacier and snow covered areas within the Kabul River basin. The HYMOD_DS is composed of hydrological process modules that repre-

sent soil moisture accounting, evapotranspiration, snow processes, glacier processes and flow routing. The model operates on a daily time step and requires daily precipitation and mean temperature as input variables. The overall model structure of the HYMOD_DS and its 15 parameters are described in Fig. 3 and Table 2, respectively. Further details are provided below.

The HYMOD conceptual watershed model has been extensively used in studies on streamflow forecasting and model calibration (Wagener et al., 2004; Vrugt et al., 2008; Kollat et al., 2012; Gharari et al., 2013; Remesan et al., 2013). The HYMOD is a soil moisture accounting model based on the probability–distributed storage capacity concept proposed by Moore (1985). This conceptualization represents a cumulative distribution of varying storage capacities (C) with the following function:

$$F(C) = 1 - \left(1 - \frac{C}{C_{\max}}\right)^B \quad 0 \leq C \leq C_{\max}, \tag{1}$$

where the exponent B is a parameter controlling the degree of spatial variability of storage capacity over the basin and $C_{\max}$ is the maximum storage capacity. The model assumes that all storages within the basin are filled up to the same critical level ($C^*(t)$), unless this amount exceeds the storage capacity of that particular location. With this assumption, the total water storage $S(t)$ contained in the basin corresponds to

$$S(t) = \frac{C_{\max}}{B+1}\left(1 - \left(1 - \frac{C^*(t)}{C_{\max}}\right)^{B+1}\right). \tag{2}$$

Consequently, two parameters are introduced for the runoff generation process with two components:

$$\mathrm{Runoff}_1 = \begin{cases} P(t) + C^*(t-1) - C_{\max} \text{ if } P(t) \\ \quad + C^*(t-1) \geq C_{\max} \\ 0 \text{ if } P(t) + C^*(t-1) < C_{\max} \end{cases}, \tag{3}$$

$$\mathrm{Runoff}_2 = \begin{cases} (P(t) - \mathrm{Runoff}_1) + (S(t) - S(t-1)) \\ \quad \text{if } P(t) - \mathrm{Runoff}_1 \geq S(t) - S(t-1) \\ 0 \text{ if } P(t) - \mathrm{Runoff}_1 < S(t) - S(t-1) \end{cases}, \tag{4}$$

where $P(t)$ is precipitation, Runoff_1 is surface runoff, and Runoff_2 is subsurface runoff. A parameter (α) is introduced to represent how much of the subsurface runoff is routed over the fast (Q_{fast}) and slow (Q_{slow}) pathway:

$$Q_{\mathrm{fast}} = \mathrm{Runoff}_1 + \alpha \cdot \mathrm{Runoff}_2, \tag{5}$$

$$Q_{\mathrm{slow}} = (1 - \alpha) \cdot \mathrm{Runoff}_2. \tag{6}$$

The potential evapotranspiration (PET) is derived based on the Hamon method (Hamon, 1961), in which daily PET in millimeters is computed as a function of daily mean temperature and hours of daylight:

$$\mathrm{PET} = \mathrm{Coeff} \cdot 29.8 \cdot L_{\mathrm{d}} \cdot \frac{0.611 \cdot \exp\left(17.27 \cdot \frac{T}{(T+273.3)}\right)}{T + 273.3}, \tag{7}$$

Table 2. HYMOD_DS parameters.

Parameter name	Description	Feasible range	
		Lower bound	Upper bound
Coeff	Hamon potential evapotranspiration coefficient	0.1	2
C_{max}	Maximum soil moisture capacity (mm)	5	1500
B	Shape for the storage capacity distribution function	0.01	1.99
α	Direct runoff and base flow split factor	0.01	0.99
K_s	Release coefficient of groundwater reservoir	0.00005	0.001
DDF_s	Degree day snowmelt factor (mm °C day^{-1})	0.001	10
T_{th}	Snowmelt temperature threshold (°C)	0	5
T_s	Snow/rain temperature threshold (°C)	0	5
r	Glacier melt rate factor	1	2
K_g	Glacier storage release coefficient	0.01	0.99
T_g	Glacier melt temperature threshold (°C)	0	5
N	Unit hydrograph shape parameter	1	99
K_q	Unit hydrograph scale parameter	0.01	0.99
Velo	Wave velocity in the channel routing (m s^{-1})	0.5	5
Diff	Diffusivity in the channel routing (m^2 s^{-1})	200	4000

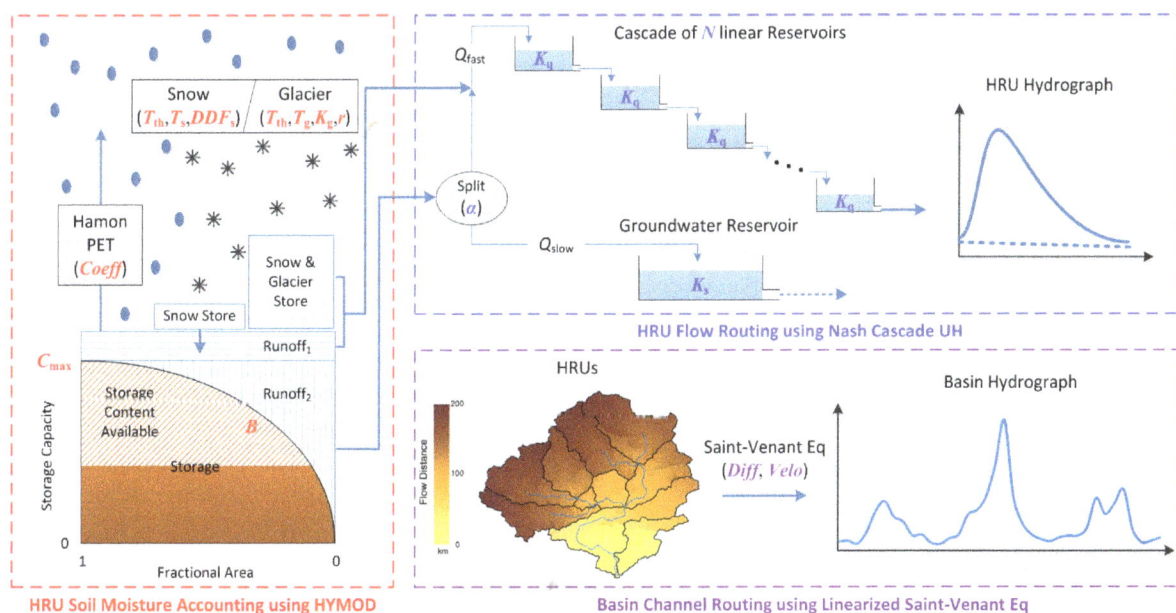

Figure 3. Distributed version of the HYMOD model (HYMOD_DS).

where L_d is the daylight hours per day, T is the daily mean air temperature (°C), and Coeff is a bias correction factor. The hours of daylight is calculated as a function of latitude and day of year based on the daylight length estimation model (CBM model) suggested by Forsythe et al. (1995).

The HYMOD_DS includes snow and glacier modules with separate runoff processes, i.e., the runoff from the glacierized area is calculated separately and added to runoff generated from the soil moisture accounting module coupled with the snow module. The implicit assumption here is that there is no interchange of water between soil layers and

glacial area and runoff from glacial areas is regarded as surface flow. The runoff from each area is weighted by its area fraction within the basin to obtain total runoff.

The time rate of change in snow and glacier volume governed by ice accumulation and ablation (melting and sublimation) is expressed by the degree day factor (DDF) mass balance model (Moore, 1993; Stahl et al., 2008). The dominant phase of precipitation (snow vs. rain) is determined by a temperature threshold (T_{th}). The snowmelt M_s and glacier melt M_g is calculated as

$$M_s = \mathrm{DDF}_s \times (T - T_s), \tag{8}$$

$$M_g = \mathrm{DDF}_g \times \left(T - T_g\right), \tag{9}$$

with DDF_s (T_s) and DDF_g (T_g) applied separately for snow and glacier modules, respectively. To account for the higher melting rate of glaciers than snow owing to the low albedo (Konz and Seibert, 2010; Kinouchi et al., 2013), we introduced a parameter $r > 1$ to constrain DDF_g to be larger than DDF_s (i.e., $\mathrm{DDF}_g = r \times \mathrm{DDF}_s$). For the rain that falls on the glacierized area, the glacier parameter K_g determines the portion of rain becoming surface runoff as a multiplier for the rainfall. The remaining rainfall is assumed to be accumulated to the glacier store.

The within-grid routing process for direct runoff is represented by an instantaneous unit hydrograph (IUH) (Nash, 1957), in which a catchment is depicted as a series of N reservoirs each having a linear relationship between storage and outflow with the storage coefficient of K_q. Mathematically, the IUH is expressed by a gamma probability distribution:

$$u(t) = \frac{K_q}{\Gamma(N)} \left(K_q t\right)^{N-1} \exp\left(-K_q t\right), \tag{10}$$

where Γ is the gamma function. The within-grid groundwater routing process is simplified as a lumped linear reservoir with the storage recession coefficient of K_s.

The transport of water in the channel system is described using the diffusive wave approximation of the Saint-Venant equation (Lohmann et al., 1998):

$$\frac{\partial Q}{\partial t} + C \frac{\partial Q}{\partial x} - D \frac{\partial^2 Q}{\partial^2 x^2} = 0, \tag{11}$$

where C and D are parameters denoting wave velocity (Velo) and diffusivity (Diff), respectively.

Similar to most other hydrological models (Efstratisdis et al., 2008), HYMOD_DS is not designed to model water abstractions for agricultural lands and dam operations within the basin. According to the World Bank (2010), water demand for agricultural use is about 2000 million cubic meters, or about 8.3 % of the total annual flow. The Naglu dam (Fig. 1) upstream of the Daronta streamflow gage forms the largest and most important reservoir in the basin, with an active storage of 379 million cubic meters. In our hydrologic modeling process, the water consumed by irrigated croplands is implicitly accounted for by the evapotranspiration module. We note that the degree of irrigation impact during the time frame used for calibration (1960–1981) is likely much smaller than the current level. We also expect that using monthly data for calibration somewhat reduces the bias from human interference, particularly the daily operations of the Naglu dam. Nevertheless, the calibration results for the gage below this dam (Daronta), and to a lesser extent the basin

outlet (Dakah), should be approached with caution. Given that a majority of the gages examined in this study are on an underdeveloped branch of the Kabul River, issues of human interference on calibration are somewhat mitigated.

4   Methods

The purpose of this study is to explore the implications of different calibration strategies and choices for a computationally expensive distributed hydrologic model. A variety of calibration experiments are conducted, with the results from preceding experiments informing choices made for subsequent ones. All calibration approaches are tested in terms of their ability to predict flows at interior site gages that were left out of the calibration process. In all cases, the genetic algorithm (GA) introduced by Wang (1991) is used as an optimization method for model parameter calibration, and the objective function is based simply on the Nash–Sutcliffe efficiency (NSE) (Nash and Sutcliff, 1970), which is by far the most utilized performance metric in hydrological model applications (Biondi et al., 2012). A multisite average of the NSE is used when evaluating performance across multiple sites. We fully recognize that the use of one objective, such as the NSE, is inferior compared to multiobjective approaches that can identify Pareto optimal solutions that provide good model performance across different components of the flow regime (Madsen, 2003; Efstratiadis and Koutsoyiannis, 2010; Li et al., 2010; Kumar et al., 2013). However, in this particular study daily hydrologic model simulations can only be compared against available monthly streamflow records, reducing the number of viable objectives against which to calibrate. That is, statistics representing peak flows, extreme low flows, and other daily flow regime characteristics often used in multiobjective optimization approaches are unavailable. We believe that the use of a monthly NSE value as a single objective, while coarse, does not inhibit our ability to provide insight into the research questions posed. In addition to the NSE, the Kling–Gupta efficiency (KGE) (Gupta et al., 2009) is adopted as an alternative model performance metric, which equally weights model mean bias, variance bias, and correlation with observations.

In this study, three levels of parameter complexity are considered: lumped, semi-distributed, and fully distributed formulations (Fig. 4). The different levels of parameter complexity are defined according to the spatial distribution of unique hydrologic model parameters. In the lumped formulation a single parameter set is applied to the entire basin. In the semi-distributed formulation, a unique parameter set is assigned to each sub-basin, defined based on the location of available streamflow gaging sites. The fully distributed parameter structure follows the spatial discretization of climate input grids, allowing for a unique parameter set for each grid cell. No matter the parameterization scheme, the model

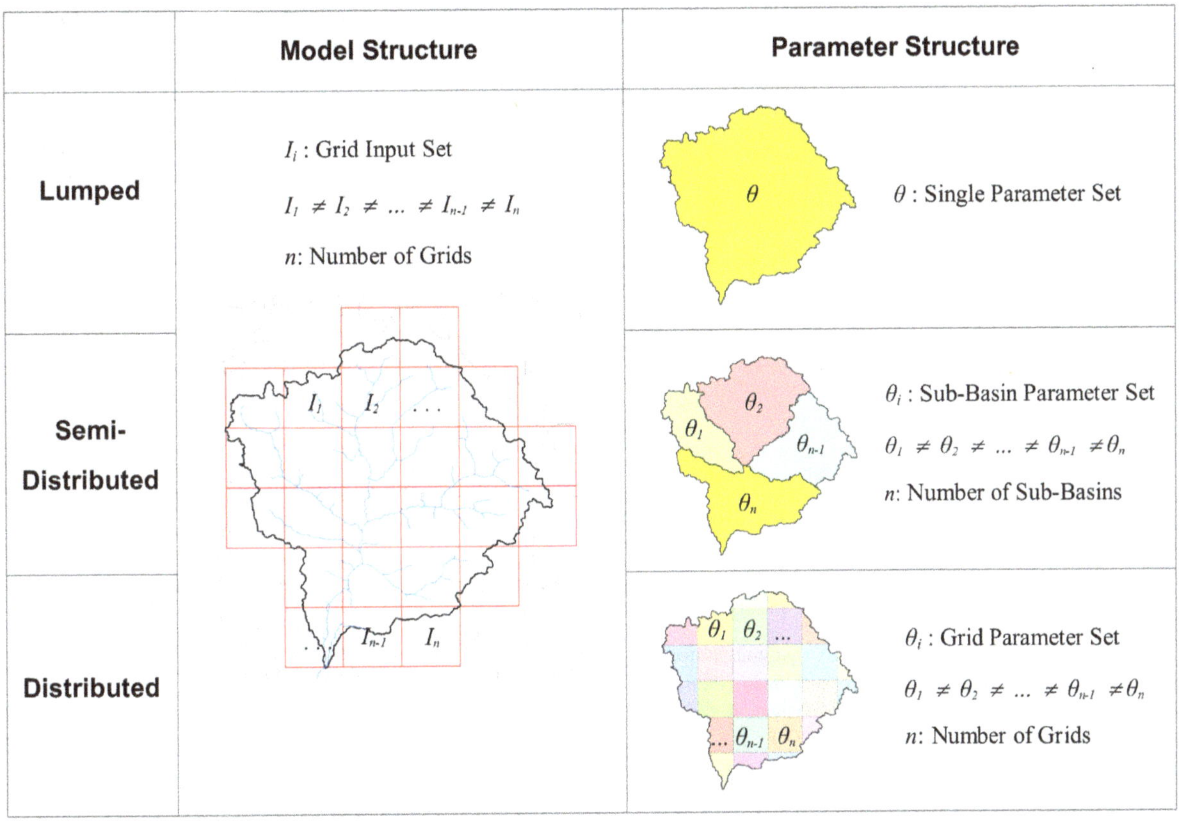

	Model Structure	**Parameter Structure**
Lumped	I_i : Grid Input Set $I_1 \neq I_2 \neq ... \neq I_{n-1} \neq I_n$ n: Number of Grids	θ θ : Single Parameter Set
Semi-Distributed	I_1 I_2 ... I_{n-1} I_n	θ_2 θ_1 θ_{n-1} θ_n θ_i : Sub-Basin Parameter Set $\theta_1 \neq \theta_2 \neq ... \neq \theta_{n-1} \neq \theta_n$ n: Number of Sub-Basins
Distributed		θ_1 θ_2 ... θ_{n-1} θ_n θ_i : Grid Parameter Set $\theta_1 \neq \theta_2 \neq ... \neq \theta_{n-1} \neq \theta_n$ n: Number of Grids

Figure 4. Model structure based on climate input grids and three different parameterization concepts.

structure follows the climate input grids; i.e., the hydrological water cycle within each grid cell is modeled separately. We note that a lumped model structure (i.e., no gridded or sub-unit structure) has often been considered as a baseline model formulation in the assessment of distributed modeling frameworks (e.g., see Smith et al., 2013). However, the focus of our study is on ungaged interior site streamflow estimation, making this formation somewhat inappropriate. Furthermore, preliminary tests comparing streamflow simulations at the basin outlet (Dakah) between a gridded and basin-averaged structure, both with a lumped parameter formulation, support the use of the distributed grid structure (Fig. S3).

The parameter complexity will vary depending on the calibration experiment being conducted but, for each experiment regardless of the parameterization, the optimization is implemented 50 times using the GA algorithm to explore calibration uncertainty. The considerably high computational cost required to perform a large number of calibrations is managed using the parallel computing power provided by the Massachusetts Green High-Performance Computing Center (MGHPCC), from which several thousands of processors are available.

In the first modeling experiment, we explore two calibration strategies for using multisite streamflow data, a stepwise and pooled approach. In the stepwise calibration, parameters are calibrated for upstream gaged sub-catchments and subsequently fixed during calibration of downstream points, while for the pooled approach, parameters are calibrated for multiple sub-catchments simultaneously. Both approaches are assessed for the semi-distributed formulation. The better of the two methods is identified for use in the second experiment, where the effects of increased parameter complexity are tested in terms of streamflow prediction accuracy and uncertainty. In the third experiment, we consider the situation where there is only data at the basin outlet for calibration. Here, the model is calibrated against the outlet gage under all levels of parameter complexity and is compared against the best combination of calibration strategy (stepwise or pooled) and parameter complexity (lumped, semi-distributed, or fully distributed) identified in the previous experiments. Finally, a subset of the calibration approaches deemed worthy of further investigation are compared in terms of their projections of future streamflow under climate change to highlight how model calibration differences can alter the results of a climate change assessment for water resources applications. These experiments are described in further detail below.

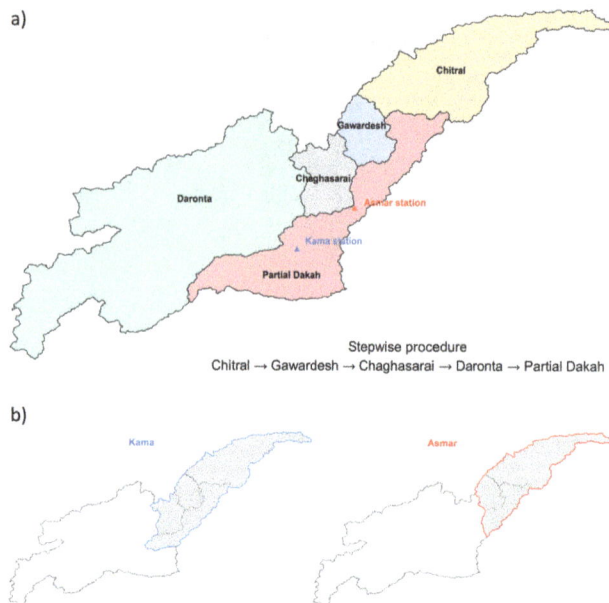

Figure 5. (a) Sub-basins corresponding to five gaging stations are used for the multisite calibrations. **(b)** Two sub-basins (Kama and Asmar) are assumed to be ungaged and used for evaluating the calibration approaches.

4.1   Multisite calibration: stepwise and pooled approaches

In the first experiment, the semi-distributed parameterization concept is compared under alternative multisite calibration strategies, the stepwise and pooled calibration approaches. To conduct the stepwise calibration, a nested class of sub-basins is defined corresponding to multiple gaging stations. In the first step of the stepwise calibration, the optimization process is carried out with nested sub-basins at the lowest level (i.e., the most upstream sites). Once parameters of nested sub-basins are determined, the parameters are fixed, and the calibration procedure proceeds with nested basins at upper levels until parameters for the entire basin are determined. In this particular application to the Kabul River basin, five gaged sub-basins were selected and the stepwise calibration procedure for those sub-basins followed this direction: Chitral → Gawardesh → Chaghasarai → Daronta → Dakah (Fig. 5). The stepwise calibration approach involves a number of GA implementations corresponding to the number of gaging sites. The GA optimization was carried out a total of 250 times in this application, with 50 optimization runs containing GA implementations for five sub-basin regions.

The pooled calibration strategy involves calibrating all parameters of the model domain simultaneously against multiple streamflow gages within the watershed. This approach aims at looking for suitable parameters that are able to produce satisfactory model results at all gaging stations in a single implementation of GA optimization. That is, the GA searches the entire parameter space at once to maximize the average NSE across all sites. This operational feature reduces the processing time spent on the GA implementation compared to the stepwise calibration strategy. To identify the better of the two multisite calibration approaches, the comparison focused on their ability to predict streamflow and calibration uncertainties at two interior site gages (Kama and Asmar) that were assumed to be ungaged (Fig. 5), as well as for validation data at the basin outlet.

It is important to note that the evaluation of these multisite calibration strategies is somewhat weakened because of the lack of overlapping data periods among most of the stations (Fig. 2). This drawback prevents the calibration methods from accounting for simultaneous information from different tributaries, which, if available, would better enable the calibration methods to account for heterogeneity of hydrological processes across the sub-basins.

4.2   Increased parameter complexity

In the second experiment, the better of the two approaches (stepwise or pooled) identified in the first experiment is further tested with respect to the three different levels of parameter complexity. In addition to the semi-distributed parameter formulation considered in the first experiment, lumped and fully distributed parameter formulations are calibrated for the selected approach to investigate the gain or loss arising from different levels of parameter complexity. Since the hydrologic model HYMOD employed in this study involves 15 parameters, the lumped version of the HYMOD_DS contains a single, 15-member parameter set applied to all model grid cells. The semi-distributed conceptualization of HYMOD_DS contains a single parameter set for each sub-basin, totaling 75 parameters. In the distributed parameterization the number of parameters increases dramatically. With 160 0.25 °C grid cells, the number of parameters requiring calibration reaches 2400. As the number of parameters increase across the parameterization schemes, calibration becomes increasingly computationally expensive. The number of model runs used in the GA optimization algorithm for the lumped, semi-distributed, and distributed parameterization schemes are 15 000 (150 populations × 100 generations), 75 000 (750 × 100), and 480 000 (2400 × 200), respectively. These population/generation sizes were supported using convergence tests for each calibration. Again, 50 separate GA optimizations were used to explore calibration uncertainties for each parameterization scheme. To give a sense of the computational burden of this experiment, we note that 50 trials of the HYMOD_DS calibration under the distributed conceptualization required 1000 processors over 7 days on the MGHPCC system.

4.3   Basin outlet calibration

The third experiment considers the situation where there is only gaged data at the basin outlet (Dakah) for calibration, a common situation when calibrating hydrologic models in data-scarce river basins. Here, we evaluate the potential of the basin outlet calibration to estimate interior watershed flows in terms of both accuracy and precision at all gaging stations. All levels of parameter complexity are considered for this calibration. The main purpose of this experiment is to compare the veracity of a distributed hydrologic model calibrated only using basin outlet data with results from multisite calibrations to better understand the degradation in model performance under data scarcity. Other than the use of an NSE objective only at the basin outlet, all other GA settings for each level of parameter complexity are identical to the settings used in the second experiment.

4.4   Climate change projections of streamflow

The fourth experiment investigates how the choice of calibration approach can alter the projections of future streamflow under climate change. To explore this question, streamflow simulations for the 2050s, defined as the 30-year period spanning from 2036 to 2065, are carried out using climate projections from the CMIP5 (Talyor et al., 2012). A total of 36 different climate models run under two future conditions of radiative forcing (RCP 4.5 and 8.5) are used. Streamflow projections are developed for the basin outlet (Dakah) and two interior gages left out of the calibration (Kama and Asmar). By using 36 different GCMs and 50 optimization trials for each calibration scheme, this analysis compares the uncertainty in future streamflow projections originating from uncertainty in different hydrologic model parameterization schemes and under alternative future climates.

Streamflow projections are considered under all three parameterization schemes (lumped, semi-distributed, and fully distributed) for both the basin outlet model and the best multisite calibration approach (stepwise or pooled). Multiple streamflow characteristics are evaluated, including monthly streamflow, wet (April–September) and dry (October–March) season flows, and daily peak flow response. The differences and uncertainty in these metrics across calibration approaches will highlight the importance of calibration strategy for evaluating future water availability and flood risk.

5   Results

For the remaining part of the paper, we introduce the following shorthand: Lump, Semi, and Dist indicate the lumped, semi-distributed, and fully distributed parameterization schemes, and Outlet, Stepwise, and Pooled correspond to basin outlet, stepwise, and pooled calibrations. The comparison between different calibration strategies is based on the model performance evaluated with the NSE, as well as an alternative metric, the KGE.

5.1   Pooled calibration vs. stepwise calibration

This section reports the results from the first experiment comparing the stepwise and pooled calibration approaches for the semi-distributed model parameterization. Figure 6 shows the comparison between the Semi-Stepwise and Semi-Pooled with box plots representing the 50 trials of calibration. Under the stepwise calibration the results for four sub-basins (Chitral, Gawardesh, Chaghasarai, and Daronta) are optimal because there is no interaction between those sub-basins. However, the calibrated parameter sets of each sub-basin act as constraints in the last step of the Semi-Stepwise resulting in the degradation of model skill at the basin outlet (Dakah) and two left-out gages (Asmar and Kama). This becomes apparent when comparing the Semi-Stepwise to the Semi-Pooled results. The model skill under the Semi-Pooled is similar to that from the Semi-Stepwise with respect to the four upstream sub-basins, but it outperforms at the verification gages. This is particularly true for the Asmar gage, which exhibits a downward bias and substantial variability in performance under the Semi-Stepwise. The Semi-Pooled results suggest that small sacrifices of model performance at certain sites can improve and stabilize basin-wide performance. Expected values of KGE from 50 calibrations are also provided (values in parenthesis in the bottom of Fig. 6) and this performance metric also leads to the same conclusion. Therefore, the Semi-Pooled was selected as the better multisite calibration strategy and is considered for further analyses in the following sections.

5.2   Pooled calibration with alternative parameterizations

Here we examine results for the three levels of parameter complexity applied to the pooled calibration approach. Figure 7 shows the comparison of the pooled calibrations. Unsurprisingly, streamflow predictions from the Lump-Pooled have the lowest accuracy and largest uncertainty at the calibration sites, particularly for the Chaghasarai and Daronta sites. This demonstrates the well-known difficulty in representing flow characteristics of a spatially variable system with a homogenous parameter set (Beven, 2012). The pooled calibration substantially improves with increasing parameter complexity at the calibration sites. Both the Semi-Pooled and Dist-Pooled produce NSE values above 0.8 for all calibration sites; however, the Dist-Pooled shows a somewhat higher performance, undoubtedly from its greater freedom to overfit to the calibration data. However, the advantage of the Dist-Pooled with respect to streamflow predictions at validation sites becomes less clear. Only the Dist-Pooled at Kama shows marginally better predictions, while the results are ambiguous at Dakah and Asmar. Overall, this likely suggests

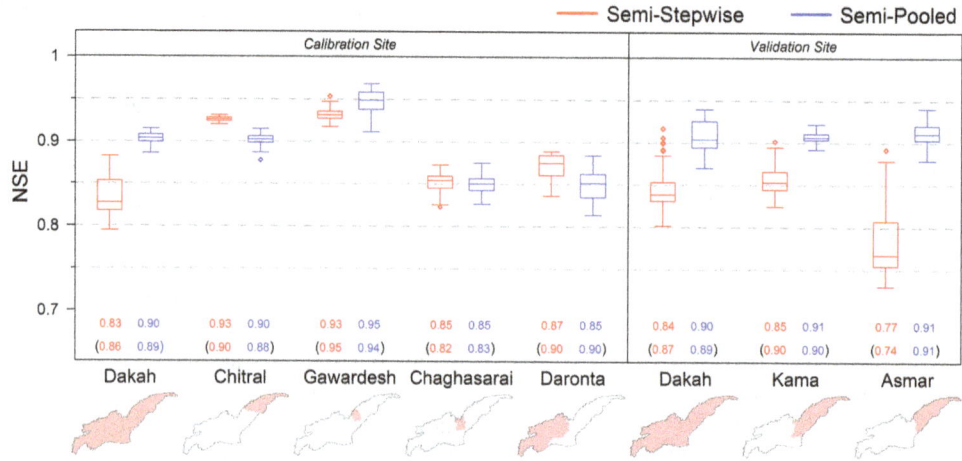

Figure 6. Comparison of the stepwise and pooled calibrations under the semi-distributed parameterization. Each calibration is conducted 50 times. Values on the bottom represent expected values of NSE (in upper row) and KGE (within parenthesis in lower row) from 50 calibrations.

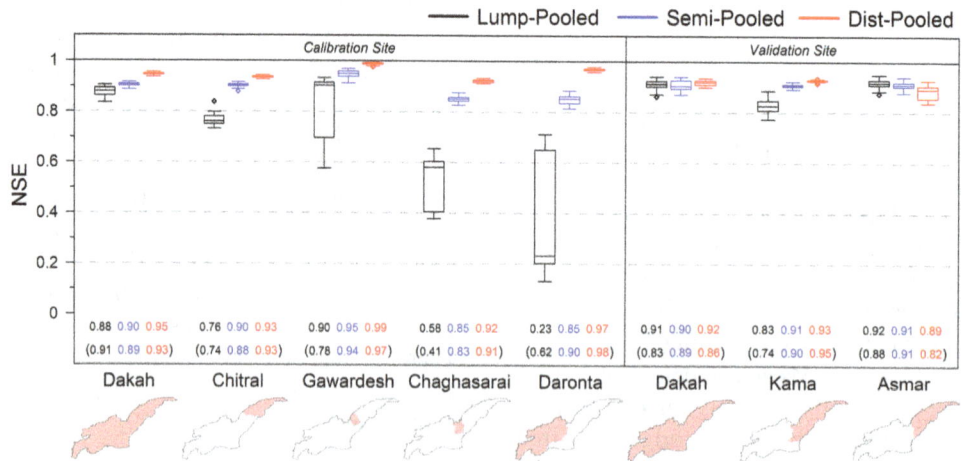

Figure 7. Comparison of the pooled calibrations for the 3 parameterizations of lumped, semi-distributed, and distributed. Each calibration is conducted 50 times. Values on the bottom represent expected values of NSE (in upper row) and KGE (within parenthesis in lower row) from 50 calibrations.

that the fully distributed conceptualization leads to overfitting of the model as compared to the Semi-Dist conceptualization. We reached the same conclusion when examining the KGE values, which rise with greater parameter complexity at calibration sites but no longer follow this pattern strictly at validation sites.

Interestingly, the Lump-Pooled performs well at the verification sites despite its poor performance at calibration sites. The Lump-Pooled does not show significant degradation in skill at Kama compared to the more complex parameterizations, and the flow prediction at Asmar actually exhibits the best performance of all three model variants. A partial reason for this unexpected result arises from different overlapping periods in the calibration and validation data (see Fig. 2). The periods used for the calibration for Chitral (1978–1981) and Gawardesh (1975–1978) have no overlapping periods with

the one for Asmar (1966–1971), which encompasses those two sub-basins. Instead, the validation at Asmar is mostly affected by the calibration to Dakah because of the overlapping 4 years (1968–1971) between those two sites. This explains the reason why the Lump-Pooled shows high skill at Asmar despite the low skill at its sub-basins. However, the low model skill at Chaghasarai from the Lump-Pooled propagates to the validation result at Kama, as these two sites have a relatively long overlapping period (8 years, from 1967 to 1974).

5.3 Limitations of the basin outlet calibration

In the third experiment the HYMOD_DS was calibrated only to data at the basin outlet under all levels of parameter complexity, and streamflow records for all six sub-basins, as well as flows at Dakah not used during calibration, are

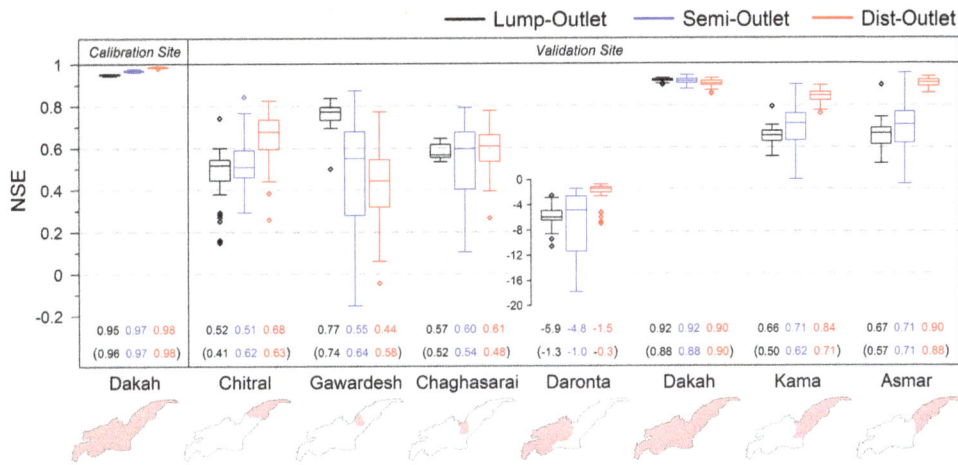

Figure 8. Comparison of the basin outlet calibrations for the three parameterizations of lumped, semi-distributed, and distributed. Each calibration is conducted 50 times. Values on the bottom represent expected values of NSE (in upper row) and KGE (within parenthesis in lower row) from 50 calibrations.

used for model validation. First, we consider the flows at Dakah. During the calibration period, all three parameterization schemes produce very accurate streamflow predictions with NSE (KGE) values above 0.95 (0.96) (Fig. 8). High accuracy holds even under the Lump-Outlet, despite the spatial heterogeneity of the basin. While NSE and KGE values at Dakah rise marginally with greater parameter complexity during calibration, this no longer holds during the validation period, suggesting no benefit with an increase in parameter complexity.

The validation results for the six sub-basins demonstrate the danger in relying on outlet data alone when calibrating a distributed model for flow prediction at interior points. Streamflow predictions at interior sites exhibit low accuracy and high uncertainty, with the worst performance at the Daronta site (all NSEs and KGEs are negative). We note that the poor performance at Daronta is likely due in part to the impacts of water abstraction and the operation of Naglu dam. Further examination (Fig. S4) showed that the HYMOD_DS significantly overestimated streamflow at Daronta and underestimated flow at three sites in the eastern part of the basin (Chitral, Gawardesh, and Chaghasarai). Model performance at Kama and Asmar is somewhat better than at the other validation sites, although improvements are not the same across all parameterizations. The Lump-Outlet predictions at these sites still have low average accuracy (average NSE < 0.7 and average KGE < 0.6), while the Semi-Outlet exhibits large uncertainty in performance across the 50 optimization trials. Surprisingly, the over-parameterized Dist-Outlet shows promising results with high expected accuracy at Kama and Asmar (mean NSE (KGE) of 0.84 (0.71) and 0.90 (0.88), respectively) and comparable performance at many of the other sites. One exception is Gawardesh, where the Lump-Outlet outperforms the other model variants, although the reason for this is not immediately clear. Overall,

the results indicate that any calibration based on basin outlet data should be used with substantial caution when predicting flows at interior basin sites.

After reviewing all of the calibration experiments, it becomes clear that the Semi-Pooled and Dist-Pooled calibrations provide more robust performance compared to the basin outlet calibrations due to their improved representation of internal hydrologic processes across the basin. To further compare these calibration strategies against one another, we evaluate the variability in optimal parameters resulting from the 50 trials of the GA algorithm. Figure 9 shows the coefficient of variation (CV) of C_{max} (a parameter for the soil moisture account module) over the basin from all combinations of calibration approaches (the outlet and pooled) and three parameterization schemes. A clear pattern of increasing variability (higher uncertainty in C_{max}) emerges as parameter complexity increases for both the outlet and pooled calibration strategies. That is, the semi- and fully distributed parameterizations lead to significantly variable parameter sets that produce similar representations of the observed basin response. Figure 9 also suggests that the equifinality can be alleviated to an extent by pooling data across sites. The pooled calibration approaches consistently show lower variability in C_{max} compared to the outlet calibration at the same level of parameter complexity. These results are relatively consistent across the remaining 14 HYMOD_DS parameters. The implications of parameter stability on streamflow projections under climate change is addressed in the next section.

5.4 Climate change projections of streamflow with uncertainty

Here we explore how projections of future water availability and flood risk under climate change are influenced by the choice of calibration approach. For the Kabul River basin,

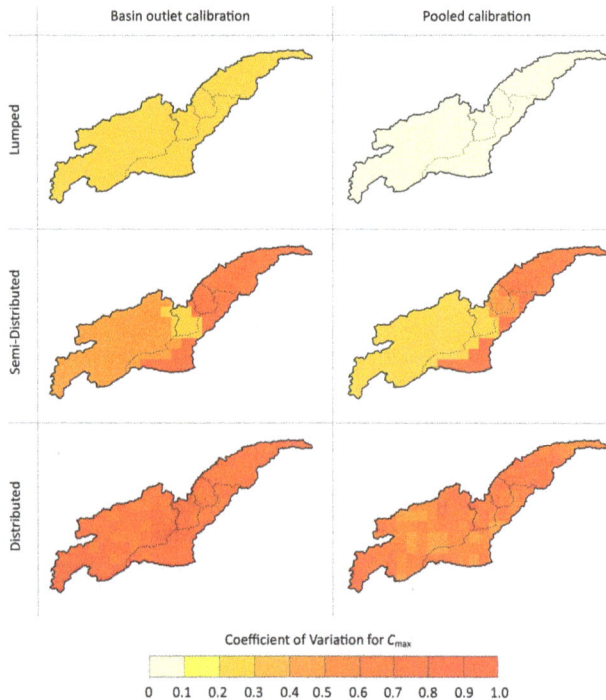

Figure 9. Coefficient of variation (CV) of 50 optimal values of $C_{\max}$ (parameter for the soil moisture accounting module in the HYMOD_DS) from the basin outlet calibrations (left panel) and the pooled calibrations (right panel).

the CMIP5 GCM projections of monthly total precipitation and mean temperature are shown in Fig. S5. According to the CMIP5 ensemble, precipitation projections show no clear trend; the average precipitation change in monthly total precipitation fluctuates between −10 and 10 mm. On the other hand, temperature clearly shows an upward trend for both radiative forcing scenarios. The average changes in annual temperature are +2.2 and +2.8 °C for RCPs 4.5 and 8.5, which, using the Hamon method, correspond to an increase in annual PET by approximately 100 and 150 mm, respectively.

We first examine average monthly streamflow estimates across four calibration strategies: the Semi-Pooled and Dist-Pooled (most promising calibration strategies), as well as the Lump-Outlet (as a baseline) and Dist-Outlet (the best outlet calibration strategy). Figure 10 shows the monthly streamflow estimates for the historical period with the whisker bars indicating the uncertainty range across the 50 calibration trials. The monthly streamflow predictions are also provided for the 2050s under the RCP 4.5 and 8.5 scenarios. For the future scenarios, the whisker bars are derived by averaging over the 36 different climate projections for each of the 50 trials. For the historical time period, all calibration schemes match the observed monthly streamflow at Dakah well, but monthly streamflow is underestimated in most months at Kama and Asmar under the basin outlet calibrations, particularly by the

Lump-Outlet. The historical monthly streamflow estimates from the outlet calibration strategies also tends to be highly uncertain for the months of June, July, August, and September, especially compared to the Semi-Pool and Dist-Pool.

Under future climate projections for the 2050s, the four calibration strategies show similar changes in monthly streamflow at Dakah, but the magnitudes of change are somewhat different. All calibration strategies suggest reduction in streamflow for June, July, and August under both RCP 4.5 and 8.5 scenarios. Also, the peak monthly flow, which occurred in June or July in the historical period, is shifted to May at Dakah. However, the Lump-Outlet predicts less reduction of flow in June and July and a greater reduction in August and September as compared to the other three calibrations. Considering that all calibration schemes had similar levels of good performance at this site for both calibration and validation periods, it is notable that they project future streamflow somewhat differently.

Future monthly streamflow predictions at Kama and Asmar vary widely between the four calibration schemes, mostly an artifact of their historic differences (Fig. 10). Streamflow projections under the outlet calibration strategies tend to show large uncertainties at these two sites, particularly the Lump-Outlet calibration. For three months, July–September, the outlet calibration and pooled calibration strategies provide substantially different insights about future water availability at Kama and Asmar. The outlet calibrations suggest less water with large uncertainties for those months as compared to the pooled calibrations. At Kama, the pooled calibrations suggest significant changes in the pattern of peak monthly flow timing under both RCP scenarios; instead of having a clear peak in July, streamflow from May to August show similar amounts of water.

To further understand the sources of uncertainty in future water availability, we evaluate the separate and joint influence of uncertainties in parameter estimation and future climate on seasonal streamflow projections across all calibration schemes. Figure 11 represents the uncertainty of wet and dry seasonal streamflow at Dakah from three sources: (1) calibration uncertainty across the 50 trials, with future climate uncertainty averaged out for each trial; (2) future climate uncertainty across the 36 projections, with calibration uncertainty averaged out across the 50 trials; and (3) the combined uncertainty across all 1800 (50 × 36) simulations. The results suggest somewhat surprisingly that uncertainty reduction can be expected as parameter complexity increases and, less surprisingly, by applying pooled calibration approaches. Another clear point is that the uncertainty resulting from different climate change scenarios substantially outweighs that from calibration uncertainty.

Up to this point, there has been little difference between the Semi-Pooled and Dist-Pooled model variants. These two versions were further analyzed with respect to extreme streamflow to see if distinguishing characteristics emerge. It has been demonstrated that clear gains in predicting peak

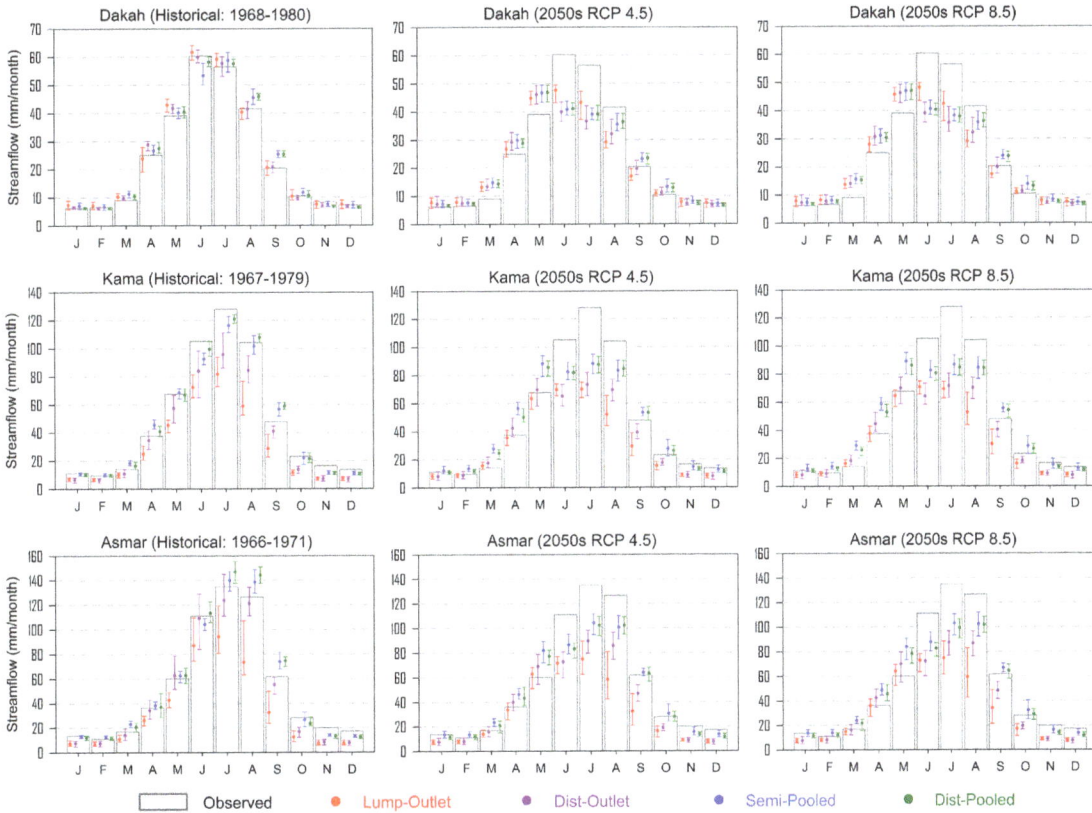

Figure 10. Historical and 2050s average monthly streamflow predictions at Dakah, Kama, and Asmar under four calibration strategies: Lump-Outlet, Dist-Outlet, Semi-Pooled, and Dist-Pooled. The error bars represent the streamflow ranges resulting from 50 trails of the HYMOD_DS calibration. For each of the 50 trials, the 2050s streamflow predictions are averaged over 36 GCM climate projections.

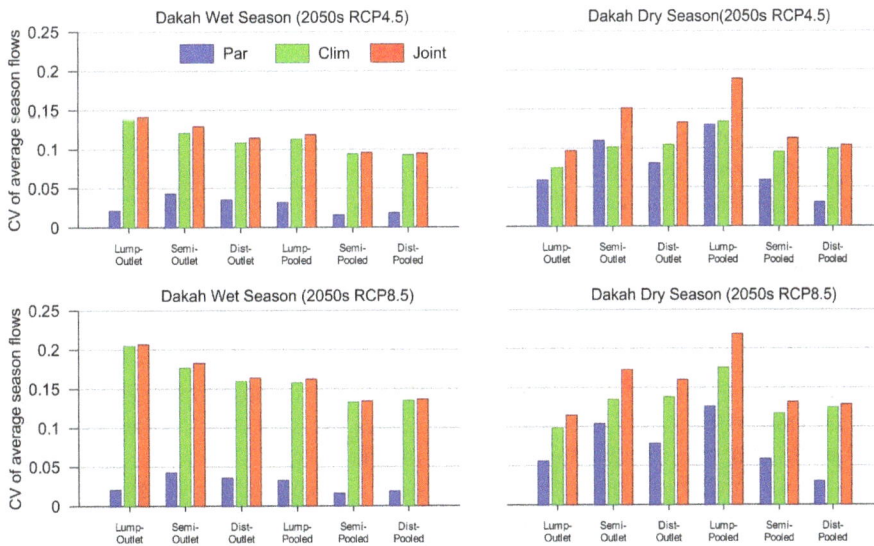

Figure 11. Uncertainties in wet and dry season average streamflow predictions for 2050s are derived from the basin outlet and pooled calibrations for Dakah. Uncertainties are evaluated by the CV of average season streamflow predictions. Three uncertainty sources are considered: calibration uncertainty across 50 calibration trials (Par), climate uncertainty across GCM projections (Clim), and combined uncertainty (Joint).

flows from distributed models are noticeable (Reed et al., 2004) and spatial variability in model parameters significantly influence the runoff behavior (Brath and Montanari, 2000; Pokhrel and Gupta, 2011). The spatial variability of optimal parameters derived from the Semi-Pooled and Dist-Pooled is shown in Fig. S6, with larger variability across all parameters for the Dist-Pooled than for the Semi-Pooled. To understand the effects of spatial variability and calibration uncertainty of parameters on extreme event estimation, the 100-year daily flood event was calculated under the Semi-Pooled and Dist-Pooled for each of the 50 historic simulations and 1800 future simulations across both RCP scenarios. Although the intermodel comparison is intended to be a useful addition that provides a distinction between the parameterization schemes in the pooled calibration approach, results from this analysis should be viewed in the context of a theoretical calibration exercise, not for decision-making purposes, because no observed daily streamflow is available against which to compare the estimated 100-year daily flood events. Projections of the 100-year daily flood, estimated using a log-Pearson type III distribution fit to annual peaks of 30 years, differ somewhat between the Semi-Pooled and Dist-Pooled (Fig. 12). At three validation sites, extreme floods are consistently larger under the Semi-Pooled than the Dist-Pooled, and the mean difference in the 100-year daily flood estimate between the two calibration approaches grows between the historic runs and the RCP 4.5 and 8.5 scenarios. This suggests that the flood-generation process is fundamentally different between the two parameterizations, with the Semi-Pooled formalization magnifying the effect of climate change on extremes. Furthermore, there is substantially more uncertainty in the 100-year daily flood estimate under the Semi-Pooled. Figure 12 shows the combined uncertainty across both climate projections and calibrations, but this uncertainty is broken down further in Fig. 13. Similar to Fig. 11, three sources of uncertainty are evaluated for the 100-year daily flood, including calibration uncertainty alone, climate projection uncertainty alone, and their combined effect. For both the Semi-Pooled and Dist-Pooled, calibration uncertainty has a smaller influence than projection uncertainties and, for all sites, the Dist-Pooled has a smaller uncertainty range than the Semi-Pooled, even for calibration uncertainty alone. This was a truly surprising result, given the parametric freedom in the Dist-Pooled model and the fact that no daily data were ever used in the calibration of either model. It appears that a lack of model parsimony does not necessarily lead to greater uncertainty in model simulations under different climate conditions, somewhat counter to what would be expected of overfit models. One possible reason for this result would be if increased parametric freedom somehow offset the effects of structural deficiencies in the model. However, further research is needed to investigate this issue.

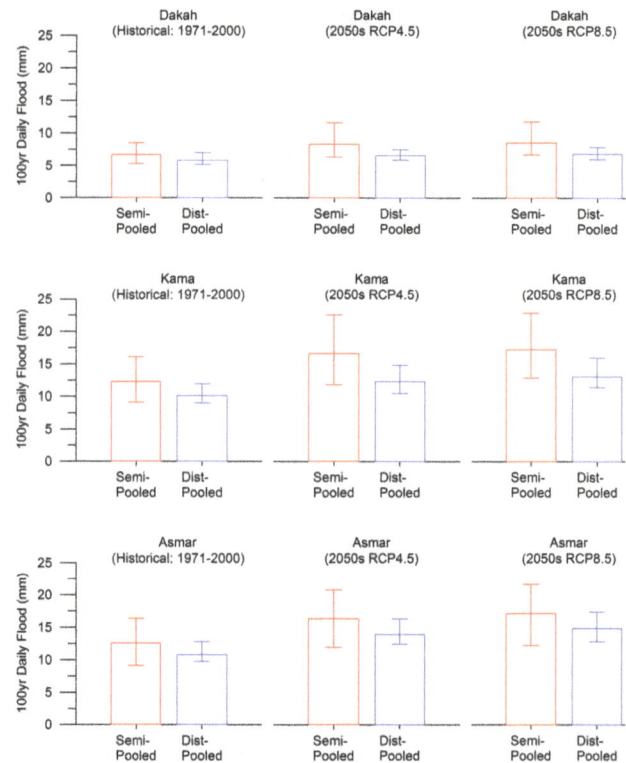

Figure 12. Comparison of GCM average 100-year daily flood events derived from the semi-distributed and distributed pooled calibrations. The uncertainty range is from 50 trials of the model calibration.

6   Discussion and conclusion

In this study we examined a variety of calibration experiments to better understand the benefits and costs associated with different calibration choices for a complex, distributed hydrologic model in a data-scarce region. The goal of these experiments was to provide insight regarding the use of multisite data in calibration, the effects of parameter complexity, and the challenges of using limited data for distributed model calibration, all in the context of projecting future streamflow under climate change.

This study tested two multisite calibration strategies, the stepwise and pooled approaches, finding that the pooled approach using all data simultaneously provides improved calibration results. This suggests that small sacrifices of model performance at certain sites can improve and stabilize basin-wide performance. The pooled calibration substantially improves with increasing parameter complexity at the calibration sites, but similar streamflow predictions at the validation sites between the semi-distributed and distributed pooled calibrations were found, suggesting overfitting of the model from the fully distributed conceptualization. It is worth noting that for the transformation of rainfall to runoff, up to five or six parameters can be identified on the basis of a single hydrograph (Wagner et al., 2001). Under this premise, the

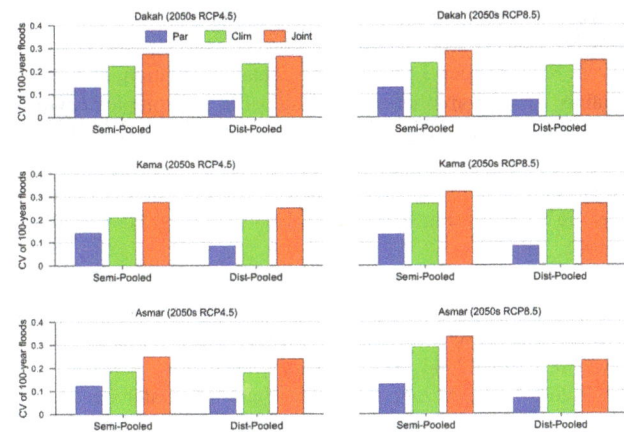

Figure 13. Uncertainties in 100-year daily flood estimates for 2050s are assessed using the Semi-Pooled and Dist-Pooled calibrations. Uncertainties are evaluated by calculating the CV of the 2050s 100-year flood estimates under three uncertainty sources: calibration uncertainty across 50 calibration trials (Par), climate uncertainty across GCM projections (Clim), and combined uncertainty (Joint).

number of the HYMOD_DS parameters being calibrated in the semi-distributed approach remains realistic, but the fully distributed parameterization scheme likely causes poor identifiability of the parameters. Thus, pursuing a parsimonious configuration (e.g., optimization for a small portion of the parameters) with an effort to increase the amount of information (e.g., multivariable/multisite) is critical in the calibration of watershed system models (Gupta et al., 1998; Efstratiadis et al., 2008). We also note the important role of experienced hydrologists in designing a parsimonious hydrologic calibration (e.g., Boyle et al., 2000). In this study, the feasible ranges of the HYMOD_DS parameters were kept wide (as is often done in automatic hydrologic calibrations) without consideration of the physical properties of the basin; the judgment of local hydrologic experts could help reduce the feasible ranges used during the calibration and thus contribute to a reduction of calibration uncertainty.

Calibration only based on data at the basin outlet is all too common in hydrologic model applications and is sometimes considered comparable to multisite calibrations even for predictions at interior gauges (Lerat et al., 2012). In contrast, others have reported improvements in interior flow predictions by using internal flow measurements (Anderson et al., 2001; Wang et al., 2012; Boscarello et al., 2013). This is in agreement with the findings from this study, demonstrating the superiority of the pooled calibration approach to the basin outlet calibration in terms of its ability to represent interior hydrologic response correctly. This study shows the danger in relying on an outlet calibration for interior flow prediction.

It was shown that caution is needed when using an outlet calibration approach for streamflow predictions under future climate conditions. This study showed that the basin outlet calibration can lead to projections of mid-21st cen-

tury streamflow that deviate substantially from projections under multisite calibration strategies. From the test of implications of the pooled calibration in the context of climate change, it was found that applying the pooled calibration with semi-distributed and distributed parameter formulations showed clear gains in reducing uncertainties in predictions of monthly and seasonal water availability as compared to the basin outlet calibrations. Surprisingly, increased parameter complexity in the calibration strategies did not increase the uncertainty in streamflow projections, even though parameter equifinality did emerge. The results suggest that increased (excessive) parameter complexity does not always lead to increased uncertainty if structural uncertainties in the model are present.

The semi-distributed pooled and distributed pooled calibrations are very similar for monthly streamflow projections, yet differ in their projections of extreme flows in part due to their differences in the spatial variability of optimal parameters, with the distributed pooled calibration showing less uncertainty for 100-year daily flood events. We evaluated the separate and joint influence of uncertainties in parameter estimation and future climate on projections of seasonal streamflow and 100-year daily flood across calibration schemes and found that the uncertainty resulting from variations in projected climate between the CMIP5 GCMs substantially outweighs the calibration uncertainty. These results agree with other studies showing the dominance of GCM uncertainty in future hydrologic projections (Chen et al., 2011; Exbrayat et al., 2014). While the GCM-based simulations still have widespread use in assessing the impacts of climate change on water resources availability, the bounds of uncertainty resulting from an ensemble of GCMs cannot be well-defined because of the low credibility with which GCMs are able to produce time series of future climate (Koutsoyiannis et al., 2008). This issue hinders a straightforward appraisal of future water availability under climate change and has motivated other efforts; e.g., performance-based selection of GCMs (Perez et al., 2014).

In addition to the uncertainties surrounding model parameters and future climate explored in this study, there is also significant uncertainty in streamflow projections stemming from structural differences between applied hydrologic models, which can be especially pertinent where robust calibration is hampered by the scarcity of data (Exbrayat et al., 2014). Furthermore, the residual error variance of hydrologic model simulations would increase the effects of hydrologic model uncertainty as compared to that of the climate projections (Steinschneider et al., 2014). These issues need to be addressed in future work for exploring a comprehensive uncertainty assessment of climate change risk for poorly monitored hydrologic systems.

Successful automatic calibration algorithms for hydrologic models are based primarily on global optimization algorithms that are computationally expensive and require a large number of function evaluations (Kuzmin et al., 2008).

Although the speed and capacity of computers have increased multifold in the past several decades, the time consumed by running hydrological models (especially complex, physically based, distributed hydrological models) is still a concern for hydrology practitioners. A single trial of parameter optimization of HYMOD_DS associated with 100 000 runs can take 28 days on a single processor (Fig. S7). Accordingly, the use of high-performance computing power was essential in this study to better understand the implications of different calibration choices and their associated uncertainty for streamflow projections. Enhanced data with high spatial and temporal resolution are increasingly available from remote sensing and satellite products. In the future, remote sensing and satellite information can be integrated into calibration approaches to develop more robust estimates of spatially distributed parameter values, enabling internal consistency of distributed hydrological modeling. Significant progress has been made toward this end (Tang et al., 2009; Khan et al., 2011; Thirel et al., 2013). Future work will consider using high-performance computing power (e.g., Laloy and Vrugt, 2012; Zhang et al., 2013) to understand how such information can enhance the hydrologic simulation at ungaged sites and reduce the calibration uncertainty of distributed hydrologic models in data-scarce regions.

Acknowledgements. The authors are grateful to Efrat Morin, Andreas Efstratiadis, and one anonymous reviewer for their constructive suggestions for improving this manuscript.

This research is funded by a World Bank grant: Hydro-Economic Modeling for Brahmaputra and Kabul River. The views expressed in this paper are those of the authors and do not necessarily reflect the views of the World Bank.

We acknowledge the use of the supercomputing facilities managed by the Research Computing department at the University of Massachusetts.

Edited by: E. Morin

References

Ahmad, S.: Towards Kabul Water Treaty: Managing Shared Water Resources – Policy Issues and Options, Karachi, Pakistan, 2010.

Ajami, N. K., Gupta, H., Wagener, T., and Sorooshian, S.: Calibration of a semi-distributed hydrologic model for streamflow estimation along a river system, J. Hydrol., 298, 112–135, 2004.

Anderson, J., Refsgaard, J. C., and Jensen, K. H.: Distributed hydrological modeling of the Senegal river basin – model construction and validation, J. Hydrol., 247, 200–214, 2001.

Bandaragoda, C., Tarboton, D. G., and Woods, R.: Application of TOPNET in the distributed model intercomparison project, J. Hydrol., 298, 178–201, 2004.

Beven, J. K.: Rainfall-Runoff Modelling: The Primer, 2nd Edition, Wiley-Blackwell, Chichester, 2012.

Beven, K. and Freer, J.: Equifinality, data assimilation, and uncertainty estimation in mechanistic modelling of complex environmental systems using the GLUE methodology, J. Hydrol., 249, 11–29, 2001.

Beven, K.: How far can we go in distributed hydrological modelling?, Hydrol. Earth Syst. Sci., 5, 1–12, doi:10.5194/hess-5-1-2001, 2001.

Biondi, D., Freni, G., Iacobellis, V., Mascaro, G., and Montanari, A.: Validation of hydrological models: conceptual basis, methodological approaches and a proposal for a code of practice, Phys. Chem. Earth, 42–44, 70–76, 2012

Boscarello, L., Ravazzani, G., and Mancini, M.: Catchment multi-site discharge measurements for hydrological model calibration, Procedia Environmental Sciences, 19, 158–167, 2013.

Boyle, D. P., Gupta, H. V., and Sorooshian, S.: Toward improved calibration of hydrologic models: Combining the stregths of manual and automatic methods, Water Resour. Res., 36, 3663–3674, 2000.

Boyle, D. P.: Multicriteria calibration of hydrologic models, Ph.D. thesis, Department of Hydrology and Water Resources Engineering, The University of Arizona, USA, 2001.

Brath, A. and Montanari, A.: The effects of the spatial variability of soil infiltration capacity in distributed flood modelling, Hydrol. Process., 14, 2779–2794, 2000.

Brath, A., Montanari, A., and Toth, E.:. Analysis of the effects of different scenarios of historical data availability on the calibration of a spatially-distributed hydrological model, J. Hydrol., 291, 232–253, 2004.

Breuer, L., Huisman J. A., Willems, P., Bormann, H., Bronstert, A., Croke, B. F. W., Frede, H. G., Gräff, T., Hubrechts, L., Jakeman, A. J., Kite, G., Lanini, J., Leavesley, G., Lettenmaier, D. P., Lindström, G., Seibert, J., Sivapalan, M., and Viney, N. R.: Assessing the impact of land use change on hydrology by ensemble modeling (LUChEM). I: Model intercomparison with current land use, Adv. Water Resour., 32, 129–146, 2009

Cao, W., Bowden, W. B., Davie, T., and Fenemor, A.: Multi-variable and multi-site calibration and validation of SWAT in a large mountainous catchment with high spatial variability, Hydrol. Process, 20, 1057–1073, 2006.

Chen, J., Brissette, F. P., Poulin, A., and Leconte, R.: Overall uncertainty study of the hydrological impacts of climate change for a Canadian watershed, Water Resour. Res., 47, W12509, doi:10.1029/2011WR010602, 2011.

Cole, S. J. and Moore, R. J.: Hydrological modelling using raingauge- and radar-based estimators of areal rainfall, J. Hydrol., 358, 159–181, 2008.

DAWN: Pakistan, Afghanistan mull over power project on Kunar River, available at: http://www.dawn.com/news/1038435 (last access: 2 January 2015), 2013.

Eckhardt, K., Fohrer, N., and Frede, H. G.: Automatic model calibration, Hydrol. Process., 19, 651–658, 2005.

Efstratiadis, A. and Koutsoyiannis, D.: One decade of multiobjective calibration approaches in hydrological modelling: a review, Hydrolog. Sci. J., 55, 58–78, 2010.

Efstratiadis, A., Nalbantis, I., Koukouvinos, A., Rozos, E., and Koutsoyiannis, D.: HYDROGEIOS: a semi-distributed GIS-based hydrological model for modified river basins, Hydrol.

Earth Syst. Sci., 12, 989–1006, doi:10.5194/hess-12-989-2008, 2008.

Exbrayat, J. F., Buytaert, W., Timbe, E., Windhorst, D., and Breuer, L.: Addressing sources of uncertainty in runoff projections for a data scarce catchment in the Ecuadorian Andes, Climatic Change, 125, 221–235, 2014.

Flugel, W. A.: Delineating Hydrological Response Units (HRU's) by GIS analysis for regional hydrological modelling using PRMS/MMS in the drainage basin of the River Brol, Germany, Hydrol. Process., 9, 423–436, 1995.

Forsythe, W. C., Rykiel Jr., E. J., Stahl, R. S., Wu, H., Schoolfield, R. M.: A model comparison for daylength as a function of latitude and day of year, Ecol. Model., 80, 87–95, 1995.

Frances, F., Velez, J. I., and Velez, J. J.: Split-parameter structure for the automatic calibration of distributed hydrological models, J. Hydrol., 332, 226–240, 2007.

Gharari, S., Hrachowitz, M., Fenicia, F., and Savenije, H. H. G.: An approach to identify time consistent model parameters: sub-period calibration, Hydrol. Earth Syst. Sci., 17, 149–161, doi:10.5194/hess-17-149-2013, 2013.

Grinsted, A.: An estimate of global glacier volume, The Cryosphere, 7, 141–151, doi:10.5194/tc-7-141-2013, 2013.

Gupta, H. V., Kling, H., Yilmaz, K. K., and Martinez, G. F.: Decomposition of the mean squared error and NSE performance criteria: Implications for improving hydrological modelling, J. Hydrol., 377, 80–91, 2009.

Gupta, H. V., Sorooshian, S., and Yapo, P. O.: Towards improved calibration of hydrologic models: Multiple and noncommensurable measures of information, Water Resour. Res., 34, 751–763, 1998.

Hamon, W. R.: Estimating potential evapotranspiration, J. Hydr. Eng. Div.-ASCE, 87, 107–120, 1961.

Hewitt, K., Wake, C. P., Young, G. J., and David, C.: Hydrologicl investigations at Biafo glacier, Karakoram Himalaya, Pakistan: An important source of water for the Indue River, Ann. Glaciol., 13, 103–108, 1989.

Immerzeel, W. W., van Beek, L. P. H., Konz, M., Shrestha, A. B., and Bierkens, M. F. P.: Hydrological response to climate change in a glacierized catchment in the Himalayas, Climatic Change, 110, 721–736, 2012.

IUCN: Towards Kabul Water Treaty: Managing Shared Water Resources – Policy Issues and Options, IUCN Pakistan, Karachi, 11 pp., 2010.

Jarvis, A., Reuter, H. I., Nelson, A., and Guevara, E.: Hole-filled seamless SRTM data V4, International Centre for Tropical Agriculture (CIAT), available at: http://srtm.csi.cgiar.org (last access: 2 January 2015), 2008.

Khakbaz, B., Imam, B., Hsu, K., and Sorooshian, S.: From lumped to distributed via semi-distributed: Calibration strategies for semi-distributed hydrologic models, J. Hydrol., 418–419, 61–77, 2012.

Khan, S. I., Yang, H., Wang, J., Yilmaz, K. K., Gourley, J. J., Adler, R. F., Brakenridge, G. R., Policell, F., Habib, S., and Irwin, D.: Satellite remote sensing and hydrologic modeling for flood inundation mapping in Lake Victoria Basin: Implications for hydrologic prediction in ungauged basins, IEEE T. Geosci. Remote, 49, 85–95, 2011.

Kinouchi, T., Liu, T., Mendoza, J., and Asaoka, Y.: Modeling glacier melt and runoff in a high-altitude headwater catchment in the

Cordillera Real, Andes, Hydrol. Earth Syst. Sci. Discuss., 10, 13093–13144, doi:10.5194/hessd-10-13093-2013, 2013.

Kollat, J. B., Reed, P. M., and Wagener, T.: When are multiobjective calibration trade-offs in hydrologic models meaningful?, Water Resour. Res., 48, W03520, doi:10.1029/2011WR011534, 2012.

Konz, M. and Seibert, J.: On the value of glacier mass balances for hydrological model calibration, J. Hydrol., 385, 238–246, 2010.

Koren, V., Reed, S., Smith, M., Zhang, Z., and Seo, D. J.: Hydrology laboratory research modeling system (HL-RMS) of the US national weather service, J. Hydrol., 291, 297–318, 2004.

Koutsoyiannis, D., Efstratiadis, A., Mamassis, N., and Christofides, A.: On the credibility of climate predictions, Hydrolog. Sci. J., 53, 671–684, 2008.

Kumar, R., Samaniego, L., and Attinger, S.: Implications of distributed hydrologic model parameterization on water fluxes at multiple scales and locations, Water Resour. Res., 49, 360–379, 2013.

Kuzmin, V., Seo D., and Koren V.: Fast and efficient optimization of hydrologic model parameters using a priori estimates and stepwise line search, J. Hydrol., 353, 109–128, 2008.

Laloy, E. and Vrugt, J. A.: High-dimensional posterior exploration of hydrologic models using multiple-try DREAM(ZS) and high-performance computing, Water Resour. Res., 48, W01526, doi:10.1029/2011WR010608, 2012.

Latham, J., Cumani, R., Rosati, I., and Bloise, M.: Global Land Cover SHARE (GLC-SHARE) database Beta-Release Version 1.0, available at: http://www.glcn.org/databases/lc_glcshare_en.jsp (last access: 2 January 2015), 2014.

Leavesley, G. H., Hay, L. E., Viger, R. J., and Markstrom, S. L.: Use of Priori Paramter-Estimation Methods to Constrain Calibration of Distributed-Parameter Models, Water. Sci. Appl., 6, 255–266, 2003.

Legates, D. R. and Willmott, C. J.: Mean seasonal and spatial variability in gauge-corrected, global precipitation, Int. J. Climatol., 10, 111–127, 1990.

Lerat, J., Andreassian V., Perrin, C., Vaze, J., Perraud J. M., Ribstein, P., and Loumagne C.: Do internal flow measurements improve the calibration of rainfall-runoff models?, Water Resour. Res., 48, W02511, doi:10.1029/2010WR010179, 2012.

Li, X., Weller, D. E., and Jordan, T. E.: Watershed model calibration using multi-objective optimization and multi-site averaging, J. Hydrol., 380, 277–288, 2010.

Lohmann, D., Raschke, R., Nijssen, B., and Lettenmaier, D. P.: Regional scale hydrology: I. Formulation of the VIC-2L model coupled to a routing model, Hydrolog. Sci. J., 43, 131–141, 1998.

Madsen, H.: Parameter estimation in distributed hydrologicl catchment modelling using automatic calibration with multiple objectives, Adv. Water Resour., 26, 205–216, 2003.

Moore, R. D.: Application of a conceptual streamflow model in a glacierized drainage basin, J. Hydrol., 150, 151–168, 1993.

Moore, R. J.: The probability-distribted principle and runoff production at point and basin scales, Hydrolog. Sci. J., 30, 273–297, 1985.

Nash, J. E. and Sutcliff, J. V.: River flow forecasting through conceptual models: Part 1. A discussion of priciples, J. Hydrol., 10, 282–290, 1970.

Nash, J. E.: The form of the instantaneous unit hydrograph, International Association of Science and Hydrology, 3, 114–121, 1957.

Olson, S. A. and Williams-Sether, T.: Streamflow characteristics at streamgages in Northern Afghanistan and selected locations, U.S. Geological Survey, Reston, Virginia, 2010.

Palazzi, E., von Hardenberg, J., and Provenzale, A.: Precipitation in the Hindu-Kush Karakoram Himalaya: Observations and future scenarios, J. Geophys. Res., 118, 85–100, 2013.

Perez, J., Menendez, M., Mendez, F. J., and Losada, I. J.: Evaluating the performance of CMIP3 and CMIP5 global climate models over the north-east Atlantic region, Clim. Dynam., 43, 2663–2680, 2014.

Pfeffer, T. W., Arendt, A. A., Bliss, A., Bolch, T., Cogley J. G., Gardner, A. S., Hagen, J. O., Hock R., Kaser, G., Kienholz, C., Miles E. S., Moholdt, G., Molg, N., Paul, F., Radic, V., Rastner, P., Raup, B. H., Rich, J., Sharp, M. J., and The Randolph Consortium: The Randolph Glacier Inventory, J. Glaciol., 60, 537–552, 2014.

Pokhrel, P. and Gupta, H. V.: On the use of spatial regularization strategies to improve calibration of distributed watershed models, Water. Resour. Res., 46, W01505, doi:10.1029/2009WR008066, 2010.

Pokhrel, P. and Gupta, H. V.: On the ability to infer spatial catchment variability using streamflow hydrographs, Water Resour. Res., 47, W08534, doi:10.1029/2010WR009873, 2011.

Radic, V., Bliss, A., Beedlow, A. C., Hock, R., Miles, E., and Cogley, J. G.: Regional and global projections of twenty-first century glacier mass changes in response to climate scenarios from global climate models, Clim. Dynam., 42, 37–58, 2014.

Reed, S., Koren, V., Smith, M., Zhang, Z., Moreda, F., Seo, D. J., and DMIP Participants: Overall distributed model intercomparison project results, J. Hydrol., 298, 27–60, 2004.

Remesan, R., Bellerby, T., and Frostick, L.: Hydrological modelling using data from monthly GCMs in a regional catchment, Hydrol. Process., 28, 3241–3263, 2013.

Safari, A., De Smedt, F., and Moreda, F.: WetSpa model application in the Distributed Model Intercomparison Project (DMIP2), J. Hydrol., 418–419, 77–89, 2012.

Shakir, A. S., Rehman, H., and Ehsan, S.: Climate change impact on river flows in Chitral watershed, Pakistan Journal of Engineering and Applied Sciences, 7, 12–23, 2010.

Smith, M. B., Koren, V., Reed, S., Zhang, Z., Zhang, Y., Moreda, F., Cui, Z., Mizukami, N., Anderson, E. A., and Cosgrove, B. A.: The distributed model intercomparison project – Phase 2: Motivation and design of the Oklahoma experiments, J. Hydrol., 418–419, 3–16, 2012.

Smith, M. B., Seo, D. J., Koren, V. I., Reed, S. M., Zhang, Z., Duan, Q., Moreda, F., and Cong, S.: The distributed model intercomparison project (DMIP): motivation and experiment design, J. Hydrol., 298, 4–26, 2004.

Smith, M., Koren, V., Zhang, Z., Moreda, F., Cui, Z., Cosgrove, B., Mizukami, N., Kitzmiller, D., Ding, F., Reed, S., Anderson, E., Schaake, J., Zhang, Y., Andreassian, V., Perrin, C., Coron, L., Valery, A., Khakbaz, B., Sorooshian, S., Behrangi, A., Imam, B., Hsu, K. L., Todini, E., Coccia, G., Mazzetti, C., Andres, E. O., Frances, F., Orozco, I., Hartman, R., Henkel, A., Fickenscher, P., and Staggs, S.: The distributed model intercomparison project – Phase 2: Experiment design and summary results of the western basin experiments, J. Hydrol., 507, 300–329, 2013.

Stahl, K., Moore, R. D., Shea, J. M., Hutchinson, D., and Cannon, A. J.: Coupled modelling of glacier and streamflow response

to future climate scenarios, Water Resour. Res., 44, W02422, doi:10.1029/2007WR005956, 2008.

Steinschneider, S., Polebitski, A., Brown, C., and Letcher, B. H.: Toward a statistical framework to quantify the uncertainties of hydrologic response under climate change, Water Resour. Res., 48, W11525, doi:10.1029/2011WR011318, 2012.

Steinschneider, S., Wi, S., and Brown, C.: The integrated effects of climate and hydrologic uncertainty on future flood risk assessments, Hydrol. Process., doi:10.1002/hyp.10409, accepted, 2014.

Talyor, K. E., Stouffer, R. J., and Meehl, G. A.: An Overview of CMIP5 and the Experiment Design, B. Am. M. Soc., 93, 485–498, 2012.

Tang, Q., Gao, H., Lu, H., and Lettenmaier, D. P.: Remote sensing: hydrology, Prog. Phys. Geog., 33, 490–509, 2009.

Thirel, G., Salamon, P., Burek, P., and Kalas, M.: Assimilation of MODIS snow cover area data in a distributed hydrological model using the particle filter, Remote Sensing, 5, 5825–5850, 2013

USDA-NRCS: FAO-UNESCO Soil Map of the World, available at: http://www.nrcs.usda.gov/wps/portal/nrcs/detail/soils/use/?cid=nrcs142p2_054013 (last access: 2 January 2015), 2005.

Vrugt, J. A., ter Braak, C. J. F., Gupta, H. V., and Robinson, B. A.: Equifinality of formal (DREAM) and informal (GLUE) Bayesian approaches in hydrologic modeling?, Stoch. Env. Res. Risk A., 23, 1011–1026, 2008.

Wagener, T., Boyle, D. P., Lees, M. J., Wheater, H. S., Gupta, H. V., and Sorooshian, S.: A framework for development and application of hydrological models, Hydrol. Earth Syst. Sci., 5, 13–26, doi:10.5194/hess-5-13-2001, 2001.

Wagener, T., Wheater, H. S., and Gupta, H. V.: Rainfall-Runoff Modelling in Gauged and Ungauged Catchments, Imperical College Press, London, 2004.

Wang, Q. J.: The Genetic Algorithm and Its Application to Calibrating Conceptual Rainfall-Runoff Models, Water Resour. Res., 27, 2467–2471, 1991.

Wang, S., Zhang, Z., Sun, G., Strauss, P., Guo, J., Tang, Y., and Yao, A.: Multi-site calibration, validation, and sensitivity analysis of the MIKE SHE Model for a large watershed in northern China, Hydrol. Earth Syst. Sci., 16, 4621–4632, doi:10.5194/hess-16-4621-2012, 2012.

Wilby, R. L.: Uncertainty in water resource model parameters used for climate change impact assessment, Hydrol. Process., 19, 3201–3219, 2005.

Wood, A. W., Leung, L. R., Sridhar, V., and Lettenmaier, D. P.: Hydrologic Implications of Dynamical and Statistical Approaches to Downscaling Climate Model Outputs, Climatic Change, 62, 189–216, 2004.

World Bank: Afghanistan – Scoping strategic options for development of the Kabul River Basin: a multisectoral decision support system approach, World Bank, Washington, D.C., 2010.

Yatagai, A., Kamiguchi, K., Arakawa, O., Hamada, A., Yasutomi, N., and Kitoh, A.: APHRODITE: Constructing a Long-Term Daily Gridded Precipitation Dataset for Asia Based on a Dense Network of Rain Gauges, B. Am. Meteorol. Soc., 93, 1401–1415, 2012.

Yu, W., Yang, Y. C. E., Savitsky, A., Alford, D., Brown, C., Wescoat, J., Debowicz, D., and Robinson, S.: The Indus Basin of Pakistan: The Impacts of Climate Risks on Water and Agriculture, World Bank, Washington DC, 2013.

Zhang, X., Beeson, P., Link, R., Manowitz, D., Izaurralde, R. C., Sadeghi, A., Thomson, A. M., Sahajpal, R., Srinivasan, R., and Arnold, J. G.: Efficient multi-objective calibration of a computationally intensive hydrologic model with parallel computing software in Python, Environ. Modell. Softw., 46, 208–218, 2013.

Zhu, C. and Lettenmaier, D. P.: Long-Term Climate and Derived Surface Hydrology and Energy Flux Data for Mexico: 1925–2004, J. Climate., 20, 1936–1946, 2007.

On the sensitivity of urban hydrodynamic modelling to rainfall spatial and temporal resolution

G. Bruni[1], R. Reinoso[2], N. C. van de Giesen[1], F. H. L. R. Clemens[1,3], and J. A. E. ten Veldhuis[1]

[1]Department of Water Management, Faculty of Civil Engineering and Geosciences, Delft University of Technology, Delft, the Netherlands
[2]Department of Geoscience and Remote Sensing, Faculty of Civil Engineering and Geosciences, Delft University of Technology, Delft, the Netherlands
[3]Deltares, Delft, the Netherlands

Correspondence to: G. Bruni (g.bruni@tudelft.nl)

Abstract. Cities are increasingly vulnerable to floods generated by intense rainfall, because of urbanisation of flood-prone areas and ongoing urban densification. Accurate information of convective storm characteristics at high spatial and temporal resolution is a crucial input for urban hydrological models to be able to simulate fast runoff processes and enhance flood prediction in cities. In this paper, a detailed study of the sensitivity of urban hydrodynamic response to high resolution radar rainfall was conducted. Rainfall rates derived from X-band dual polarimetric weather radar were used as input into a detailed hydrodynamic sewer model for an urban catchment in the city of Rotterdam, the Netherlands. The aim was to characterise how the effect of space and time aggregation on rainfall structure affects hydrodynamic modelling of urban catchments, for resolutions ranging from 100 to 2000 m and from 1 to 10 min. Dimensionless parameters were derived to compare results between different storm conditions and to describe the effect of rainfall spatial resolution in relation to storm characteristics and hydrodynamic model properties: rainfall sampling number (rainfall resolution vs. storm size), catchment sampling number (rainfall resolution vs. catchment size), runoff and sewer sampling number (rainfall resolution vs. runoff and sewer model resolution respectively).

Results show that for rainfall resolution lower than half the catchment size, rainfall volumes mean and standard deviations decrease as a result of smoothing of rainfall gradients. Moreover, deviations in maximum water depths, from 10 to 30 % depending on the storm, occurred for rainfall res-

olution close to storm size, as a result of rainfall aggregation. Model results also showed that modelled runoff peaks are more sensitive to rainfall resolution than maximum in-sewer water depths as flow routing has a damping effect on in-sewer water level variations. Temporal resolution aggregation of rainfall inputs led to increase in de-correlation lengths and resulted in time shift in modelled flow peaks by several minutes. Sensitivity to temporal resolution of rainfall inputs was low compared to spatial resolution, for the storms analysed in this study.

1 Introduction

Rainfall is a key input to hydrological models and a crucial issue for hydrologists is to find the importance of the spatial structure of rainfall in relation to flood generation (Segond et al., 2007). Many studies conducted in large natural catchments have shown that spatial variability of rainfall is important in determining both timing and volume of rainfall transformed into runoff (Obled et al., 1994) and thus timing of simulated basin response and magnitude of the response peak (Dawdy and Bergman, 1969; Krajewski et al., 1991; Seliga et al., 1992). It has been suggested, with much less evidence, that this is also true for small catchments with shorter response times, such as urban catchments (Blanchet et al., 1992; Obled et al., 1994). Urban catchments are characterised by a high percentage of imperviousness, which leads to a high proportion of the rainfall producing runoff.

It is therefore expected that the effect of spatial rainfall variability on water flows is greater in urban catchments than in rural ones, where local variation of rainfall input is smoothed and delayed within the soil as a result of infiltration in pervious areas (Obled et al., 1994, among others). Previous studies have shown that urban catchments, characterised by a fast hydrological response due to both low interception and infiltration, are highly sensitive to small-scale spatial and temporal variability of the precipitation field (Bell and Moore, 2000; Einfalt et al., 2004; Gires et al., 2013) In the past, a lot of studies have addressed requirements and approaches for flood modelling (Schmitt et al., 2004; Balmforth and Dibben, 2006; Parker et al., 2011; Pathirana et al., 2011; Priest et al., 2011; Neal et al., 2012; Ozdemir et al., 2013). More recently, studies have shown the impact of rainfall variability on hydrodynamic models outputs (Gires et al., 2012; Liguori et al., 2012; Vieux and Imgarten, 2012).

As resolutions of available data and models have increased, rainfall variability information at high resolution has become a critical component to study hydrological response in urban drainage systems using hydrological models. Weather radars are more suitable for this purpose than rain gauge networks as they have better spatial coverage. Weather radars, such as S-band and C-band radars, are already used by meteorological institutes worldwide in order to (indirectly) measure and predict precipitation at national and regional scales. Nonetheless, several studies have shown that the spatial resolution of operational radar network measurements is insufficient to meet the scale of urban hydrodynamics (Berne et al., 2004; Emmanuel et al., 2011; Schellart et al., 2012). Because of their relatively low cost and small size, X-band radars are ideally suited for local rainfall estimation. These radars measure at high resolutions, both in space and time, and much closer to the ground than S- or C-band radars, which for operational purposes, cover large distances and thus point higher especially at locations several tens of kilometres away from the radar sites. X-band radars have been tested locally and show better performances in catching the rapidly changing characteristics of intense rainfall than rain gauges (Jensen and Pedersen, 2005). This is particularly the case when the distance between rain gauges is larger than 3 to 4 km (Wood et al., 2000).

The effects of radar spatial resolution on hydrological model outputs were addressed by Ogden and Julien (1994) by using length scales to characterise rainfall data and catchments, such as storm de-correlation length, grid size of rainfall data, characteristic catchment length and grid size of the distributed runoff model. In their study, Ogden and Julien aimed to explain variability in hydrological responses based on rainfall and catchment characteristics, for two catchments of 30 and $100\,km^2$, using fully distributed rainfall–runoff models. They recommended rainfall spatial resolution of 0.4 the square root of the watershed area, in order to avoid deviations in runoff flows. This corresponds to 1 km resolution for a $10\,km^2$ watershed, and 4 km resolution for a $100\,km^2$ wa-

tershed, as was also found by Segond et al. (2007). Several other studies on natural catchments also found that the influence of rainfall resolution is directly related to the spatial variability of the storm and of the catchment that transforms rainfall into runoff (Krajewski et al., 1991; Winchell et al., 1998; Koren et al., 1999, among others).

The purpose of this paper was to analyse the sensitivity of urban hydrodynamic model outputs to spatial and temporal resolutions of rainfall inputs derived from weather radar data at intra-urban scale. Sensitivity was analysed according to spatial characteristics of rainfall and urban catchment properties as well as model topology. Sensitivity was quantified using dimensionless parameters that describe relationships between rainfall resolution and spatial characteristics of the urban catchment, storm cells and model topology. Some of them were chosen according to their previous use by Odgen and Julien (1994). In this study rainfall estimates were used derived from dual-polarimetric X-band radar (IDRA), operated by Delft University of Technology (TU Delft) and located at CESAR, Cabauw Experimental Site for Atmospheric Research (Leijnse et al., 2010; Otto and Russchenberg, 2011). A detailed urban hydrodynamic model for a catchment in the city of Rotterdam was chosen as a pilot case. Catchment conditions are representative of urban districts in lowland areas, especially delta cities, where almost half of the world population lives. Lowland catchments are characterised by flat terrain, therefore the mechanism dominating sewer flow is different from sloped terrain, where flow is driven by gravitation. This study aims at analysing the sensitivity of this urban hydrodynamic model to changes in rainfall spatial and temporal resolution. The study's focus is on model uncertainty related to rainfall input; model performance is not tested here, since storms were virtually applied to the catchment, which did not allow a proper model validation based on water level and flow observations. However, model geometry was strictly checked and model parameters were estimated based on literature values and experts opinion, so that the model is considered to be a reliable representation of local pluvial response.

Results were used to address the following questions:

- Does small-scale precipitation variability affect hydrological response and can a highly detailed semi-distributed model properly describe such a response?

- Is high-resolution rainfall information required when a storm does not present pronounced space–time variability?

- Does sensitivity of small-sized urban catchments to spatial and temporal variability of precipitation depend on catchment scale?

The findings have relevance for the use of high-resolution radar data in flood forecasting and flood protection in cities, at intra-urban scale. It provides a contribution to the debate on radar spatial resolution requirements for urban drainage modelling of small-scale urban catchments at district level, i.e. up to 3 km^2.

The paper is organised as follows. Section 2 presents the case study, hydrodynamic modelling approach and provides an analysis and description of rainfall fields used to conduct the sensitivity analysis. In Sect. 3 scale lengths are defined and then used to obtain a set of dimensionless parameters that will characterise relationships between rainfall fields, spatial resolution of rainfall and catchment characteristics. In Sect. 4 results of the scale analysis are shown and discussed. Lastly, conclusions are presented in Sect. 5.

2 Presentation of the case study and data sets

2.1 Case study and model description

This paper focuses on the central district of Rotterdam, the Netherlands (Fig. 1). The district is densely populated and includes mainly residential areas with approximately 30 000 inhabitants, as well as businesses and shopping centres. The district has a size of 3.4 km^2. Two green areas are located in the southern part of the district, sized 6 and 24 ha. The southern border of the district is formed by the Meuse River. The district belongs to a polder area below sea level. As a result, the area is nearly flat and there is not a dominant flow direction. During rainfall, excess storm water needs to be pumped out into the river system or temporarily stored elsewhere. Meanwhile, net rainfall fills sewer systems and storage basins up to the level of external weirs, where overflows to surface water take place if rainfall continues. An underground storage facility with a capacity of 10 000 m^2 has been built in the district to reduce flood risk during heavy rainfall events.

A hydrodynamic urban drainage model has been built for the catchment area using SOBEK-Urban software (Deltares, 2014). Although fully distributed models best describe the effect of rainfall variability on a catchment, the use of a highly detailed semi-distributed model with runoff areas of approximately the same size or smaller than the highest rainfall input resolution of 100 m × 100 m, is a close alternative. The combined sewer system was modelled in 1-D and consists of around 3000 manhole nodes (most of them are with runoff) and 11 external weirs, which serve as outflow points. The model contains four pressurised pipes interconnecting parts of the sewer system. Two external pumping stations transport water to the waste water treatment plant and to the river. Rainfall–runoff processes are modelled in SOBEK-RR (Deltares, 2014). The main components in this model are surface water storage, evaporation, infiltration and delay of surface runoff before entering the sewer system. Surface water storage occurs when rainwater forms puddles. When the water level exceeds the given maximum street storage, runoff is generated. Infiltration is computed on pervious surfaces by the Horton equation. Runoff to the sewer system is computed

Table 1. Surface characteristics of the central district catchment in Rotterdam used for hydrodynamic modelling: percentage, runoff coefficient and storage coefficient.

Type of area	Overall percentage (%)	Runoff coefficient (min^{-1})	Storage coefficient (mm)
Open paved flat	40	0.2	0.5
Closed paved flat	14	0.2	0.5
Roof flat	16	0.2	2
Roof sloped (slope larger than 4 %)	30	0.5	0

Table 2. Specification of the X-band radar of CESAR.

Dual polarimetric X-band radar	
Radar type	FMCW
Polarisation	Dual polarisation
Frequency	9.475 GHz
Highest range resolution	30 m
Min range	230 m
Max range	< 122 km
Max unambiguous radial velocity	19 m s^{-1}
Temporal resolution	1 min
Beamwidth	1.8°
Elevation	0.5°

as a function of net rainfall and runoff factors, which depend on length, roughness, slope and percentage of imperviousness of the areas. According to Dutch guidelines (Stichting RIONED, 2004), four different area types were used with different sets of runoff parameter values (Table 1): closed paved, open paved, roof flat and roof sloped (with slope larger than 4 %) areas. The open paved area type represents paved streets with bricks, which allow water to infiltrate and to be retained within the road surface. Green areas are not taken into account by the model, as they are assumed to be disconnected from the sewer system. The rainfall–runoff module is lumped and its basic unit is the "runoff area". Each runoff area contains different types of surface, the runoff of which enters the sewer system through the manhole nodes. Further details of the software package used in this study are provided in the Appendix.

2.2 Rainfall data

Rainfall data were obtained from CESAR (Leijnse et al., 2010) which provides data from a dual-polarimetric X-band radar collected at 30 m range resolution and a maximum unambiguous range of 15 km approximately. Other specifications on the new generation X-band radar device can be found in Table 2. Aggregations were made from radar rainfall rates at 30 m polar pixels based on reflectivity for values smaller than 30 dBZ, differential phase otherwise (Otto and

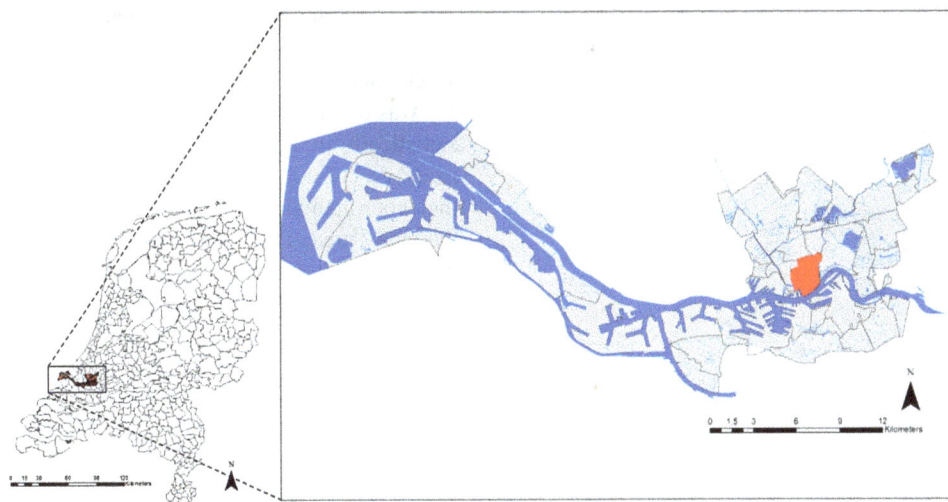

Figure 1. Localisation of Centrum district (in red in the right panel), situated in Rotterdam urban area (right panel and in red in the left panel), the Netherlands (left panel).

Russchenberg, 2011). The X-band radar has been operational intermittently since 27 June 2008. From the available data sets provided by CESAR, four rainfall storms could be selected for analysis based on a minimum mean rainfall volume of 3 mm over the area size of the studied catchment, the size of which is $3.4 \, \text{km}^2$. Lower rainfall volumes produce insufficient runoff to allow proper hydrodynamic analysis. According to the classification adopted by Emmanuel et al. (2012), events are grouped as follows:

- Event 1 and Event 2: storm organised in rain bands;

- Event 3: storm less organised;

- Event 4: light rain.

In Event 1, a long-lived squall line was measured on 3 January 2012. The convective storm moved eastward with a velocity of $20 \, \text{m s}^{-1}$ approximately. A squall line is a line of convective cells that forms along a cold front with a predominately trailing stratiform precipitation (Storm et al., 2007). Squall lines are typically associated with a moderate shear between 10 and $20 \, \text{m s}^{-1}$ and strong updraft (Weisman and Rotunno, 2004). If winds increase rapidly with height ahead of a strong front, thunderstorms triggered along the boundary may organise into severe storms called supercell storms. The X-band radar was able to capture storm features associated with supercell. The overall duration of the event was short, 1 h in total, but the most intense peak lasted 10 min at the end of the storm, and with rainfall intensities higher than $100 \, \text{mm h}^{-1}$. The most affected part of the catchment was the central and the northwestern part, while the southern part was affected by light rain. Event 2, occurring on 10 September 2011, can be characterised as a cluster of convective and organised storm cells that moved in a northeast direction. The

storm moved northeastward with a velocity of $16 \, \text{m s}^{-1}$ approximately. The storm system showed a convective spread area larger than the first event and with slower shift. The storm lasted 2 h, between 18:00–20:00 UTC, with the most intense part concentrated between 19:00 and 20:00 UTC. Intensities ranged between 30 and $60 \, \text{mm h}^{-1}$, and the whole central part, from south to north of the catchment was affected, while east and west bands were less exposed. In Event 3, which occurred on 28 June 2011 from 22:00 to 24:00 UTC, mesoscale observations showed a non-organised squall line moving northeast, with a speed of $15 \, \text{m s}^{-1}$ approximately and containing rainfall rate cores of at least $10 \, \text{mm h}^{-1}$. Rainfall rate values of $50 \, \text{mm h}^{-1}$ were founded over small areas during 22:00–23:00 UTC, travelling from southwest toward northeast and affecting all the catchment. Lastly, Event 4, which occurred on 29 October 2012, was a stratiform precipitation moving eastward at $13 \, \text{m s}^{-1}$ approximately and showing uniform rainfall rates. Rainfall retrieval was based on reflectivity only, of about $8 \, \text{mm h}^{-1}$. Storms motions and directions were estimated based on a centroid-based storm association algorithm, inspired by Johnson et al. (1998). For each event, total rainfall volumes in terms of minimum, maximum and mean value of all pixels affecting the area can be found in Fig. 2, as well as their standard deviation, giving a first insight into the variability of the event. Figure 3 presents radar images showing the maximum intensity minute of each one of the selected rainfall events, as well as the location of the catchment with respect to them and the main direction of the storms.

3 Methods

In this study, effects of radar spatial resolution on hydrological model outputs were analysed by means of length

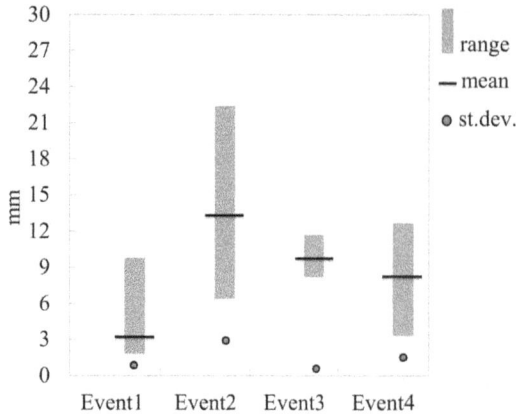

Figure 2. Characteristics of the four selected storm events: rainfall volume range (maximum and minimum for all $100\,\mathrm{m} \times 100\,\mathrm{m}$ pixels over the catchment area), mean and standard deviations.

scales. Building upon the approach introduced by Ogden and Julien (1994), length scales were developed for urban catchments and adjusted and extended for application to hydrodynamic urban drainage models (Table 3, Fig. 4). A scale dependency between storm, catchment and model topology for small-scale urban catchments, was studied based on rainfall fields derived from polarimetric radar, using spatial resolutions of 100, 500, 1000 and 2000 m, obtained by upscaling the original resolution. The finest spatial resolution, namely 100 m, was chosen for being the highest resolution at which radar rainfall data were provided. The 1000 m resolution was selected for being the resolution at which most of the national weather radar networks work, the 500 m was chosen as an intermediate resolution between X-band radar and C-band national radar network resolutions. The 2000 m resolution was used to represent uniform rainfall conditions over the catchment. Results were analysed to investigate the effect of different spatial and temporal rainfall data resolutions on rainfall volumes, peak runoff and in-sewer water depths at locations inside the catchment, according to dimensionless parameters specified.

3.1 Scale lengths

3.1.1 Rainfall lengths

Rainfall length L_R was defined as the rainfall resolutions used as input into the hydrodynamic model to observe the response of the catchment. Rainfall data were spatially aggregated from the original resolution (30 m near the radar, 100 m elsewhere) to 500, 1000 and 2000 m. In this work storms where captured at distances from radar such that the finest grid resolution was $100\,\mathrm{m} \times 100\,\mathrm{m}$.

3.1.2 Storm and catchment lengths

To characterise storm size, de-correlation length of the storm L_D was defined as the distance from which rainfall rates are statistically independent. For each of the four storms under study, de-correlation lengths were determined as the range of the experimental anisotropic semi-variogram computed over the study area. The semi-variogram function was originally defined by Matherson (1963) as half the average squared difference between points separated by a distance h (Eq. 1). It is calculated as

$$\gamma(h) = \frac{1}{2m(h)} \sum_i [(Z(x_{i+h}) - Z(x_i))^2], \qquad (1)$$

where $m(h)$ is the set of all pairwise Euclidean distances h, and Z are the rainfall values at spatial locations. Storm de-correlation length was defined as the range of the semi-variogram, i.e. the distance at which the sill is first reached; the sill is defined as the limit of the semi-variogram tending to infinite lag distances (see Fig. 5). Besides the value of the lag distance, in this paper the direction is also taken into account: we computed the anisotropic semi-variogram (Goovaerts, 2000; Haberlandt, 2007; Emmanuel et al., 2012), in four directions, spaced 45°. Since the limiting length is the minimum storm length, the minimum of the four ranges was taken as storm length for the study.

Storm de-correlation length was compared to pixel size of radar rainfall estimates L_R and to catchment length L_C, computed as the square root of the catchment size.

3.1.3 Model lengths

Characteristic lengths of the model topology are a result of modeller's choices based on available data, options of applied software and acceptable computational effort. Runoff length L_{RA} characterises the spatial resolution of the runoff model and was defined as the square root of the averaged runoff areas' size. Runoff length quantifies the size of the grid over which runoff is generated: if $L_{RA} \ll L_C$, the catchment is divided into sufficiently small elements to describe the spatial variability of the catchment characteristics. Moreover, spatial variability in rainfall rates can be properly captured by the runoff model if $L_R < L_{RA}$. If $L_R > L_{RA}$, rainfall rates can no longer be correctly attributed to associated runoff areas, which may distort the hydrological response pattern (Ogden and Julien, 1994).

Sewer length L_S characterises the inter-pipe distance; it is roughly the urban equivalent of drainage density for natural catchments. L_S was defined as the ratio between catchment size and the total length of the piped system. Similar to L_{RA}, the condition $L_R \ll L_S$ guarantees that the sewer pipe system routes the correct rainfall volume, previously transformed in runoff over the corresponding runoff area.

Table 3. Scale lengths related to catchment, runoff areas and sewer density, for the total catchment as well as length scale ranges for the 10 subcatchments.

Length scales (m)	Code	Event 1	Event 2	Event 3	Event 4
Storm de-correlation length	L_D	950	1000	1480	1600
Runoff length: mean (median)	L_{RA}		28 (23)		
Sewer length	L_S		43		
Catchment length	L_C		2.024		
Subcatchment runoff length (range)	(Sub) L_{RA}		21–59		
Subcatchment sewer length (range)	(Sub) L_S		33–78		
Subcatchment length (range)	(Sub) L_C		429–2.024		

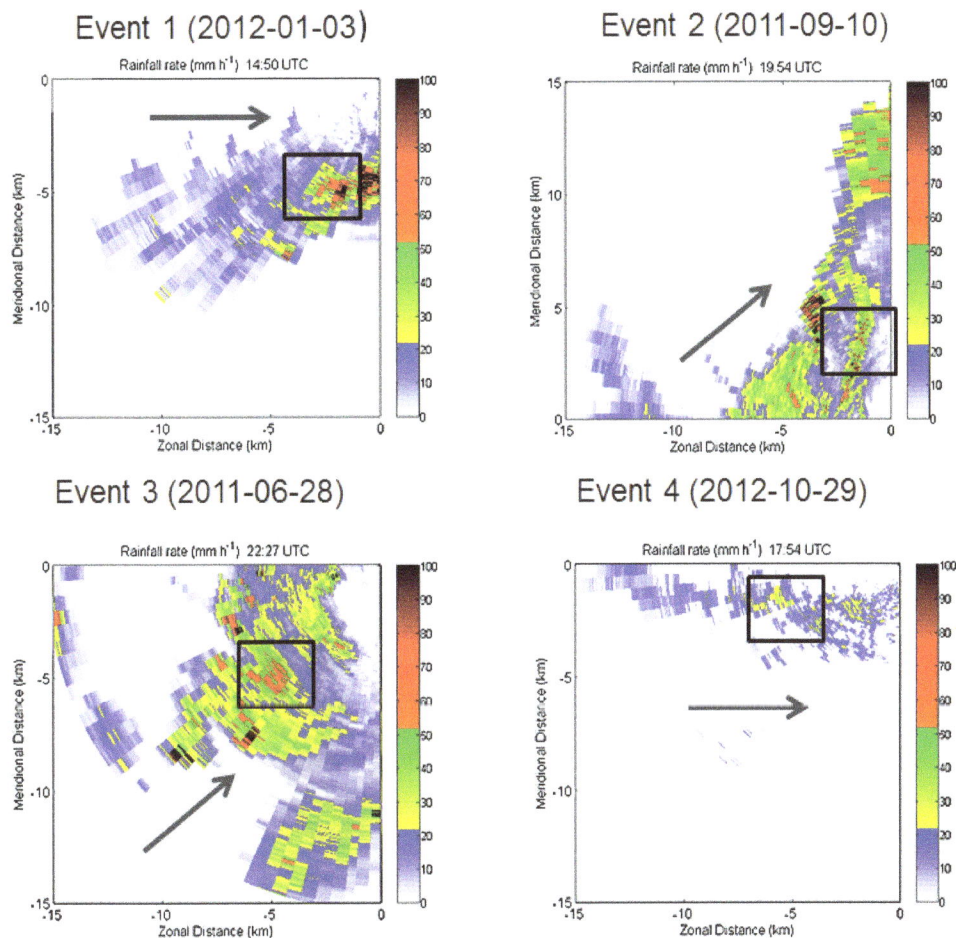

Figure 3. Plots of the maximum intensity time step for the four storm events, main direction of the storm (grey arrow), and virtual position of the catchment with respect to storm movement (black square). Zonal distances in east–west and north–south direction from X-band radar position. The latter is at (0, 0) and the maximum range is 15 km. Events 1, 3 and 4 were detected in the southwestern quadrant of the radar coverage, while Event 2 was detected in the northwestern quadrant.

3.1.4 Definition of subcatchments

The analysis involving model lengths was conducted at subcatchment scale to compare results for different model lengths: the district was divided into 11 subcatchments (Fig. 4). In lowland areas, drainage systems are often inter-linked and looped and flow direction changes over the course of a storm event as the system first fills and then starts routing the storm water. This implies that flow directions and subcatchment boundaries are changeable and cannot be defined based on topography or network configuration. For this reason, in order to define subcatchment boundaries, we performed the following steps (according to a previous work of ten Veldhuis and Skovgård Olsen, 2012):

Figure 4. Storm de-correlation length (L_D) and rainfall resolution (L_R) in left panel. Catchment length (L_C), runoff length (L_{RA}) in right panel; the catchment is divided into 11 independent subcatchments. Red arrows represent main flow directions. Runoff areas are also displayed, their average size is reported in Table 2.

1. We ran simulations under long-lasting uniform storms;

2. We made sure no overflow towards surface water bodies occurred (in that case, a direction change would affect the sewer flow);

3. We detected sewer pipes with $Q = 0$;

4. We delineate subcatchments as if the latter were removed;

5. We compared flows at outlets of the 11 subcatchments in "looped" conditions (the original model) and "branched" conditions (model after the removal of cross boundary conduits); we found high agreement between the two results; therefore we accepted the catchments' delineation as a satisfactory approximation.

A visual inspection of the sewer network helped to understand the direction of flow: since no overflows occurred for the events used in this study, the system drains received water toward the main pumping station. Under this condition the main sewer conduits collect all water from peripheral conduits. We could therefore observe the flow direction in the main conduit.

3.2   Dimensionless parameters

Using the length scales, dimensionless parameters were computed to analyse relationships between spatial characteristics of rainfall, catchment and its hydrological response.

3.2.1   Rainfall sampling number (L_R/L_D)

"Rainfall sampling number" (L_R/L_D) was defined as the ratio between rainfall length (L_R) and storm de-correlation length (L_D) in order to study rainfall gradient smoothing

in terms of the relationship between the estimated rainfall field and the storm inherent structure. This parameter is similar to the "storm smearing" effect defined by Ogden and Julien (1994); it accounts for the deformation of the storm structure caused by rainfall measurements of coarser resolution than the storm length. For instance, rainfall intensities in storm cells with sizes smaller than applied rainfall spatial resolution will be averaged out, leading to an underestimation in rainfall rates in the area affected by the storm cells and a overestimation in the area surrounding the cells.

In other words, as L_R tends to L_D, rain rates in high intensity regions tend to decrease, and conversely rainfall intensities in adjacent regions tend to increase. The overall effect is a reduction of rainfall gradients. Dimensionless rainfall sampling number quantifies this effect.

3.2.2   Catchment sampling number (L_R/L_C)

The second dimensionless parameter, "catchment sampling number" (L_R/L_C), also referred to as "watershed smearing" by Ogden and Julien (1994), was defined as the ratio between rainfall length (L_R) and catchment length (L_C). It accounts for rainfall transfer across catchment boundaries, as the rainfall spatial resolution approaches the size of the catchment. When the parameter exceeds 1, location of rainfall cells with respect to the catchment becomes uncertain and rainfall variability is not properly captured by the catchment. In other words, when dealing with small size storms, the position of the storm with respect to the catchment is no longer properly represented for rainfall resolutions approaching or exceeding catchment length. This affects the hydrological response: a storm moving near the boundaries of the catchment is averaged across the catchment boundary, so rainfall is artificially transferred outside the catchment. This effect is quantified by the catchment sampling number, relating the size of the catchment to the size of the radar pixel.

3.2.3   Runoff sampling number (L_R/L_{RA})

The third parameter is called "runoff sampling number" (L_R/L_{RA}), which is the ratio between rainfall length (L_R) and runoff area length (L_{RA}). This, similar to catchment sampling number, quantifies the correct assignment of rainfall values to the corresponding runoff area. The higher this ratio, the less precise is the rainfall assignment to the correct runoff area, but also the lower this ratio, the more unable is the model to capture rainfall variability, as the model resolution is coarser than the rainfall resolution. This parameter relates to the rainfall–runoff module of the model, which has rainfall as input and runoff discharge into one of the nodes of the sewer network as output. Runoff sampling number relates model input data resolution to runoff model resolution, and intends to measure the "smearing" of runoff flows induced by low rainfall resolution compared to runoff area resolution.

3.2.4 Sewer sampling number (L_R/L_S)

The fourth dimensionless parameter is the "sewer sampling number" (L_R/L_S), defined as the ratio between rainfall length (L_R), and intra-sewer length (L_S), which is computed as the average length of conduits in the system. The lower the sewer sampling number, the less sensitive is the drainage network to rainfall variability: a low "sewer sampling number" means that the inter-pipe distance is larger than the rainfall pixel size, so the sewer system cannot catch rainfall variability. Conversely, for higher sewer sampling numbers rainfall input is too coarse compared to the sewer network density and this may result in lack of accuracy of modelled water levels and sewer overflows. The "smearing effect" for sewer flows is related to the runoff smearing effect, quantified by the runoff sampling number, but they differ in this respect: the latter focuses on runoff model output, namely discharge towards the sewer network, while the sewer index represents the routing within the piped system and so it quantifies the smearing effect for in-sewer water levels. Water levels in pipes are affected by runoff discharge but also by upstream sewer inflows. As it is not possible to isolate the effect at the level of individual pipes, it is analysed at the outlet of each independent subcatchment.

3.2.5 Normalisation of model output results

To compare results between rainfall resolutions and between storms, model results were normalised with respect to results related to the highest rainfall spatial resolution: total rainfall volumes, runoff peaks and maximum in-sewer water depths were normalised according Eqs. (2), (3) and (4) respectively:

$$V_{norm}\left(L_{R_i}\right) = \frac{V(L_{R_i})}{V(L_{R100})} \tag{2}$$

$$Q_{norm}\left(L_{R_i}\right) = \frac{Q(L_{R_i})}{Q(L_{R100})} \tag{3}$$

$$WD_{norm}\left(L_{R_i}\right) = \frac{WD(L_{R_i})}{WD(L_{R100})}, \tag{4}$$

where L_{R_i} represents parameter values at the rainfall resolution under consideration (100, 500, 1000 or 2000 m) and L_{R100} represents values at 100 m rainfall resolution, used as a reference for normalisation.

3.3 Temporal resolution analysis

While the focus of this paper is on spatial scales, a preliminary investigation of the effect of temporal resolution on model outcomes was conducted to see how temporal resolution interrelates with spatial resolution. To this end, rainfall data were aggregated to 5 and 10 min temporal resolutions.

The temporal aggregation was performed by averaging out 5 (10) consecutive 1 min rainfall values at a time, obtaining temporal resolution of 5 (10) min. Semi-variograms were computed for these resolutions to study the relation be-

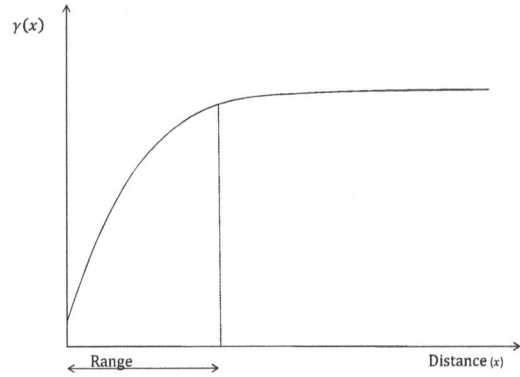

Figure 5. Sketch of semi-variogram: the range is the distance (x) from the origin beyond which the semi-variogram $\gamma(x)$ tends to infinity.

tween temporal resolution and the spatial structure of rainfall. The effect of the variation in rainfall temporal resolution on model outputs was quantified through the comparison of time to maximum water depths. Combined time–space resolutions were studied for Events 3 and 4: both events were simulated at two spatial and two temporal resolutions, namely 100 m, 1000 m, 5 min and 10 min, composing four different spatio-temporal rainfall scenarios.

4 Results and discussion

Results of length scale calculations are presented in Table 3, dimensionless parameter values are shown in Table 4. Storm de-correlation lengths vary between 950 and 1600 m for the four storm events. Subcatchment lengths vary from 429 to 2024 m, while runoff and sewer lengths in the hydrodynamic model vary between about 20 and 80 m, representing the model's high spatial resolution. Dimensionless parameter values show that rainfall sampling numbers vary from 0.06 for Event 4 to 0.11 for Event 1 at 100 m rainfall resolution and increase to 1.25 and 2.11 respectively at 2000 m rainfall resolution. Catchment sampling number increases from 0.05 to 0.99 for 100 and 2000 m, while runoff and sewer sampling numbers vary from 1.9 to 4.7 at 100 m resolution to 25.5 to 93.3 at 2000 m resolution, runoff sampling numbers being slightly higher than sewer sampling numbers.

Model results of the four storm events were compared against dimensionless parameters to identify trends and variability as a function of storm characteristic, radar resolution and model resolution.

Table 4. Dimensionless parameters values derived from scale length values, for the four different rainfall resolutions used in the study. Values presented for runoff sampling and sewer sampling numbers, represent value ranges for the 10 subcatchments (outlined in Fig. 4).

Rainfall resolution	Rainfall sampling number L_R/L_D				Catchment sampling number	Runoff sampling number	Sewer density sampling number
(m)	Event 1	Event 2	Event 3	Event 4	L_R/L_C	L_R/L_{RA}	L_R/L_S
100	0.11	0.10	0.07	0.06	0.05	2.6–4.7	1.9–3.8
500	0.53	0.50	0.34	0.31	0.25	13.1–23.3	6.4–19.1
1000	1.05	1.00	0.68	0.63	0.49	26.1–46.7	12.8–38.3
2000	2.11	2.00	1.35	1.25	0.99	52.3–93.3	25.5–76.5

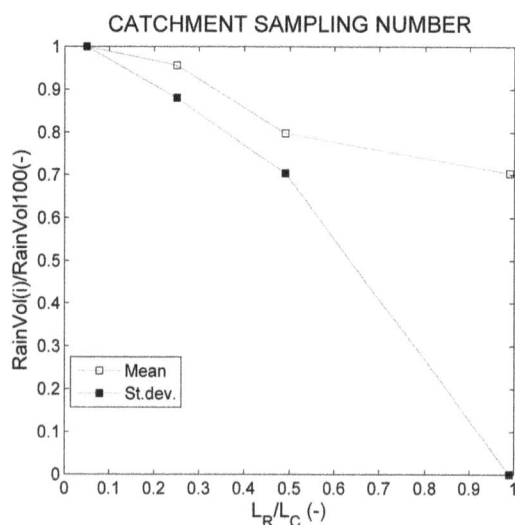

Figure 6. Normalised rainfall volumes versus catchment sampling number (L_R/L_C): mean and standard deviation of normalised rainfall volumes computed over the catchment, for the four events.

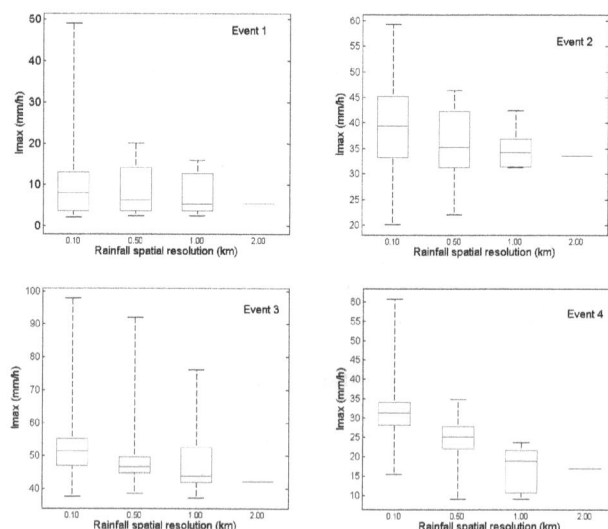

Figure 7. Boxplots of maximum rainfall intensity (mm h^{-1}) among all pixels covering the catchment area, for the four spatial resolutions (the 2000 m shows a unique value corresponding to rainfall uniformly distributed over the catchment), for the four events analysed.

4.1 Effect of spatial resolution

4.1.1 Total rainfall volumes versus catchment sampling number

Figure 6 shows mean and standard deviation of normalised rainfall volumes (according to Eq. 2) computed over the catchment, versus catchment sampling number. This result was obtained only analysing various rainfall resolutions and no hydrological modelling was used.

The results show that mean normalised rainfall volumes decrease by 5, 20 and 30 % with respect to the 100 m resolution reference, for L_R/L_C 0.2, 0.5 and 1 respectively. Standard deviations decrease by 2, 30 and 100 % respectively. Mean and standard deviation decrease progressively for catchment sampling number values above 0.2. This means that rainfall gradients decrease as rainfall values are smoothed at coarser resolution and that rainfall volumes decrease as smoothing of rainfall values at the catchment boundaries artificially transfers rainfall across the boundary. According to the findings of Ogden and Julien (1994), this effect, called by them "catchment smearing", occurs for catchment sampling numbers greater than 0.4. In contrast, results of the present study show that this effect already occurs at smaller sampling numbers, namely 0.2, and becomes stronger for values greater than 0.2. Figure 7 presents boxplots for maximum rainfall intensity values per pixel, over the studied catchment as a function of rainfall spatial resolution. The median of maximum intensity values shows a mild decrease for coarser rainfall resolutions. The smoothing effect is more pronounced for Event 3 and Event 4, where convective cells move closer to catchment boundaries. This results in storm cells being smoothed across catchment boundaries.

Event 1 is characterised by a 1 km wide storm line passing over the catchment very rapidly, resulting in steep rainfall gradients that are strongly smoothed when rainfall input resolution is reduced. When resolution is reduced from 100 to 500 m, spatial structure of the storm line is decomposed, leading to a reduction in maximum rainfall intensities (Fig. 7) in the area affected by the storm. As resolution is reduced from 1000 to 2000 m resolution, storm structure is lost and rainfall becomes uniform over the catchment. Storm cells in Event 2 are characterised by steeper spatial gradients in rainfall intensities compared to Event 1 and as a result maximum rainfall intensity values are more strongly affected by changes in rainfall resolution: the upper 25 % values decrease as a result of rainfall gradient smoothing, especially as resolution is reduced from 100 to 500 m. The lowest 25 % values increase as a result of gradient smoothing and storm structure decomposition, especially as resolution is reduced from 500 to 1000 m, where the variation between first and third quartile values is reduced from about 10 to 5 mm h^{-1}. Events 3 and 4 present a clear reduction of the median as a result of rainfall aggregation across the catchment boundary. The variation between first and third quartile values is larger at 1000 m resolution than at 100 and 500 m resolution. For Event 3, this is due to the non-organised structure of rainfall cells: local rainfall cells found at 100 m resolution are smoothed out at 500 m resolution, while at 1000 m resolution the most active convective area affects two out of nine pixels covering the catchment, i.e. the lowest 25 % values are relatively high. Event 4 is characterised by stratiform precipitation showing uniform rainfall rates. Upper quartile values decrease from 34 to 22 mm h^{-1}; lower quartile values reduce from 28 mm h^{-1} to about 10 mm h^{-1} at 1000 m resolution. This is a result of rainfall gradient smoothing and storm cells spreading southward due to spatial aggregation, while the core of the storm remains within the catchment boundaries. The strongest effect of rainfall coarsening in this case is found in a strong reduction of rainfall gradients. As a general conclusion, spatial aggregation leads to smoothing of rainfall gradients, while the effect on rainfall intensities' distribution strongly depends on spatial dimensions of storm cells and the movement of storm cells relative to the catchment boundaries.

4.1.2 Normalised maximum in-sewer water depths and runoff peaks versus rainfall resolution

Figure 8 summarises the effect of rainfall spatial resolution coarsening on semi-distributed hydrodynamic model outputs in terms of maximum computed water depths and maximum runoff flows in all nodes, per storm event. The in-sewer maximum water depths and runoff peaks at every node of the model are normalised using Eqs. (3) and (4). Results presented in the boxplots show that normalised runoff peaks are more strongly affected by changing spatial resolution of rainfall inputs compared to normalised maximum water depths. The largest effect of spatial aggregation is found for Event 4 (Fig. 8 last column), where upper and lower quartile values of runoff peaks are reduced by 40 to 60 % at

2000 m resolution with respect to the reference at 100 m resolution. Normalised maximum water depths are less strongly affected; upper quartile values remain almost unchanged, while lower quartile values decrease by up to 30 %. Event 4 has a pronounced spatial structure that is strongly affected by rainfall resolution coarsening and this directly translates into stronger changes in runoff volumes compared to the other events. Largest changes in normalised maximum water depths are found for Event 1, where upper and lower quartile values change by up to 40 % as a result of spatial redistribution of rainfall due to resolution coarsening. This event is characterised by small total rainfall volumes, resulting in small flows and water depth variations, which in turn translate into large relative differences.

Smaller changes in water depths compared to runoff flows are explained by the fact that water depths are influenced by rainfall–runoff inputs as well as by sewer routing and by storage being dominant over flow in drainage systems characterised by small gradients.

For Events 1, 2 and 3, changes in normalised water depths and runoff flows are of the same order of magnitude at 500 and 1000 m resolutions, which indicates that the effect of rainfall resolution coarsening from 100 to 500 m is not further amplified as resolution is reduced to 1000 m. When resolution is further reduced to 2000 m, corresponding to uniform rainfall over the catchment, values above 3rd quartile tend to increase as areas previously affected by low rainfall receive higher rainfall as a result of gradient smoothing.

4.1.3 Spatial structure of rainfall: anisotropic semi-variogram

Figure 9 shows experimental multi-directional spatial semi-variograms for each of the four storm events. For each storm and each time step, the semi-variogram was computed in four directions, from 0 to 180°, starting from north and going clockwise at an angle step of 45° (directions at 0 and 180° are the same, thus plots coincide). To obtain a unique semi-variogram representative of overall storm duration, for each direction, a weighted average of all semi-variograms was computed, assigning a higher weight to those of higher variance. This criterion was chosen to focus the study on more pronounced spatial rainfall structures, without losing information on the temporal evolution of the storm. Rainfall data used for the calculation are those estimated at the highest temporal and spatial resolution of IDRA radar, 1 min and 100 m respectively, in order to analyse rainfall structure at its most accurate description. The semi-variogram of Event 1 (Fig. 9 top left) presents a unique structure with a range of 1200 m in three out of four directions, while at 90° direction the range is smaller, reaching a de-correlation distance at 950 m. This is quite expected since Event 1 is a squall line moving from west to east, thus the gradient at 90° is steeper than at 180°. All four semi-variograms show a fast rise, al-

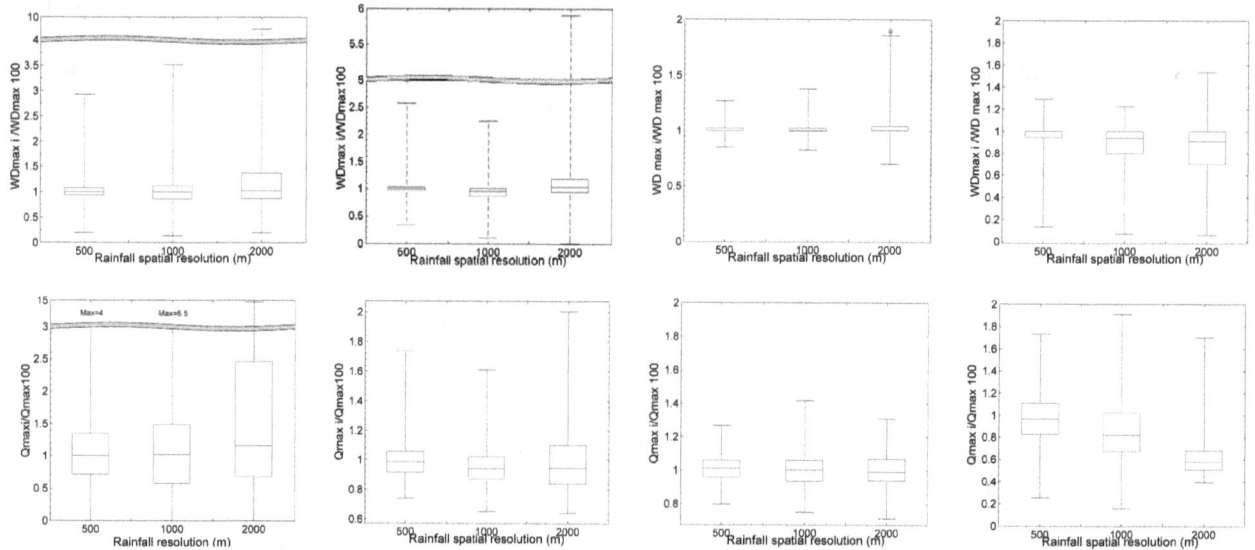

Figure 8. Boxplots of the normalised maximum water depths (top panel) and runoff peaks (bottom panel) computed for all nodes in the model, for Events 1 (left) to 4 (right).

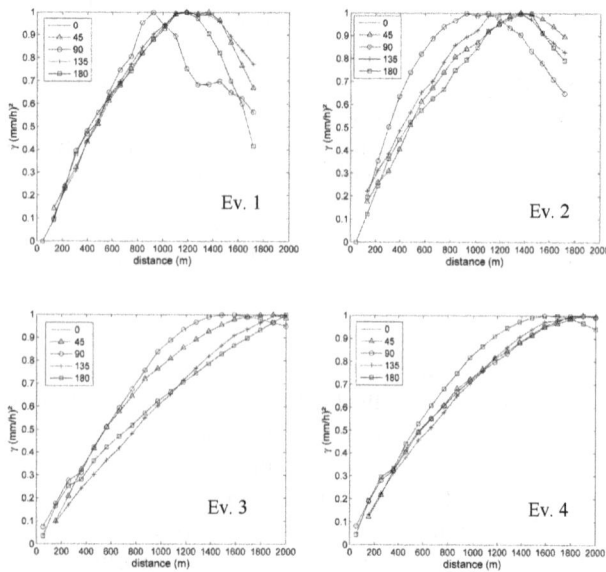

Figure 9. Instantaneous experimental multi-directional spatial semi-variogram of non-zero rainfall for each of the four storms.

Table 5. Range derived from experimental semi-variograms for different temporal aggregations, for all four events.

Rainfall	Range (m)		
	$\Delta t = 1\,\text{min}$	$\Delta t = 5\,\text{min}$	$\Delta t = 10\,\text{min}$
Event 1	950	960	970
Event 2	1000	1200	1450
Event 3	1480	> 2000	> 2000
Event 4	1600	1500	1500

Semi-variograms of Events 3 and 4 show a milder rise compared to Events 1 and 2. They are characterised by a different type of rainfall structure: Event 3 is a non-organised storm band, it seems to have a more defined structure in the 45 and 90° directions, the range of which is 1480 m (see also Table 5). The curves at 135 and 180° directions do not reach stability, meaning that the de-correlation distance exceeds the catchment size for which the semi-variogram was calculated. The rainfall structure of Event 4 shows a more isotropic behaviour. This is an expected result, since light rain storms are characterised by low and uniform rainfall rates. The de-correlation distance is 1600 m, highest among the four events, found in the 180° direction. The de-correlation distances found by means of this geostatistical approach were used to compute the rainfall sampling numbers discussed in the next section.

though the shape of the one at 90° diverts considerably from the rest.

The same results are found for Event 2: the directional semi-variogram at 90° shows a faster rise compared to the other directions, thus the storm structure is clearly oriented. The de-correlation distance is 1000 m. No explanation was found to interpret the pronounced decrease in the semi-variograms of Events 1 and 2. We can only report that the same behaviour was found in storms belonging to the same rainfall group defined by Emmanuel et al. (2012).

4.1.4 Normalised in-sewer maximum water depths versus rainfall sampling number

The rainfall sampling number is a measure for what Ogden and Julien (1994) referred to as "storm smearing": rainfall rates in convective regions tend to decrease while rain rates in low-intensity regions tend to increase as a result of spatial aggregation. The overall effect is thus a flattening of rainfall gradients. This happens when the resolution of the volume unit measured by the weather radar approaches or exceeds the rainfall de-correlation length, thus the rainfall sampling number exceeds 1. This effect is also due to "catchment smearing" addressed in Sect. 4.1.1. The effect of rainfall sampling number on in-sewer water depths was analysed for all four rainfall events. In-sewer depths were analysed at the outlets of the 10 subcatchments (Fig. 4) to study the effect of storm smearing in relation to catchment characteristics and in-sewer flow routing. Maximum water depth values were normalised with respect to values at 100 m resolution. Figure 10 shows normalised maximum water depths against rainfall sampling number, at the outlet nodes of the 10 subcatchments and the outlet node of the whole catchment (catchment number 11 in Fig. 10). For all events, deviations in normalised water depth increase for L_R/L_D increasing to 0.5 and 1. For Event 1, when L_R/L_D exceeds 1, deviations slightly reduce in 5 of out 11 catchments while slightly increasing for 6 subcatchments, depending on local re-distribution of rainfall. Subcatchment 2 shows highest deviation at $L_R/L_D = 0.5$, followed by a decrease for coarser resolutions. This is because the subcatchment is located at the boundary of the storm, where at 500 m spatial resolution rainfall gradients increase, while at 1000 m resolution gradients are reduced due to averaging over a larger region not affected by the storm. This directly affects the maximum water depth in underlying subcatchments. The opposite situation occurs in subcatchment 5, which is located in the southern part of the catchment with the closest node at 1.2 km from the convective region, beyond the de-correlation length. The storm only affects this southern region when rainfall data are aggregated to the 2000 m resolution, so the storm "virtually" extends from the northern part of the catchment to the whole catchment. A similar effect is noticed at the same subcatchment for Event 2. Results suggest that for most subcatchments, storm smearing occurs for L_R/L_D ratio above 0.5. Figure 9 shows that while storm smearing already affects water depths at values of L_R/L_D below 0.5, the effect becomes a lot stronger for values between 0.5 and 2. Results are in agreement with the findings of Ogden and Julien (1994), who found for their catchments that "storm smearing" occurred for $L_R/L_D > 0.8$. This implies that for the storm events used in this study, with de-correlation lengths of 0.95 to 1.6 km, the current resolution of operational weather radars (1000 m) is insufficient to have a proper estimation of intra-urban hydrodynamics.

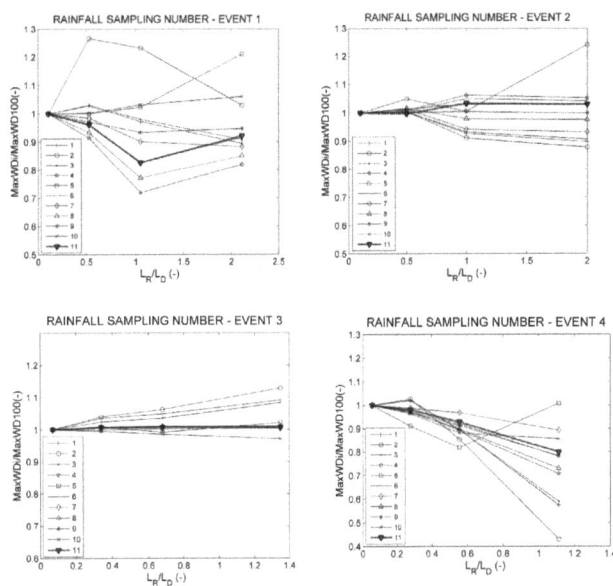

Figure 10. Normalised maximum in-sewer water depths versus rainfall sampling number (L_R/L_D): results at the outlet of the 10 subcatchments (numbered 1 to 10) and of the whole catchment (no. 11).

4.1.5 Normalised runoff peaks versus runoff sampling number

Normalised maximum runoff flows of all runoff areas were averaged within each of the 11 (sub)catchments for all four events and plotted versus corresponding runoff sampling numbers (Fig. 11) to study effects of rainfall smoothing on runoff inputs at (sub)catchment level. Deviations from 100 m simulation results remain between 0.9 and 1.1 for $L_R/L_{RA} < 20$, while higher deviations up to almost 50 % occur for $L_R/L_{RA} > 20$. At the original rainfall input resolution of 100 m, L_R/L_{RA} is below 10, so rainfall pixel size used to feed the urban hydrological model is up to 10 times larger than runoff model resolution. As L_R/L_{RA} grows larger, computed maximum runoff flows increasingly deviate as a result of rainfall smoothing and of catchment smearing, discussed in Sect. 4.1.1.

4.1.6 Normalised maximum water depths versus sewer sampling number

As presented in Sect. 3, sewer sampling number represents a measure of the ability of the sewer system to capture rainfall variability. For the model used in this study, intra-sewer pipe distances are quite small, ranging from 33 to 78 m: this means that there are 700 to 900 m of sewer pipes per 100 m × 1100 m of catchment area. The idea here is to analyse the combination effect of rainfall resolution and sewer model resolution. Figure 12 presents normalised maximum water depths as a function of sewer sampling numbers aver-

RUNOFF SAMPLING NUMBER

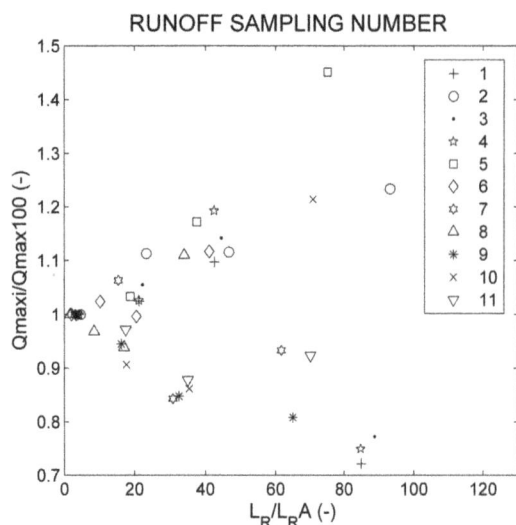

Figure 11. Normalised runoff peaks versus runoff sampling number (R_R/L_{RA}): results averaged over each of the 10 subcatchments (numbered 1 to 10) and over the whole catchment (no. 11).

SEWER SAMPLING NUMBER

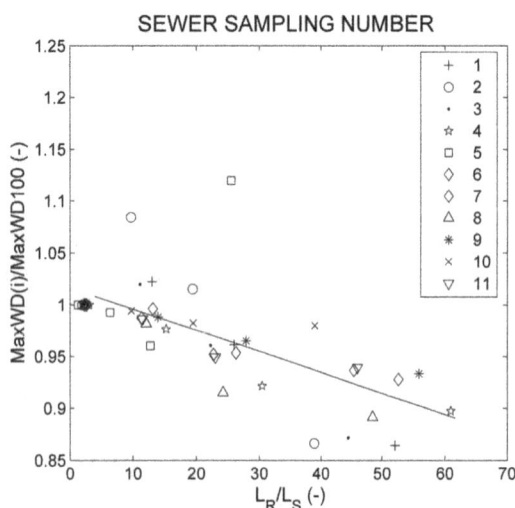

Figure 12. Normalised maximum water depths versus sewer sampling number (L_R/L_S): results at the outlet of the 10 catchments and of the whole catchment (no. 11).

aged per subcatchment, for all four events. Results show that maximum water depths tend to decrease for increasing sewer sampling numbers. In general, deviations from the reference case are smaller for in-sewer water depths, ranging from 0.87 to 1.13, than for runoff peaks, which are in the range 0.7–1.5. This is due to the smoothing effect of flow routing through the pipe system on in-sewer water depths.

4.2 Effects of temporal resolution

4.2.1 Changes in spatial structure of rainfall due to time aggregation

X-band radar images are obtained at 1 min temporal scale: the radar completes radar scans in 1 min. In order to analyse the effect of temporal resolution on rainfall spatial anisotropic semi-variograms, raw rainfall data were aggregated by averaging the original radar images to 5 and 10 min resolutions. The anisotropic experimental semi-variogram was then computed based on the aggregated values (Fig. 13). Anisotropic semi-variograms for these time resolutions were used to examine the interrelationship between temporal resolution and spatial structure of rainfall. Results show that the semi-variograms change in shape more strongly when aggregating from 1 min to 5 min compared to aggregating from 5 to 10 min resolutions. The range derived from the semi-variograms increases for lower temporal resolutions. This is especially clear for Event 3, where at 5 and 10 min the storm structure within the catchment boundaries is lost as the semi-variograms become monotonic in any of the four directions considered. In Event 4, the range expands until the catchment limits for three out of four directions, while in the 90° direction the semi-variogram range decreases. Events 1 and 2 seem less affected by changes in temporal resolution; the shape of the curves changes but the range expands only few tens of metres. Table 5 summarises ranges for all rainfall events as a function of time resolution.

4.2.2 Effect of temporal resolution on timing of maximum water depths

The effect of changes in rainfall temporal resolution on model outputs was quantified in terms of the time shift of maximum water depths with respect to the reference case. Figure 14 shows time shifts of maximum water depths between the reference simulation (100 m, 1 min) and 5 and 10 min simulations (both at 100 m spatial resolution) at the outlets of the 11 (sub)catchments, for Event 1. Results show that the timing of maximum water depths shifts by up to 4 min for aggregation to 5 min resolution and by up to 10 min for 10 min resolution. Time shifts were also calculated for all 3000 nodes of the catchment model (results not shown here). At 5 min resolution, the time shift of maximum water depths with respect to the reference case is less than 5 min for 99.4 % of all nodes. At 10 min resolution, time shifts of more than 5 min occur in 0.86 % of the nodes; time shifts in all other nodes are less than 5 min.

Figure 15 shows time shifts of maximum water depths with respect to two reference cases: 100 m, 1 min and 1000 m, 1 min. For Event 3, results show that the model is most sensitive to temporal aggregation to 10 min, at 100 m spatial resolution. The time delay of maximum water depths compared to the reference case is between 8 and 16 min. At

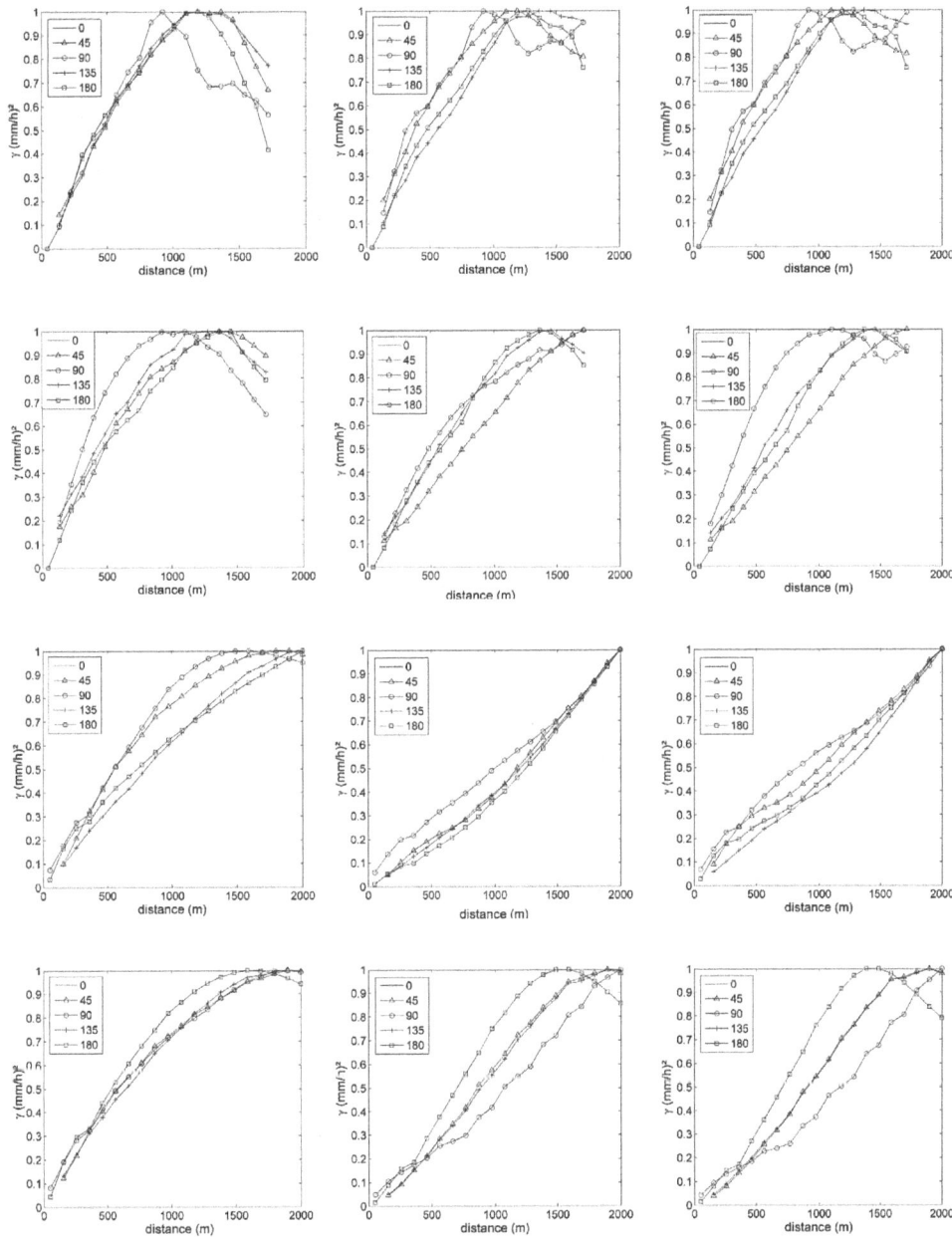

Figure 13. Anisotropic experimental semi-variograms for the four rainfall events (in rows) and different temporal resolutions: 1, 5 and 10 min (left, central and right column respectively).

the 1000 m resolution, the effect of temporal aggregation on timing of maximum water depths is comparatively smaller. The relatively high impact of 100 m and 10 min resolution simulation is explained by the change in rainfall structure induced by temporal aggregation. As shown in Fig. 13, third row, de-correlation length becomes larger than the catchment size. This effect already occurs at 5 min aggregation, but is more pronounced at 10 min aggregation. In both cases time aggregation results in enlargement of the area affected by convective storm cells, in a smoothing of rainfall peaks, and in a change in timing of rainfall peaks. This results in

delay or anticipation of maximum water depths, depending on the relative position of a node with respect to the storm and also depending on the temporal position of rainfall peak values, and therefore on the temporal sampling process (for instance, if peak values are within the same 5 or 10 min sampling interval, time to peak will be hardly shifted). If peak values are averaged out with previous or following no-peak values, this will result in an anticipation or delay of sampled rainfall values and consequently anticipation/delay in hydrological response. A possible explanation for why this effect is noticeable only at 10 min is because the concentration time

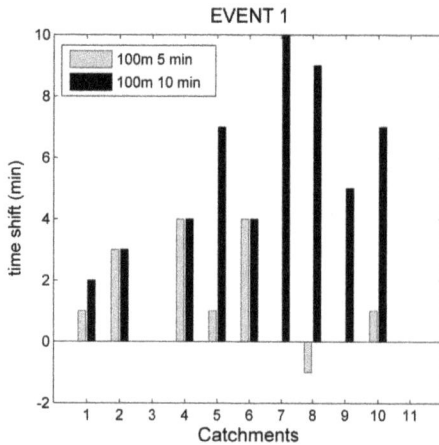

Figure 14. Time shift between maximum water depths of reference case (100 m spatial resolution, 1 min temporal resolution), and 5 and 10 min simulation, at the outlets of the 10 subcatchments and of the whole catchment (no. 11).

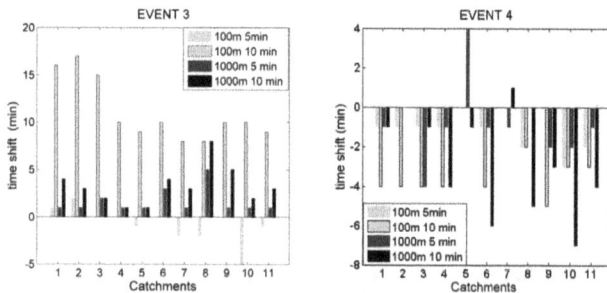

Figure 15. Differences in time to maximum water depth at the outlets of the 10 subcatchments and of the whole catchment (no. 11) for Events 3 and 4. Simulations at the highest spatial and temporal resolutions (100 m/1000 m 1 min respectively) are taken as reference.

of the 11 nodes is lower than 10 min. In order to notice an impact on model output, the time step of rainfall input must be smaller than the concentration time of the catchment at the outlet (Vaes et al., 2001) (being the concentration time the time rainfall needs to travel from the furthest place in the catchment to the chosen outlet of the sewer system). For Event 4, temporal aggregation results in anticipation of maximum water by 1 to 7 min at most of the catchment outlets.

Moreover, the effects of time aggregation on model performance have been analysed through the comparison in maximum water depths between simulations. Deviations in maximum water depths with respect to the reference case were below 0.05 m. This shows that the effect of rainfall spatial aggregation is much more important than that of temporal aggregation, in this specific case study and under these rainfall scenarios. Due to the low deviations found, results have not been reported here.

5 Conclusions

The sensitivity of an urban hydrodynamic model to spatial and temporal resolutions of weather radar data was investigated in this paper. Analyses are based on a densely populated urban catchment in Rotterdam, the Netherlands and four rainfall events that were derived from polarimetric X-band radar data. Rainfall and catchment properties were characterised using various length scales: catchment size and storm de-correlation length, which depend on the specific site and storm; rainfall data resolution, which depends on rainfall measurement resolution; and runoff resolution and sewer density, which are modeller's choices. Sensitivity of model outputs to rainfall spatial resolution was analysed in relation to: catchment size, through catchment sampling number (L_R/L_C); storm length, by means of rainfall sampling number (L_R/L_D); runoff resolution of the model, through runoff sampling number (L_R/L_{RA}); and sewer density, with the sewer sampling number (L_R/L_S). The first parameter is responsible for the uncertainty of rainfall location with respect to watershed boundaries; the second parameter describes smoothing of rain rate gradients; the third and fourth parameters describe the ability of the model (the runoff model and the sewer model respectively) to capture the rainfall structure. Storm length was been computed as the range of anisotropic experimental semi-variograms. Four rainfall spatial resolutions (100, 500, 1000 and 2000 m) and three temporal resolutions (1, 5 and 10 min) were analysed. Results obtained in this study show:

– As the ratio L_R/L_C increases (in this particular case for $L_R/L_C > 0.2$), there is a progressive decrease of both rainfall volume mean and standard deviation. Rainfall gradients decrease due to smoothing induced by rainfall resolution coarsening; mean rainfall over the catchment decreases as smoothed storm core cells extend beyond the catchment boundaries. The effect of spatial resolution coarsening on rainfall values strongly depends on the movement of storm cells relative to the catchment.

– As the ratio L_R/L_D increases (in this particular case for $L_R/L_D > 0.9$), "rainfall smearing" occurs, inducing deviations in maximum modelled in-sewer water depths. The magnitude of deviations depends on the spatial structure of the storm and variability in rainfall gradients which determines how much the rainfall field is destructured by resolution coarsening. Results are in line with what was found by Ogden and Julien (1994).

– As the ratio L_R/L_{RA} increases, deviations in runoff peaks occur. For $L_R/L_{RA} > 20$, deviations in runoff peaks are above 10 % with respect to the reference case (at 100 m rainfall resolution). This implies that, when operational weather radar products (1000 m spatial resolution) are used to feed a hydrodynamic model, runoff model outputs are not correctly represented by the model at runoff area resolutions lower than 50 m.

- As the ratio L_R/L_S increases, deviations from the reference case (100 m resolution) occur: these are smaller for in-sewer water depths, ranging from 0.87 to 1.13, than for runoff peaks, which are in the range 0.7–1.5. This is due to the smoothing effect of flow routing through the pipe system on in-sewer water depths.

Additionally, an analysis of the change in spatial structure of rainfall due to time aggregation was conducted. To this end, the impact on model results was quantified in terms of time shift of maximum water depths with respect to the reference case at 1 min temporal resolution. Experimental anisotropic semi-variograms temporal aggregations at 5 and 10 min show that rainfall field structure changes due to temporal resolution coarsening. Rainfall correlation length increases by several 100 m due to time aggregation (up to 45 % of original de-correlation length). For all rainfall events, smoothing of rainfall fields induced by temporal aggregation results in peak time shifts up to 6 min. Model outputs are most strongly affected when rainfall temporal aggregation leads to complete distortion of the rain field, which happened for one of the four events in this study.

This study was a first attempt to characterise how the effect of space and time aggregation on rainfall structure affects hydrodynamic modelling of urban catchments, for resolutions ranging from 100 to 2000 m and from 1 to 10 min. It was investigated how rainfall change in resolution is absorbed by the model, giving an indication of scale relationships between: storm structure, its representation, catchment size and model structure. In this study four storm events were used that could be derived from an experimental polarimetric X-band radar.

The findings of this study helped to provide initial insights into how small-scale precipitation variability affects hydrological response and to what extent an urban drainage model can properly describe such a response. The outcomes showed that critical thresholds are to be expected in terms of the relationship between rainfall resolution and model scales. This study points out that scale relationships are relevant in determining model output sensitivities. To give a more robust meaning to these sampling numbers, more storm events should be analysed and more catchments should be tested to confirm the findings of this study. Additionally, model sensitivity to rainfall input resolution should be analysed in relation to other sources of uncertainty, such as those related to model structure and model parameter estimation. This requires installation of a polarimetric radar in the city, which is planned for the near future. This will enable model validation according to locally observed rainfall and sewer observations and analysis of different aspects of model uncertainty under different rainfall resolution scenarios.

Such an extension of the study would allow giving reliable recommendations on what should be the model and rainfall resolution in order to prioritise either the improvement on rainfall estimation or catchment hydrological characterisation.

Appendix A: SOBEK software description

Sobek 212 is a semi-distributed hydrodynamic model from Deltares. It accounts for two modules: the rainfall–runoff module and the routing module. In the rainfall–runoff module four different types of surfaces are used depending on the runoff factor and slope: closed paved, open paved, flat roof and sloped roofs (with a slope greater than 4 %). These four categories show different runoff factor and storage coefficient. The resulting runoff is calculated based on the "rational method", where the runoff Q is given by

$$Q\left(\text{mmh}^{-1}\right) = c\left(h^{-1}\right) * p(\text{mm}), \tag{A1}$$

where p is the net rainfall and c is a runoff factor which accounts for the delay of the rainfall as overland flow to the entry point of the sewer system. The runoff factor is a function of the length, roughness and slope of the surface (Sobek, 2012). The runoff coefficient is defined as a number between 0 and 1. A coefficient of 0.5 will mean that 50 % of the runoff volume will reach the sewer entry point in 1 min. The runoff factor moves the centre of mass of the resulting hydrograph, thereby increasing the lag time. The runoff formula is applied to each one of the runoff areas connected to the node of the sewer. In semi-distributed models, the whole catchment is split into a number of subcatchments (runoff areas), each of which is treated as a lumped model (i.e. within each subcatchment rainfall input and hydrologic responses are assumed to be uniform; their spatial variability is not accounted for). Rainfall is input uniformly within each subcatchment and based on the subcatchment's characteristics; the total runoff is estimated and routed to the outlet point, which is a node of the sewer system.

Once the water enters the sewer, the routing is computed by means of the complete 1-D Saint Venant equations.

Acknowledgements. This work has been funded by the EU INTER-REG IVB RainGain Project. The authors would like to thank the RainGain Project for supporting this research (www.raingain.eu). We also thank the reviewers and editor for their valuable comments and suggestions which greatly helped to improve the quality of the paper.

Edited by: P. Molnar

References

Balmforth, D. J. and Dibben, P.: A modelling tool for assessing flood risk, Water Practice & Technology, 1, 1, doi:10.2166/WPT.2006008, 2006.

Bell, V. A. and Moore, R. J.: The sensitivity of catchment runoff models to rainfall data at different spatial scales, Hydrol. Earth Syst. Sci., 4, 653–667, doi:10.5194/hess-4-653-2000, 2000.

Berne, A., Delrieu, G., Creutin, J.-D., and Obled, C.: Temporal and spatial resolution of rainfall measurements required for urban hydrology, J. Hydrol., 299, 166-179, doi:10.1016/j.jhydrol.2004.08.002, 2004.

Blanchet, F., Brunelle, D., and Guillon, A.: "Influence of the spatial heterogeneity of precipitation upon the hydrologic response of an urban catchment", 2nd Int. Symp. Hydrological Applications of Weather Radar, Hannover, p. 8, Preprint, 1992.

Dawdy, D. R. and Bergmann, J. M.: Effect of rainfall variability on streamflow simulation, Water Resour. Res., 5, 958–966, 1969.

Deltares: Sobek, Hydrodynamics, Rainfall Runoff and Real Time Control, User Manual. Version: 1.00.34157, 3 June 2014. Copyright© 2014 Deltares, Delft, NL, 2014.

Einfalt, T., Arnbjerg-Nielsen, K., Golz, C., Jensen, N.-E., Quirmbach, M., Vaes, G., and Vieux, B.: Towards a roadmap for use of radar rainfall data in urban drainage, J. Hydrol., 299, 186–202, doi:10.1016/j.jhydrol.2004.08.004, 2004.

Emmanuel, I., Andrieu, H., Leblois, E., and Flahaut, B.: Temporal and spatial variability of rainfall at the urban hydrological scale, J. Hydrol., 430–431, 162–172, doi:10.1016/j.jhydrol.2012.02.013, 2012.

Gires, A., Onof, C., Maksimovic, C., Schertzer, D., Tchiguirinskaia, I., and Simoes, N.: Quantifying the impact of small scale unmeasured rainfall variability on urban runoff through multifractal downscaling: A case study, J. Hydrol., 442–443, 117–128, doi:10.1016/j.jhydrol.2012.04.005, 2012.

Gires, A., Tchiguirinskaia, I., Schertzer, D., and Lovejoy, S.: Multifractal analysis of a semi-distributed urban hydrological model, Urban Water J., 10, 195–208, doi:10.1080/1573062X.2012.716447, 2013.

Goovaerts, P.: Geostatistical approaches for incorporating elevation into the spatial interpolation of rainfall, J. Hydrol., 228, 113–129, doi:10.1016/S0022-1694(00)00144-X, 2000.

Haberlandt, U.: Geostatistical interpolation of hourly precipitation from rain gauges and radar for a large-scale extreme rainfall event, J. Hydrol., 332, 144–157, 2007.

Jensen, N. E. and Pedersen, L.: Spatial variability of rainfall: Variations within a single radar pixel, Atmos. Res., 77, 269–277, doi:10.1016/j.atmosres.2004.10.029, 2005.

Johnson, J. T., MacKeen, P. L., Witt, A., Mitchell, E. D. W., Stumpf, G. J., Eilts, M. D., and Thomas, K. W.: The storm cell identification and tracking algorithm: An enhanced WSR-88D algorithm, Weather Forecast., 13, 263–276, 1998.

Koren, V., Finnerty, B., Schaake, J., Smith, M., Seo, D.-J., and Duan, Q.-Y.: Scale dependencies of hydrologic models to spatial variability of precipitation, J. Hydrol., 217, 285–302, doi:10.1016/S0022-1694(98)00231-5, 1999.

Krajewski, W. F., Lakshmi, V., Georgakakos, K. P., and Jain, S. C.: A Monte Carlo study of rainfall sampling effect on a distributed catchment model, Water Resour. Res., 27, 119–128, 1991.

Leijnse, H., Uijlenhoet, R., van de Beek, C. Z., Overeem, A., Otto, T., Unal, C. M. H., Dufournet, Y., Russchenberg, H. W. J., Figueras i Ventura, J., Klein Baltink, H., and Holleman, I.: Precipitation Measurement at CESAR, the Netherlands, J. Hydrometeorol., 11, 1322–1329, doi:10.1175/2010JHM1245.1, 2010.

Liguori, S., Rico-Ramirez, M. A., Schellart, A. N. A., and Saul, A. J.: Using probabilistic radar rainfall nowcasts and NWP forecasts for flow prediction in urban catchments, Atmos. Res., 103, 80–95, doi:10.1016/j.atmosres.2011.05.004, 2012.

Matheron, G.: Principles of geostatistics, Econom. Geol., 58, 1246–1266, 1963.

Neal, J., Villanueva, I., Wright, N., Willis, T., Fewtrell, T., and Bates, P.: How much physical complexity is needed to model flood inundation?, Hydrol. Process., 26, 2264–2282, doi:10.1002/hyp.8339, 2012.

Obled, C., Wendling, J., and Beven, K.: The sensitivity of hydrological models to spatial rainfall patterns: an evaluation using observed data, J. Hydrol., 159, 305–333, 1994.

Ogden, F. L. and Julien, P. Y.: Runoff model sensitivity to radar rainfall resolution, J. Hydrol., 158, 1–18, 1994.

Otto, T. and Russchenberg, H. W. J.: Estimation of Specific Differential Phase and Differential Backscatter Phase From Polarimetric Weather Radar Measurements of Rain, IEEE Geosci. Remote Sens. Lett., 8, 988–992, doi:10.1109/LGRS.2011.2145354, 2011.

Ozdemir, H., Sampson, C. C., de Almeida, G. A. M., and Bates, P. D.: Evaluating scale and roughness effects in urban flood modelling using terrestrial LIDAR data, Hydrol. Earth Syst. Sci., 17, 4015–4030, doi:10.5194/hess-17-4015-2013, 2013.

Parker, D. J., Priest, S. J., and McCarthy, S. S.: Surface water flood warnings requirements and potential in England and Wales, Appl. Geogr., 31, 891–900, doi:10.1016/j.apgeog.2011.01.002, 2011.

Pathirana, A., Tsegaye, S., Gersonius, B., and Vairavamoorthy, K.: A simple 2-D inundation model for incorporating flood damage in urban drainage planning, Hydrol. Earth Syst. Sci., 15, 2747–2761, doi:10.5194/hess-15-2747-2011, 2011.

Priest, S. J., Parker, D. J., Hurford, A. P., Walker, J., and Evans, K.: Assessing options for the development of surface water flood warning in England and Wales, J. Environ. Manage., 92, 3038–3048, doi:10.1016/j.jenvman.2011.06.041, 2011.

Schellart, A. N. A., Shepherd, W. J., and Saul, A. J.: Influence of rainfall estimation error and spatial variability on sewer flow prediction at a small urban scale, Adv. Water Resour., 45, 65–75, doi:10.1016/j.advwatres.2011.10.012, 2012.

Schmitt, T. G., Thomas, M., and Ettrich, N.: Analysis and modeling of flooding in urban drainage systems, J. Hydrol., 299, 300–311, doi:10.1016/j.jhydrol.2004.08.012, 2004.

Segond, M.-L., Wheater, H. S., and Onof, C.: The significance of spatial rainfall representation for flood runoff estimation: A numerical evaluation based on the Lee catchment, UK, J. Hydrol., 347, 116–131, doi:10.1016/j.jhydrol.2007.09.040, 2007.

Seliga, T. A., Aron, G., Aydin, K., and White, E.: Storm runoff simulation using radar-estimated rainfall rates and a Unit Hydrograph model (SYN-HYD) applied to the GREVE watershed, in: Am. Meteorol. Soc., 25th Int. Conf. on Radar Hydrology, 587–590, preprint, 1992.

Stichting RIONED: Rioleringsberekeningen, hydraulisch functioneren, Leidraad Riolering (Dutch Guidelines for sewer systems computations and hydraulic functioning), Stichting RIONED – National centre of expertise in sewer management and urban drainage in the Netherlands, The Netherlands, doi:10.1016/j.advwatres.2011.10.012, 2004 (in Dutch).

Storm, B. A., Parker, M. D., and Jorgensen, D. P.: A convective line with leading stratiform precipitation from BAMEX, Mon. Weather Rev., 135, 1769–1785, 2007.

Vaes, G., Willems, P., and Berlamont, J.: Rainfall input requirements for hydrological calculations, Urban Water, 3, 107–112, doi:10.1016/S1462-0758(01)00020-6, 2001.

Veldhuis ten, J. A. E. and Skovgård Olsen, A. Hydrological response times in lowland urban catchments characterised by looped drainage systems, Urban Rain 14, 9th International Workshop on Precipitation in Urban Areas, 6–9 December, St. Moritz, Switzerland, 2012.

Vieux, B. E. and Imgarten, J. M.: On the scale-dependent propagation of hydrologic uncertainty using high-resolution X-band radar rainfall estimates, Atmos. Res., 103, 96–105, doi:10.1016/j.atmosres.2011.06.009, 2012.

Weisman, M. L. and Rotunno, R.: "A theory for strong long-lived squall lines" revisited, J. Atmos. Sci., 61, 361–382, 2004.

Winchell, M., Gupta, H. V., and Sorooshian, S.: On the simulation of infiltration and saturation excess runoff using radar based rainfall estimates: Effects of algorithm uncertainty and pixel aggregation, Water Resour. Res., 34, 2655–2670, 1998.

Wood, S. J., Jones, D. A., and Moore, R. J.: Static and dynamic calibration of radar data for hydrological use, Hydrol. Earth Syst. Sci., 4, 545–554, doi:10.5194/hess-4-545-2000, 2000.

7

Assessing the impact of different sources of topographic data on 1-D hydraulic modelling of floods

A. Md Ali[1,2]**, D. P. Solomatine**[1,3]**, and G. Di Baldassarre**[4]

[1]Department of Integrated Water System and Knowledge Management, UNESCO-IHE Institute for Water Education, Delft, the Netherlands
[2]Department of Irrigation and Drainage, Kuala Lumpur, Malaysia
[3]Water Resource Section, Delft University of Technology, the Netherlands
[4]Department of Earth Sciences, Uppsala University, Sweden

Correspondence to: A. Md Ali (a.ali@unesco-ihe.org)

Abstract. Topographic data, such as digital elevation models (DEMs), are essential input in flood inundation modelling. DEMs can be derived from several sources either through remote sensing techniques (spaceborne or airborne imagery) or from traditional methods (ground survey). The Advanced Spaceborne Thermal Emission and Reflection Radiometer (ASTER), the Shuttle Radar Topography Mission (SRTM), the light detection and ranging (lidar), and topographic contour maps are some of the most commonly used sources of data for DEMs. These DEMs are characterized by different precision and accuracy. On the one hand, the spatial resolution of low-cost DEMs from satellite imagery, such as ASTER and SRTM, is rather coarse (around 30 to 90 m). On the other hand, the lidar technique is able to produce high-resolution DEMs (at around 1 m), but at a much higher cost. Lastly, contour mapping based on ground survey is time consuming, particularly for higher scales, and may not be possible for some remote areas. The use of these different sources of DEM obviously affects the results of flood inundation models. This paper shows and compares a number of 1-D hydraulic models developed using HEC-RAS as model code and the aforementioned sources of DEM as geometric input. To test model selection, the outcomes of the 1-D models were also compared, in terms of flood water levels, to the results of 2-D models (LISFLOOD-FP). The study was carried out on a reach of the Johor River, in Malaysia. The effect of the different sources of DEMs (and different resolutions) was investigated by considering the performance of the hydraulic models in simulating flood water levels as well as inundation maps. The outcomes of our study show that the use of different DEMs has serious implications to the results of hydraulic models. The outcomes also indicate that the loss of model accuracy due to re-sampling the highest resolution DEM (i.e. lidar 1 m) to lower resolution is much less than the loss of model accuracy due to the use of low-cost DEM that have not only a lower resolution, but also a lower quality. Lastly, to better explore the sensitivity of the 1-D hydraulic models to different DEMs, we performed an uncertainty analysis based on the GLUE methodology.

1 Introduction

In hydraulic modelling of floods, one of the most fundamental input data is the geometric description of the floodplains and river channels often provided in the form of digital elevation models (DEMs). During the past decades, there has been a significant change in data collection for topographic mapping techniques, from conventional ground survey to remote sensing techniques (i.e. radar wave and laser altimetry; e.g. Mark and Bates, 2000; Castellarin et al., 2009). This shift has a number of advantages in terms of processing efficiency, cost effectiveness and accuracy (Bates, 2012; Di Baldassarre and Uhlenbrook, 2012).

DEMs can be acquired from many sources of topographic information ranging from the high-resolution and accurate, but costly, lidar (Light Detection and Ranging) obtained from lower altitude, to low-cost, and coarse resolution, spaceborne data, such as ASTER (Advanced Spaceborne Thermal Emission and Reflection Radiometer) and SRTM (Shuttle Radar Topography Mission). DEMs can also be developed from traditional ground surveying (e.g. topographic contour maps) by interpolating a number of elevation points.

DEM horizontal resolution, vertical precision and accuracy differ considerably. This diversity is caused by the types of equipment and methods used in obtaining the topographic data. When used as an input to hydraulic modelling, the differences in the quality of each DEM subsequently result in differences in model output performance. In addition, resampling processes of raster data via a geographic information system (GIS) may also deteriorate the accuracy of the DEMs. The usefulness of diverse topographic data in supporting hydraulic modelling of floods is subject to the availability of DEMs, economic factors and geographical conditions of survey area (Cobby and Mason, 1999; Casas et al., 2006; Schumann et al., 2008).

To date, a number of studies have been carried out with the aim of evaluating the impact of accuracy and precision of the topographic data on the results of hydraulic models (e.g. Table 1).

Werner (2001) investigated the effect of varying grid element size on flood extent estimation from a 1-D model approach based on a lidar DEM. The study found that the flood extent estimation increased as the resolution of the DEM becomes coarser.

Horrit and Bates (2001) demonstrated the effects of spatial resolution on a raster-based flood model simulation. Simulation tests were performed at resolution sizes of 10, 20, 50, 100, 250, 500 and 1000 m and the predictions were compared with satellite observations of inundated area and ground measurements of flood-wave travel times. They found that the model reached a maximum performance at resolution of 100 m when calibrated against the observed inundated area. The resolution of 500 m proved to be adequate for the prediction of water levels. They also highlighted that the predicted flood-wave travel times are strongly dependent on the model resolution used.

Wilson and Atkinson (2005) set up a 2-D model, LISFLOOD-FP, using three different DEMs – contour data set, synthetic-aperture radar (SAR) data set, and differential global positioning system (DGPS) – used to predict flood inundation for 1998 flood event in the United Kingdom. The results showed that the contour data sets resulted in a substantial difference in the timing and the extent of flood inundation when compared to the DGPS data set. Although the SAR data set also showed differences in the timing and the extent, it was not as massive as for the contour data set. Nevertheless, the authors also highlighted a potential problem with the use of satellite remotely sensed topographic data in flood hazard assessment over small areas.

Casas et al. (2006) investigated the effects of the topographic data sources and resolution on 1-D hydraulic modelling of floods. They found that the contour-based digital terrain model (DTM) was the least accurate in the determination of the water level and inundated area of the floodplain; however, the GPS-based DTM led to a more realistic estimate of the water surface elevation and of the flooded area. The lidar-based model produced the most acceptable results in terms of water surface elevation and inundated flooded area compared to the reference data. The authors also pointed out that the different grid sizes used in lidar data have no significant effect on the determination of the water surface elevation. In addition, from an analysis of the time–cost ratio for each DEM used, they concluded that the most cost-effective technique for developing a DEM by means of an acceptable accuracy is from laser altimetry survey (lidar), especially for large areas.

Schumann et al. (2008) demonstrated the effects of DEMs on deriving the water stage and inundation area. Three DEMs at three different resolutions from three sources (lidar, contour and SRTM DEM) were used for a study area in Luxembourg. By using the HEC-RAS 1-D hydraulic model to simulate the flood propagation, the result shows that the lidar DEM derived water stages by displaying the lowest RMSE, followed by the contour DEM and lastly the SRTM. Considering the performance of the SRTM (it was relatively good with RMSE of 1.07 m), they suggested that the SRTM DEM is a valuable source for initial vital flood information extraction in large, homogeneous floodplains.

For the large flood-prone area, the availability of DEM from the public domain (e.g. ASTER, SRTM) makes it easier to conduct a study. Patro et al. (2009) selected a study area in India and demonstrated the usefulness of using SRTM DEM to derive river cross-section for the use in hydraulic modelling. They found that the calibration and validation results from the hydraulic model performed quite satisfactorily in simulating the river flow. Furthermore, the model performed quite well in simulating the peak flow, which is important in flood modelling. The study by Tarekegn et al. (2010) carried out on a study area in Ethiopia used a DEM which was generated from ASTER image. Integration between remote sensing and GIS technique were needed to construct the floodplain terrain and channel bathymetry. From the results obtained, they concluded that the ASTER DEM is able to simulate the observed flooding pattern and inundated area extents with reasonable accuracy. Nevertheless, they also highlighted the need for advanced GIS processing knowledge when developing a digital representation of the floodplain and channel terrain.

Schumann et al. (2010) demonstrate that near real-time coarse-resolution radar imagery of a particular flood event on the River Po (Italy) combined with SRTM terrain height data leads to a water slope remarkably similar to that derived by combining the radar image with highly accurate airborne laser altimetry. Moreover, they showed that this spaceborne flood wave approximation compares well to a hydraulic model thus allowing the performance of the latter,

Table 1. Summary of studies assessing the impact of topographic input data on the results of flood inundation models.

Author(s)	Numerical modelling (1-D*/1-D2-D**/2-D***)	Calibration[+]/ validation[++] data	Source of DEMs	Type of assessment	Study area
Horrit and Bates (2001)	LISFLOOD-FP**/NCFS**	SAR flood imagery[+]	Lidar	Precision	River Severn, UK
Werner (2001)	HEC-RAS*	n/a	Laser altimetry data	Precision	River Saar, Germany
Wilson and Atkinson (2005)	LISFLOOD-FP**	SAR flood imagery[++]	InSAR, topography & GPS	Accuracy	River Nene, UK
Casas et al. (2006)	HEC-RAS*	n/a	GPS, bathymetry, lidar & topography	Accuracy & precision	River Ter, Spain
Schumann et al. (2008)	REFIX*** & HEC-RAS*	Field data[+]/1-D model output[++]	Lidar, SRTM topography	Accuracy	River Alzette, Luxembourg
Schumann et al. (2010)	HEC-RAS*	Field data[+]/lidar derived water levels[++]	Lidar & SRTM	Accuracy	River Po, Italy
Yan et al. (2013)	HEC-RAS*	Field data[+]/SAR flood imagery[++]	Lidar & SRTM	Accuracy	River Po, Italy

calibrated on a previous event, to be assessed when applied to an event of different magnitude in near-real time.

Paiva et al. (2011) demonstrated the use of SRTM DEM in a large-scale hydrologic model with a full 1-D hydrodynamic module to calculate flow propagation on a complex river network. The study was conducted on one of the major tributaries of the Amazon, the Purus River basin. They found that a model validation using discharge and water level data is capable of reproducing the main hydrological features of the Purus River basin. Furthermore, realistic floodplain inundation maps were derived from the results of the model. The authors concluded that it is possible to employ full hydrodynamic models within large-scale hydrological models even when using limited data for river geometry and floodplain characterization.

Moya Quiroga et al. (2013) used Monte Carlo simulation sampling SRTM DEM elevation, and found a considerable influence of the SRTM uncertainty on the inundation area (the HEC-RAS hydraulic model of the Timis-Bega Basin in Romania was employed).

Most recently, Yan et al. (2013) made a comparison between a hydraulic model based on lidar and SRTM DEM. Besides the DEM inaccuracy, they also introduced uncertainty analysis by considering parameter and inflow uncertainty. The results of this study showed that the differences between the lidar-based model and the SRTM-based model are significant, but within the accuracy that is typically associated with large-scale flood studies.

Yet, the aforementioned studies explored the impact of topographic input data on the results flood inundation models by considering either the accuracy (or quality) or the precision (or resolution) of the DEMs (Table 1). When both accuracy and precision were considered (Casas, 2006), model results were not compared to observations via calibration and validation exercises.

This paper continues the presented line of research and deals with the assessment of the effects of using different DEM data source and resolution in a 1-D hydraulic modelling of floods. The novelty of our study is that both accuracy and precision of the DEM are explicitly considered and their impacts on hydraulic model results is evaluated in terms of both water surface elevation and inundation area. Furthermore, we compare model results via independent calibration and validation exercises and by explicitly considering parameter uncertainty and its potential compensation of inaccuracy of topographic data.

Hence, the goal of our paper is not to validate a specific approach for producing flood inundation maps, but rather to contribute to the existing literature with an original approach assessing the impact of topographic input data on hydraulic modelling of floods.

2 Study area and available data

2.1 Study area

The study area is located within the Johor River basin in the State of Johor, Malaysia. The river basin has a total area of 2690 km². The test site is a 30 km reach of the Johor River. The Johor River channel has a bankfull depth between 5 and 8 m and average slope around 0.03 %. The river reach under study is characterized by a stable main channel from 50 to 250 m wide. The study area consists of agricultural land, residential and commercial areas (see Fig. 1). As reported by Department of Irrigation and Drainage, Malaysia (DID, 2009), this test site has been experiencing some major historical flood events since 1948. The most recent ones happened in December 2006 and January 2007 when more than 3000 families were evacuated.

2.2 Hydraulic modelling

Flood inundation modelling was carried out by using the model code HEC-RAS, which was developed by the Hydrologic Engineering Center (HEC) of the United States Army Corps of Engineers (USACE, 2010). HEC-RAS is a 1-D model that can simulate both steady and unsteady flow conditions. In this study, all simulations were performed under unsteady flow conditions. To simulate open channel flows, HEC-RAS numerically solves the full 1-D Saint-Venant equations. The HEC-RAS model was set up using 32 cross-sections, whose topography is derived by different DEMs (see below). The observed flow hydrograph at an hourly time step was used as upstream boundary condition, while the friction slope was used as downstream boundary condition. The next section reports the different sources of topographic data used to define the geometric input. To develop flood inundation maps, the results were post-processed by using HEC-GeoRAS, an ArcGIS extension.

1-D hydraulic modelling does not properly simulate river hydraulics and floodplain flows. However, while 2-D models tend to schematize better flood inundation processes, they do not necessarily perform better when applied to real-world case studies because, besides model structure, many other sources of uncertainty affect model results (Werner, 2001; Bates et al., 2003; Pappenberger et al., 2005; Merwade et al., 2008; Di Baldassarre et al., 2009, 2010). A number of authors have carried out comparative studies and showed that the performance of 1-D models is often very close to that of 2-D models (e.g. Horrit and Bates, 2002; Castellarin et al., 2009; Cook and Merwade, 2009). Also, 1-D models are typically more efficient than 2-D models from a computation viewpoint, allowing for numerous simulations and uncertainty analysis to be carried out. In our case study, for a given flow, topography, river reach and a number of simulations, a HEC-RAS simulation (excluding post-processing GIS) took only 4 h to predict inundated area, whereas LISFLOOD-FP took around 26 h.

Anyhow, to properly test our model selection, we carried out a number of additional experiments (see Sect. 4.2) and compared the results of 1-D models to the results obtained with a 2-D model (LISFLOOD-FP; e.g. Hunter et al., 2006; Bates et al., 2010; Neal et al., 2012; Coulthard et al., 2013).

2.3 Digital elevation model

The required input data for the HEC-RAS include the geometry of the floodplain and the river, which is provided by a number of cross-sections. We identified several sources of DEM data for our study area (details are given below) with different spatial resolution and accuracy (Fig. 2):

Table 2. Information about the eight digital elevation models used as topographical input.

Model name	DEM type	Resolution (m)
Jhr L1	lidar	1 m
Jhr L2	(rescaled from lidar)	2
Jhr L20	(rescaled from lidar)	20
Jhr L30	(rescaled from lidar)	30
Jhr L90	(rescaled from lidar)	90
Jhr T20	Contour maps	20
Jhr A30	ASTER	30
Jhr S90	SRTM	90

1. DEMs derived from an original 1 m lidar data set (obtained from DID).

2. 20 m resolution DEM generated from the vectorial 1 : 25 000 cartography map obtained from DID with a permission of the Department of Survey and Mapping, Malaysia (DSMP).

3. 30 m resolution DEM derived from the globally and freely available ASTER data retrieved from the United States Geological Survey (USGS, http://earthexplorer.usgs.gov)

4. 90 m resolution DEM derived from the globally and freely available SRTM data retrieved from a Consortium for Spatial Information (CGIAR-CSI, www.cgiar-csi.org).

To analyse the influence of spatial resolution and separate it out from the impact of different accuracy, four additional DEMs were obtained by rescaling the original lidar DEM (1 m resolution) to the spatial resolutions of the DEMs derived from vectorial cartography (20 m), ASTER (30 m) and SRTM (90 m). Hence, a total of eight DEMs were used (see Table 2) to explore the impact of different topographic information on the hydraulic modelling of floods.

Given that the laser/radar waves used in the remote sensing techniques are not capable of penetrating the water surface and capture the river bed elevations, all the DEMs were integrated with river cross-section data derived from traditional ground survey. The ground survey of the river cross-sections within the study area was systematically carried out at about 1000 m intervals. Then, the flood simulation results across different data sets were compared to evaluate the effects of data spatial resolutions and data source differences.

3 Methodology

3.1 Evaluating DEM quality

First, the vertical error of each DEM was evaluated through comparison between the topographic data and 164 GPS

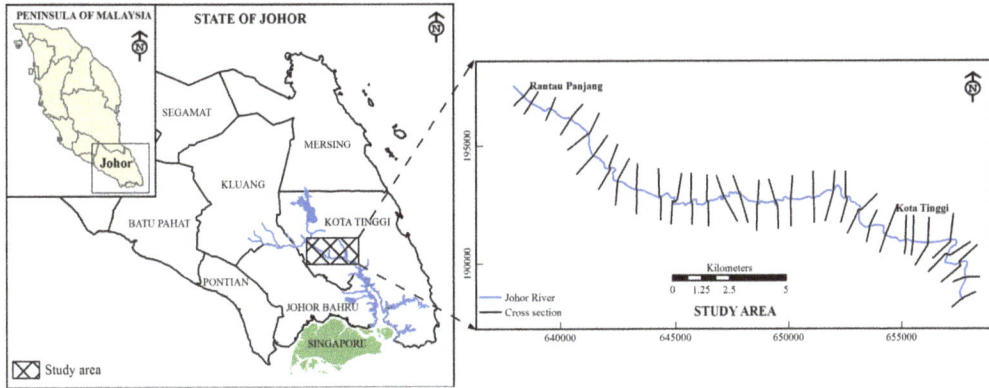

Figure 1. Layout map of study area: Johor River, Malaysia.

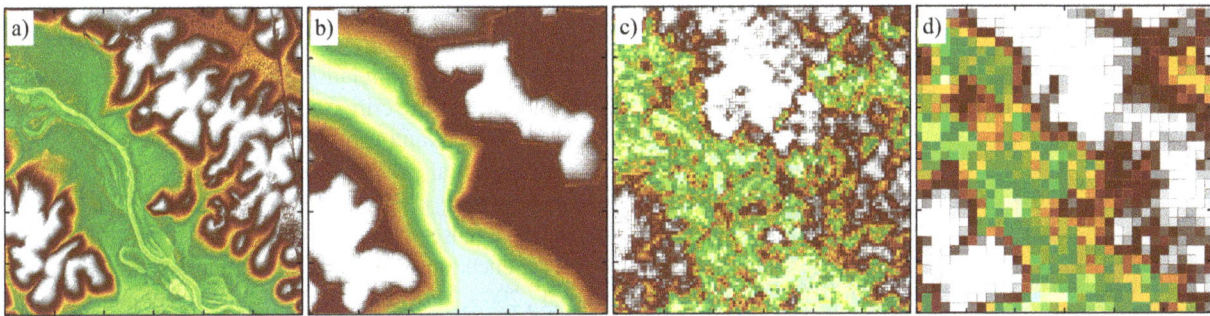

Figure 2. Original DEMs used in this study, based on (**a**) lidar data, (**b**) contour map, (**c**) ASTER data and (**d**) SRTM data.

ground points taken at random positions within the study area. The value of each reference elevation point was extracted from the study area using GPS survey equipment. The quality of each DEM is assessed by the root mean square error (RMSE$_{\text{DEM}}$) and mean error (ME$_{\text{DEM}}$). Thus

$$\text{RMSE}_{\text{DEM}} = \sqrt{\frac{\sum_{i=1}^{n} (\text{Elev}_{\text{GPS}} - \text{Elev}_{\text{DEM}})^2}{n}}, \quad (1)$$

where Elev$_{\text{GPS}}$ is the reference elevation (m) derived from GPS, Elev$_{\text{DEM}}$ is the corresponding value derived from each DEM, and n corresponds to the total numbers of points.

3.2 Model calibration and validation

Then, data from two recent major flood events that occurred along the Johor River in 2006 and 2007 were used for independent calibration and validation of the models. The estimated peak flow of the 2006 event is approximately 375 m^3 s^{-1}, while the one of the 2007 event is around 595 m^3 s^{-1}. Both discharge data were measured and recorded at Rantau Panjang hydrological station. The 2006 flood data were used for the calibration exercise, while the 2007 flood data were used for model validation.

To assess the sensitivity of the different models to the model parameters, the Manning n roughness coefficients

for all the models were sampled uniformly from 0.02 to 0.08 m$^{-1/3}$s for the river channel, and between 0.03 and 0.10 m$^{-1/3}$s for the floodplain, by steps of 0.0025 m$^{-1/3}$s. The performance of the hydraulic models in producing the observed water levels was assessed by means of the mean absolute error (MAE):

$$\text{MAE} = \frac{1}{T} \sum_{t=1}^{T} |O_t - S_t|, \quad (2)$$

where T is the number of steps in time series, O_t is the observed water level at time t, and S_t is the simulated water level at time t.

3.3 Quantifying the effect of the topographic data source on the water surface elevation and inundation area (sensitivity analysis)

The effects of DEM source and spatial resolution were further investigated by examining the sensitivity of model results in terms of maximum water surface elevation (WSE), inundation area and floodplain boundaries. For this additional analysis, the model results obtained with the most accurate and precise DEM source (lidar at 1 m resolution) were used as a reference. For WSE analysis, each model was compared to the reference model (Jhr L1, see Table 1) by means

Table 3. Statistics of errors (m) of each DEMs with respect to the GPS control points.

Model name	Min. error (m)	Max. error (m)	RMSE (m)
Jhr L1	−0.59	1.00	0.58
Jhr L2	−0.64	1.38	0.58
Jhr L20	−0.83	1.83	0.68
Jhr L30	−0.93	3.98	0.79
Jhr L90	−5.46	3.73	1.27
Jhr T20	−15.38	10.55	4.66
Jhr A30	−33.37	7.58	7.01
Jhr S90	−3.59	4.32	6.47

Table 4. Reported vertical accuracies of SRTM data.

Reference	Average height accuracy (m)		Continent
Rabus et al. (2003)		6.00	European
Sun et al. (2003)		11.20	European
SRTM mission specification (Rodríguez et al., 2005)		16.00	Global
Berry et al. (2007)	2.54	3.60	Eurasia Global
Farr et al. (2007)		6.20	Eurasia
Wang et al. (2011)		13.80	Eurasia

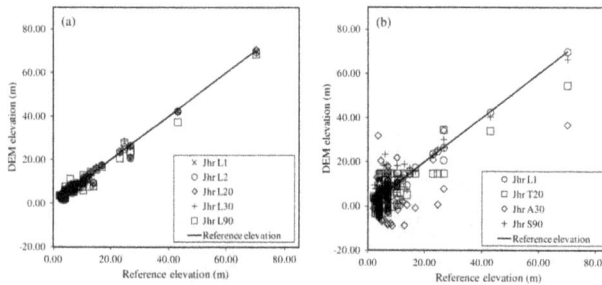

Figure 3. Comparison between GPS point elevations and elevations derived by the different DEMs: (**a**) lidar DEM at different resolution and (**b**) different sources of DEMs.

of the following measures:

$$\mathrm{MAD_{WSE}} = \frac{1}{x} \sum_{x=1}^{x} |\mathrm{WSE_{Ref}} - \mathrm{WSE_{DEM}}|, \tag{3}$$

where $\mathrm{WSE_{Ref}}$ denotes the WSE simulated by the reference model (Jhr L1), $\mathrm{WSE_{DEM}}$ the WSE estimated by the models based on DEMs of lower resolution or different source (Table 1), and x corresponds to the total number of cross-sections where models results where compared.

To analyse the sensitivity to different topographic input in terms of simulated flood extent, we used the following measure of fit:

$$F(\%) = \frac{M_1 \cap M_2}{M_1 \cup M_2}.100, \tag{4}$$

where M_1 and M_2 are the simulated and observed (i.e. simulated by the reference model) inundation areas, and $\cup$ and $\cap$ are the union and intersection GIS operations respectively. F equal to 100 % indicates that the two areas are completely coincidental.

3.4 Uncertainty estimation – GLUE analysis

In hydraulic modelling, multiple sources of uncertainty can emerge from several factors, such as model structure, topography and friction coefficients (Aronica et al., 2002; Trigg

et al., 2009; Brandimarte and Di Baldassarre, 2012; Dottori et al., 2013). A methodological approach to estimate the uncertainty is the generalized likelihood uncertainty estimation (GLUE) methodology (Beven and Binley, 1992), a variant of Monte Carlo simulation. Although some aspects of this methodology are criticized in several papers (e.g. Hunter et al., 2005; Mantovan and Todini, 2006; Montanari, 2005; Stedinger et al., 2008), it is still widely used in hydrological modelling because of its ease of implementation and a common-sense approach to use only a set of the "best" models for uncertainty analysis (e.g. Hunter et al., 2005; Shrestha et al., 2009; Vázquez et al., 2009; Krueger et al., 2010; Jung and Merwade, 2012; Brandimarte and Woldeyes, 2013).

According to the GLUE framework (Beven and Binley, 1992), each simulation, i, is associated with the (generalized) likelihood weight, W_i, ranging from 0 to 1. The weight, W_i is expressed as a function of the measure fit, ε_i, of the behavioural models:

$$W_i = \frac{\varepsilon_{\max} - \varepsilon_i}{\varepsilon_{\max} - \varepsilon_{\min}}, \tag{5}$$

where $\varepsilon_{\max}$ and $\varepsilon_{\min}$ are the maximum and minimum value of MAE of behavioural models. To identify the behaviour of the models, a threshold value (rejection criterion) has been set as follows:

1. simulations associated with MAE larger than 1.0 m and

2. Manning's n roughness coefficient of the floodplain smaller than the Manning's n roughness coefficient of the channel.

Then, the likelihood weights are the cumulative sum of 1 and the weighted 5th, 50th and 95th percentiles. The likelihood weights were calculated as follows:

$$L_i = \frac{W_i}{\sum_{i=1}^{n} W_i}. \tag{6}$$

For this study, the applications of uncertainty analysis considered only the parameter uncertainty and implemented for all DEMs.

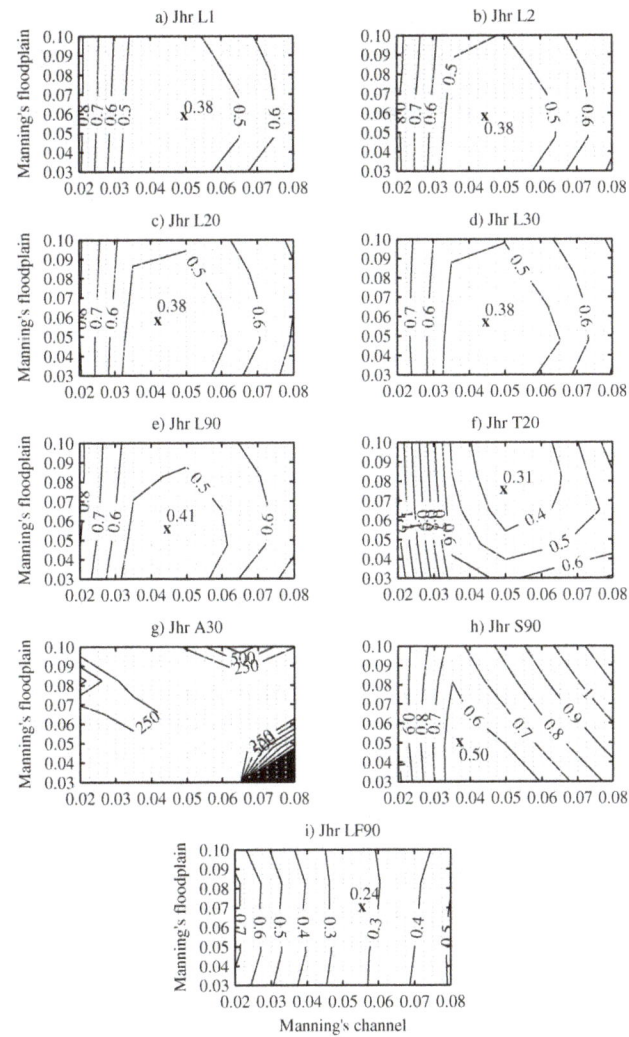

Figure 4. Model calibration: contour maps of MAE across the parameter space for (**a–h**) eight different 1-D models (HEC-RAS) and (**i**) for the 2-D model (LISFLOOD-FP).

Table 5. Model validation results.

Model name	Calibrated Manning's n roughness coefficient		MAE (m) (validation)
	Channel	Floodplain	
Jhr L1	0.0500	0.0575	0.40
Jhr L2	0.0450	0.0575	0.38
Jhr L20	0.0425	0.0575	0.37
Jhr L30	0.0450	0.0575	0.38
Jhr L90	0.0450	0.0550	0.39
Jhr T20	0.0500	0.0750	0.60
Jhr S90	0.0375	0.0500	0.60
Jhr LF90	0.0550	0.0700	0.52

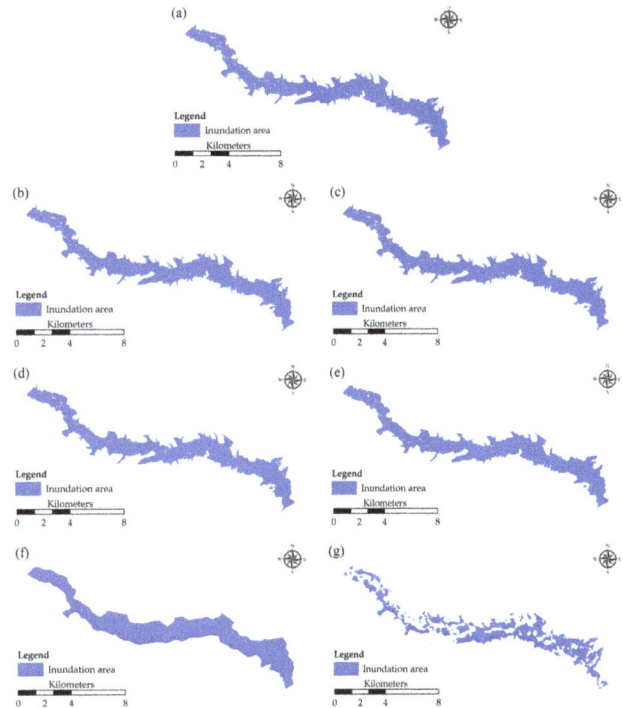

Figure 5. Effect of DEMs on Johor River. Inundation map resulting from (**a**) Jhr L1, (**b**) Jhr L2, (**c**) Jhr L20, (**d**) Jhr L30, (**e**) Jhr L90, (**f**) Jhr T20 and (**g**) Jhr S90.

4 Results and discussion

4.1 Quality of DEMs compared with the reference points

Table 3 shows the calculated statistical vertical errors for each different DEM for the same study area. As anticipated, lidar is not only the most precise DEM because of its highest resolution, but also the most accurate. The RMSE of each lidar DEM increased from 0.58 m (Jhr L1) to 1.27 m (Jhr L90) as the resolution of the DEMs reduced from 1 m (original resolution) to 90 m.

Overall, the terrain is considered well defined under the lidar DEMs even though the calculated errors are higher compared to the vertical accuracy reported in product specification (around 0.15 m). Figure 3 shows the distribution of each DEM compared to the GPS ground elevation.

Although lidar DEM gives the lowest error, it is useful to note that this type of DEM has a number of limitations as highlighted in the several papers (see Sun et al., 2003; Casas et al., 2006; Schumann et al., 2008):

1. it provides only discrete surface height samples and not continuous coverage,

2. its availability is very much limited by economic constraint,

Table 6. Effects of DEMs (source and resolution) on HEC-RAS simulations.

Model name	Inundation area (km^2)	Area difference (%)	F (%)	F (%)[+]
Jhr L1	25.86	–	–	–
Jhr L2	25.78	−0.3	96.6	–
Jhr L20	25.96	0.4	92.9	–
Jhr L30	26.18	1.2	92.2	–
Jhr L90	25.84	−0.1	89.4	–
Jhr T20	29.23	13.0	73.7	74.2
Jhr S90	16.58	−35.9	48.9	49.6

[+] Overlap-fit percentage F (%) of the floodplain inundated area with those from lidar DEMs of the same resolutions (Jhr L20, Jhr L90).

Table 7. Summary of mean absolute difference (MAD) in terms of water surface elevation simulated by the models.

Model name	MAD$_{WSE}$(m)
Jhr L1	–
Jhr L2	0.06
Jhr L20	0.05
Jhr L30	0.05
Jhr L90	0.08
Jhr T20	1.12
Jhr S90	0.76

3. it is unable to capture the river bed elevations due to the fact the laser does not penetrate the water surface, and

4. it is incapable of penetrating the ground surface in densely vegetated areas especially for the tropical region.

The RMSE value of the other DEMs is 4.66 m for contour maps, 7.01 m for ASTER and 6.47 m for SRTM. It is also noticeable that the RMSE of the SRTM DEM for this particular study area is within the average height accuracy found in other SRTM literature – either global or at particular continent (see Table 4). Whatever the case, it is proven that this type of DEM gives an acceptable result when used in large-scale flood modelling (e.g. Patro et al., 2009; Paiva et al., 2012; Yan et al., 2013).

Despite having the lowest vertical accuracies, the ASTER and contour DEMs are still widely used in the field of hydraulic flood research as they are globally available and free (e.g. Tarekegn et al., 2010; Wang et al., 2011; Gichamo et al., 2012). The differences in the vertical accuracies may be partly due to the lack of information in topographical flats areas such as floodplains. However, the further use of each DEM in this study is subject to its performance in the hydraulic flood modelling during the calibration and validation stages, which are described in the following section.

4.2 Model calibration and validation

Panels (a)–(h) of Fig. 4 show the model responses in terms of MAE provided by the eight HEC-RAS models in simulating the 2006 flood event. The models were built using the eight DEMs with different accuracy and precision (Table 2) as topographic input.

In general, all models (Fig. 4a–h) are seen to be more sensitive to the changing of Manning's n roughness coefficient of main channel than the Manning's n roughness coefficient of floodplain areas. The results of the calibration showed that the best-fit models based on lidar DEM with different resolutions (Jhr L2, Jhr L20, Jhr L30 and Jhr L90) generally gave good performances with only slight variations in

the MAE value from 0.38 to 0.41 m. The optimum channel and floodplain Manning's n roughness coefficient are centred on similar values at $n_{channel} = 0.0425$ to 0.0500 and $n_{floodplain} = 0.0575$ for Jhr L1, Jhr L2, Jhr L20, Jhr L30 and Jhr L90. The best-fit models based on topographic map and SRTM also performed well with MAE of 0.31 and 0.50 m. On the other hand, ASTER-based model completely failed (the exceptionally high values of MAE in Fig. 4g are due to model instabilities) and was therefore eliminated from further analysis.

Panel (i) of Fig. 4 shows the outcome of the additional experiment we carried out to test the appropriateness of selecting a 1-D model. In particular, a LISFLOOD-FP model was built using the lidar topography rescaled at 90 m, called here Jhr LF90. The specific topographic input was chosen as a trade-off between computational times and the need for an as-accurate-as-possible DEM for a proper comparison between 1-D and 2-D modelling. By comparing the calibration results of the LISFLOOD-FP model (Fig. 4i) to the corresponding (i.e. using the same topography) ones of the HEC-RAS model (Fig. 4e), one can observe that differences are not significant. Lastly, Fig. 4i shows that LISFLOOD-FP is also more sensitive to the main channel roughness coefficient than to the floodplain one.

The best-fit models, using the optimum Manning n roughness coefficients (Table 5), were then used to simulate the January 2007 flood event for model validation. This was carried out for all models except the ASTER-based model due to its poor performance (see Fig. 4g). Table 5 presents the MAE of each model obtained during model validation. It is noted that the MAE values for all lidar-based models (first five rows) with different resolutions remained almost the same with the difference within +0.03 m. The MAE values for the models based on topographic contour maps and SRTM DEM both provide MAE of 0.60 m.

The model validation exercise also supports the use of 1-D hydraulic models for this river reach. In particular, Table 5 shows that the LISFLOOD-FP model (Jhr LF90) provided a MAE of 0.52 m, while the corresponding HEC-RAS model (Jhr L90) provided a MAE of 0.39 m. Thus, the 1-D model performed even (slightly) better than the 2-D model.

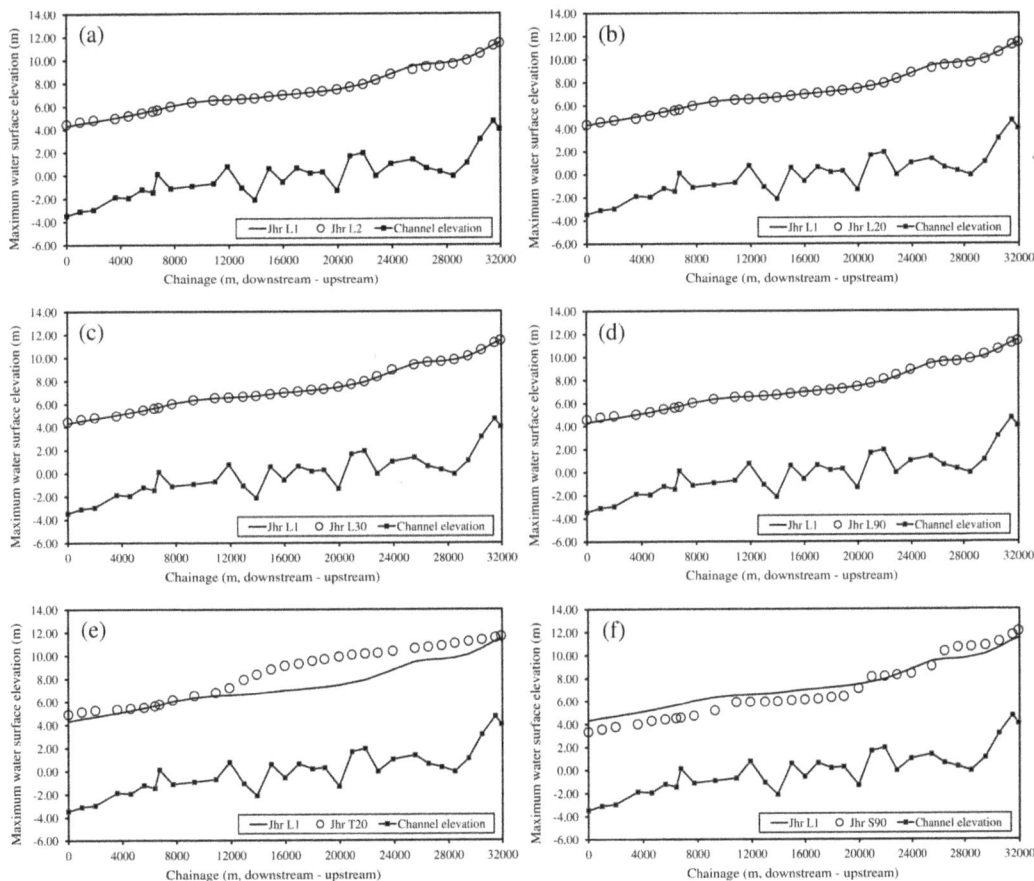

Figure 6. Maximum water surface elevation along the Johor River for the six hydraulic models compared to that simulated by the reference model.

The results of this first analysis suggest that the reduction in the resolution of lidar DEMs (from 1 to 90 m) does not significantly affect the model performance. However, the use of topographic contour maps (Jhr T20) and SRTM (Jhr S90) DEMs as geometric input to the hydraulic model produces a slight increase of model errors. For instance, Jhr L90 and Jhr S90 have the same resolution (90 m), but the different accuracy results in increased (tough not remarkably) errors in model validation (from 0.39 to 0.60 m). This limited degradation of model performance (Table 5), in spite of the much lower accuracy of topographic input (Table 2), can be attributed to the fact that models are compared to water levels observed in two cross-sections. A spatially distributed analysis (comparing the simulated flood extent and flood water profile along the river) might show more significant differences (see Sect. 4.3).

4.3 Quantifying the effect of the topographic data source on the water surface elevation and inundation area on 1-D model

4.3.1 Inundation area (sensitivity analysis)

This section reports an additional analysis aiming to better explore the sensitivity of model results to different topographic data (see Sect. 3.3). Figure 5 shows the simulated flood extent maps obtained from the seven different topographic input data. The floodplain areas simulated by the five lidar-based models (Jhr L1, Jhr L2, Jhr L20, Jhr L30 and Jhr L90) are very similar. In contrast, the floodplain areas simulated by the models based on topographic contour maps (Jhr T20) and SRTM DEM (Jhr S90) are substantially different (see Fig. 5 and Table 6).

Table 6 shows the comparison between the different models in terms of simulating flood extent. The aforementioned measure of fit F was found to decrease for both decreasing resolution and lowering accuracy. This sensitivity analysis also shows that the results of flood inundation models are more affected by the accuracy of the DEM used as topographic input than its resolution.

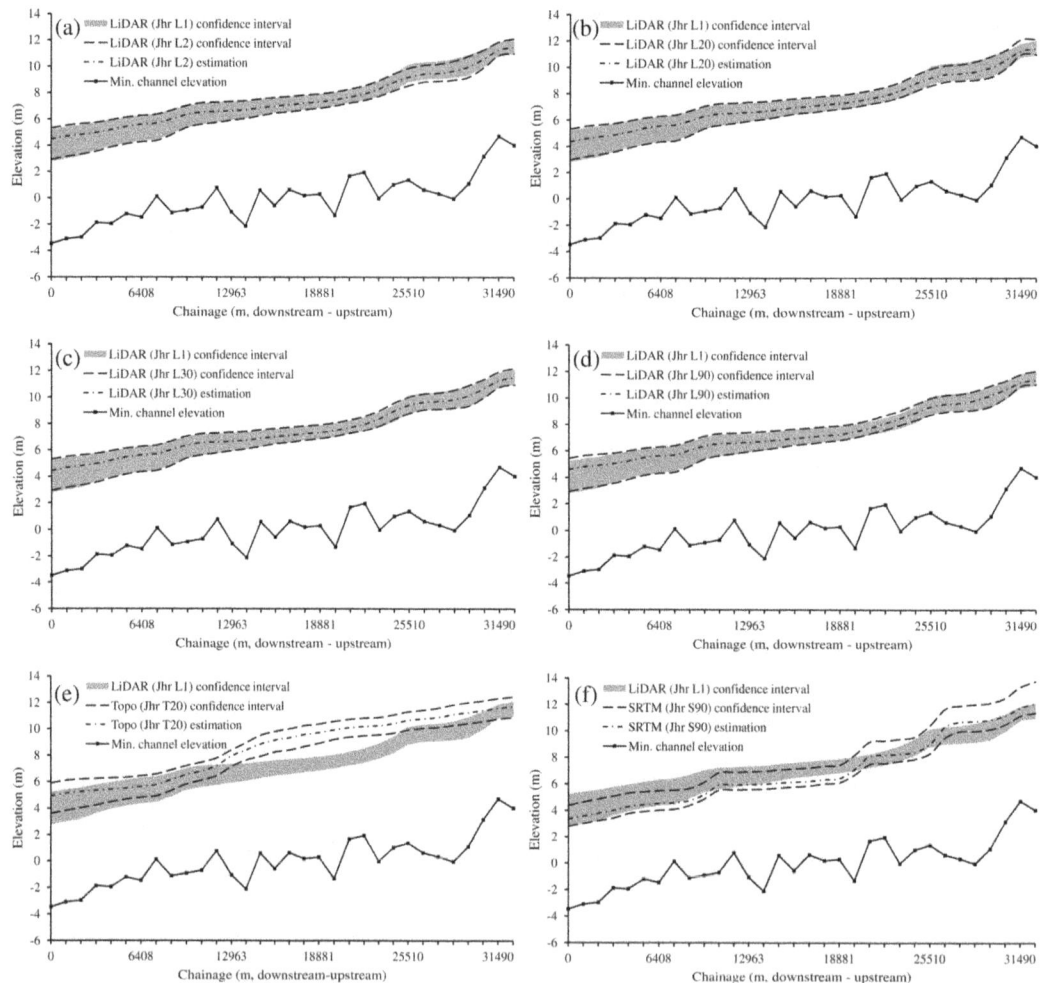

Figure 7. Comparison of uncertainty bounds (5th, 50th and 95th percentiles by considering parameter uncertainty only) between the reference model and the other models. The reference model uncertainty bounds are shown as grey areas, while the uncertainty bounds of the other six models are shown as dashed lines.

4.3.2 Water surface elevation

Figure 6 compares the flood water profiles simulated by the reference model (Jhr L1) with the flood water profiles (WSE) obtained from the other six models (Jhr L2, Jhr L20, Jhr L30, Jhr L90, Jhr T20 and Jhr S90). All these flood water profiles were obtained by simulating the 2007 flood event. Despite having different resolutions, the flood water profiles simulated from all lidar-based models portray similar flood water profiles to the reference model (see Fig. 6a to d). This is consistent with the findings about the inundation area (Fig. 5), whereas flood water profiles simulated by the models based on topographic contour maps and SRTM DEMs (see Fig. 6e and f) are rather different.

The discrepancies between the reference model (Jhr L1) and the other models as shown in Fig. 6 are quantified in terms of mean absolute difference (MAD). This shows that the re-sampled lidar data (Jhr L2, Jhr L20, Jhr L30 and Jhr L90) all have a low MAD: between 0.05 to 0.08 m. Higher

discrepancies are found with the models based on SRTM DEM (0.76 m) and contour maps (1.12 m). The great differences obtained using the topographic contour maps may be partly due to the way that the DEM height is sampled. For instance, the contour DEMs in this study were based on topographic contours at 20 m intervals and required an interpolation technique to generate a DEM. Table 7 shows the MAD in terms of water surface elevation simulated by the models.

4.3.3 Uncertainty in flood profiles obtained from different DEMs by considering parameter uncertainty

To better interpret the differences that have emerged in comparing the results of models based on different topographic data, we carried out a set of numerical experiments to explore the uncertainty in model parameters. As mentioned, we varied Manning's n roughness coefficient between 0.02 and $0.08\,\mathrm{m}^{-1/3}\mathrm{s}$ for the river channel, and from 0.03 to

$0.10\,\mathrm{m}^{-1/3}\mathrm{s}$ for the floodplain, in steps of $0.0025\,\mathrm{m}^{-1/3}\mathrm{s}$. Then a number of simulations were rejected as described in Sect. 3.4. Figure 7 shows the uncertainty bounds for the different models. The width of these uncertainty bounds was found to be between 1.5 and 1.6 m for all models (only parameter uncertainty is considered here). Nevertheless, the model based on contour maps led to significant differences from the lidar-based model, even when the uncertainty induced by model parameters is explicitly accounted for (see Fig. 7e).

5 Conclusions

This study assessed how different DEMs (derived by various sources of topographic information or diverse resolutions) affect the output of hydraulic modelling. A reach of the Johor River, Malaysia, was used as the test site. The study was performed using a 1-D model (HEC-RAS), which was found to perform as well as a 2-D model (LISFLOOD-FP) in this case study. The sources of DEMs were lidar at 1 m resolution, topographic contour maps at 20 m resolution, ASTER data at 30 m resolution, and SRTM data at 90 m resolution. The lidar DEM was also re-sampled from its original resolution data set to 2, 20, 30 and 90 m cell size. Different models were built by using them as geometric input data.

The performance of the five lidar-based models (characterized by different resolutions ranging from 1 to 90 m; see Table 5) did not show significant differences – neither in the exercise of independent calibration and validation based on water level observations in an internal cross-section, nor in the sensitivity analysis of simulated flood profiles and inundation areas. Another striking result of our study is that the model based on ASTER data completely failed because of major inaccuracies of the DEM.

In contrast, the models based on SRTM data and topographic contour maps did relatively well in the validation exercise as they provided a mean absolute error of 0.6 m, which is only slightly higher than the ones obtained with lidar-based models (all around 0.4 m). However, this outcome could be attributed to the fact that validation could only be performed by using the water level observed in a two internal cross-sections. As a matter of fact, higher discrepancies emerged when lidar-based models were compared to the models based on SRTM data or topographic contour maps in terms of inundation areas or flood water profiles. These differences were found to be relevant even when parameter uncertainty was accounted for.

The study also showed that, to support flood inundation models, the quality and accuracy of the DEM is more relevant than the resolution and precision of the DEM. For instance, the model based on the 90 m DEM obtained by re-sampling the lidar data performed better than the model based on the 90 m DEM obtained from SRTM data. These outcomes are unavoidably associated with the specific test site, but the methodology proposed here can allow a comprehensive assessment of the impact of diverse topographic data on hydraulic modelling of floods for different rivers around the world.

Acknowledgements. The authors would like to thank to the Department of Irrigation and Drainage, Malaysia (DID) for providing useful input data used in this study. We also acknowledge the Public Service Department, Malaysia for providing a PhD Fellowship funding and study leave for the first author. We thank the Editor and the two anonymous reviewers as well as Fiona, Nagendra, Micah and Yan for their constructive comments that helped to improve the manuscript.

Edited by: H. Cloke

References

Aronica, G., Bates, P. D., and Horritt, M. S.: Assessing the uncertainty in distributed model predictions using observed binary pattern information within GLUE, Hydrol. Process., 16, 2001–2016, doi:10.1002/hyp.398, 2002.

Bates, P. D., Marks, K. J., and Horritt, M. S.: Optimal use of high-resolution topographic data in flood inundation models, Hydrol. Process., 17, 5237–5257, 2003.

Bates, P. D., Horritt, M. S., and Fewtrell, T. J.: A simple inertial formulation of the shallow water equations for efficient two-dimensional flood inundation modelling, J. Hydrol., 387, 33–45, doi:10.1016/j.jhydrol.2010.03.027, 2010.

Bates, P. D.: Integrating remote sensing data with flood inundation models: how far have we got?, Hydrol. Process., 26, 2515–2521, doi:10.1002/hyp.9374, 2012.

Berry, P. A. M., Garlick, J. D., and Smith, R. G.: Near-global validation of the SRTM DEM using satellite radar altimetry, Remote Sens. Environ., 106, 17–27, doi:10.1016/j.rse.2006.07.011, 2007.

Beven, K. and Binley, A.: The future of distributed models – model calibration and uncertainty prediction, Hydrol. Process., 6, 279–298, doi:10.1002/hyp.3360060305, 1992.

Brandimarte, L. and Di Baldassarre, G.: Uncertainty in design flood profiles derived by hydraulic modelling, Hydrol. Res., 43, 753–761, doi:10.2166/nh.2011.086, 2012.

Brandimarte, L. and Woldeyes, M. K.: Uncertainty in the estimation of backwater effects at bridge crossings, Hydrol. Process., 27, 1292–1300, doi:10.1002/hyp.9350, 2013.

Casas, A., Benito, G., Thorndycraft, V. R., and Rico, M.: The topographic data source of digital terrain models as a key element in the accuracy of hydraulic flood modelling, Earth Surf. Process. Landforms, 31, 444–456, doi:10.1002/esp.1278, 2006.

Castellarin, A., Di Baldassarre, G., Bates, P. D., and Brath, A.: Optimal cross-section spacing in Preissmann scheme 1-D hydrodynamic models, J. Hydraul. Eng.-ASCE, 135, 96–105, doi:10.1061/(ASCE)0733-9429(2009)135:2(96), 2009.

Cobby, D. M. and Mason, D. C.: Image processing of airborne scanning laser altimetry for improved river flood modelling, ISPRS J. Photogramm. Remote Sens., 56, 121–138, 1999.

Cook, A. and Merwade, V.: Effect of topographic data, geometric configuration and modeling approach on flood inundation mapping, J. Hydrol., 377, 131–142, 2009.

Coulthard, T. J., Neal, J. C., Bates, P. D., Ramirez, J., de Almeida, G. A. M., and Hancock, G. R.: Integrating the LISFLOOD-FP 2D hydrodynamic model with the CAESAR model: implications for modelling landscape evolution, Earth Surf. Process. Landforms, 38, 1897–1906, doi:10.1002/esp.3478, 2013.

Department of Irrigation and Drainage, Malaysia (DID): Master plan study on flood mitigation for Johor River basin, Malaysia, 2009.

Di Baldassarre, G., Schumann, G., and Bates, P. D.: A technique for the calibration of hydraulic models using uncertain satellite observations of flood extent, J. Hydrol., 367, 276–282, 2009.

Di Baldassarre, G., Schumann, G., Bates, P. D., Freer, J. E., and Beven, K. J.: Floodplain mapping: a critical discussion on deterministic and probabilistic approaches, Hydrolog. Sci. J., 55, 364–376, 2010.

Di Baldassarre, G. and Uhlenbrook, S.: Is the current flood of data enough? A treatise on research needs for the improvement of flood modelling, Hydrol. Process., 26, 153–158, doi:10.1002/hyp.8226, 2012.

Dottori, F., Di Baldassarre, G., and Todini, E.: Detailed data is welcome, but with a pinch of salt: Accuracy, precision, and uncertainty in flood inundation modelling, Water Resour. Res., 49, 6079–6085, doi:10.1002/wrcr.20406, 2013.

Farr, T. G., Rosen, P. A., Caro, E., Crippen, R., Duren, R., Hensley, S., Kobrick, M., Paller, M., Rodriguez, E., Roth, L., Seal, D., Shaffer, S., Shimada, J., Umland, J., Werner, M., Oskin, M., Burbank, D., and Alsdorf, D.: The shuttle radar topography mission, Rev. Geophys., 45, RG2004, doi:10.1029/2005RG000183, 2007.

Gichamo, T. Z., Popescu, I., Jonoski, A., and Solomatine, D.: River cross-section extraction from the ASTER global DEM for flood modelling, Environ. Modell. Softw., 31, 37–46, doi:10.1016/j.envsoft.2011.12.003, 2012.

Horrit, M. S. and Bates, P. D.: Effects of spatial resolution on a raster based model of flood flow, J. Hydrol., 253, 239–249, 2001.

Horrit, M. S. and Bates, P. D.: Evaluation of 1-D and 2-D models for predicting river flood inundation, J. Hydrol., 180, 87–99, 2002.

Hunter, N. M., Bates, P. D., Horritt, M. S., De Roo, A. P. J., and Werner, M. G. F.: Utility of different data types for calibrating flood inundation models within a GLUE framework, Hydrol. Earth Syst. Sci., 9, 412–430, doi:10.5194/hess-9-412-2005, 2005.

Hunter, N. M., Bates, P. D., Horritt, M. S., and Wilson, M. D.: Improved simulation of flood flows using storage cell models, P. I. Civil Eng.-Wat. M., 159, 9–18, 2006.

Jung, Y. and Merwade, V.: Uncertainty quantification in flood inundation mapping using generalized likelihood uncertainty estimate and sensitivity analysis, J. Hydrol. Eng., 17, 507–520, 2012.

Krueger, T., Freer, J., Quinton, J. N., Macleod, C. J. A., Bilotta, G. S., Brazier, R. E., Butler, P., and Haygarth, P. M.: Ensemble evaluation of hydrological model hypotheses, Water Resour. Res., 46, W07516, doi:10.1029/2009WR007845, 2010.

Mantovan, P. and Todini, E.: Hydrological forecasting uncertainty assessment: incoherence of the GLUE methodology, J. Hydrol., 330, 368–381, doi:10.1016/j.jhydrol.2006.04.046, 2006.

Marks, K. and Bates, P. D.: Integration of high resolution topographic data with floodplain flow models, Hydrol. Process., 14, 2109–2122, 2000.

Merwade, V., Olivera, F., Arabi, M., and Edleman, S.: Uncertainty in flood inundation mapping: current issues and future directions, J. Hydrol. Eng., 13, 608–620, 2008.

Montanari, A.: Large sample behaviors of the generalized likelihood uncertainty estimation (GLUE) in assessing the uncertainty of rainfall-runoff simulations, Water Resour. Res., 41, W08406, doi:10.1029/2004WR003826, 2005.

Moya Quiroga, V., Popescu, I., Solomatine, D. P., and Bociort, L.: Cloud and cluster computing in uncertainty analysis of integrated flood models, J. Hydroinf., 15, 55–69, doi:10.2166/hydro.2012.017, 2013.

Neal, J., Schumann, G., and Bates, P.: A subgrid channel model for simulating river hydraulics and floodplain inundation over large and data sparse areas, Water Resour. Res., 48, W11506, doi:10.1029/2012WR012514, 2012.

Paiva, R. C. D., Collischonn, E., and Tucci, C. E. M.: Large scale hydrologic and hydrodynamic modeling using limited data and a GIS based approach, J. Hydrol., 406, 170–181, doi:10.1016/j.jhydrol.2011.06.007, 2011.

Pappenberger, F., Beven, K. J., Horritt, M., and Blazkova, S.: Uncertainty in the calibration of effective roughness parameters in HEC-RAS using inundation and downstream level observations, J. Hydrol., 302, 46–69, 2005.

Patro, S., Chatterjee, C., Singh, R., and Raghuwanshi, N. S.: Hydrodynamic modelling of a large flood-prone system in India with limited data, Hydrol. Process., 23, 2774–2791, doi:10.1002/hyp.7375, 2009.

Rabus, B., Eineder, M., Roth, A., and Bamler, R.: The shuttle radar topography mission – a new class of digital elevation models acquired by spaceborne radar, ISPRS J. Photogramm. Remote Sens., 57, 241–262. doi:10.1016/S0924-2716(02)00124-7, 2003.

Rodríguez, E., Morris, C. S., Belz, J. E., Chapin, E. C., Martin, J. M., Daffer, W., and Hensley, S.: An assessment of the SRTM topographic products, Technical Report JPL D-31639, Jet Propulsion Laboratory, Pasadena, California, 143 pp., http://www2.jplnasa.gov/srtm/srtmBibliography.html (last access: 16 December 2013), 2005.

Schumann, G., Matgen, P., Cutler, M. E. J., Black, A., Hoffmann, L., and Pfister, L.: Comparison of remotely sensed water stages from LiDAR, topographic contours and SRTM, ISPRS J. Photogramm. Remote Sens., 63, 283–296, 2008.

Schumann, G., Di Baldassarre, G., Alsdorf, D., and Bates, P. D.: Near real-time flood wave approximation on large rivers from space: application to the River Po, Northern Italy, Water Resour. Res., 46, W05601, doi:10.1029/2008WR007672, 2010.

Shrestha, D. L., Kayastha, N., and Solomatine, D. P.: A novel approach to parameter uncertainty analysis of hydrological models using neural networks, Hydrol. Earth Syst. Sci., 13, 1235–1248, doi:10.5194/hess-13-1235-2009, 2009.

Stedinger, J. R., Vogel, R. M., Lee, S. U., and Batchelder, R.: Appraisal of the generalized likelihood uncertainty estimation (GLUE) method, Water Resour. Res., 44, W00B06, doi:10.1029/2008WR006822, 2008.

Sun, G., Ranson, K. J., Kharuk, V. I., and Kovacs, K.: Validation of surface height from shuttle radar topography mission using shuttle laser altimetry, Remote Sens. Environ., 88, 401–411. doi:10.1016/j.rse.2003.09.001, 2003.

Tarekegn, T. H., Haile, A. T., Rientjes, T., Reggiani, P., and Alkema, D.: Assessment of an ASTER generated DEM for 2D flood modelling, Int. J. Appl. Earth Obs. Geoinf., 12, 457–465. doi:10.1016/j.jag.2010.05.007, 2010.

Trigg, M. A., Wilson, M. D., Bates, P. D., Horritt, M. S., Alsdorf, D. E., Forsberg, B. R., and Vega, M. C.: Amazon flood wave hydraulics, J. Hydrol., 374, 92–105, 2009.

USACE: HEC-RAS River Analysis System User's Manual. Version 4.1, Hydrologic Engineering Center, Davis, California, 2010.

Vázquez, R. F., Beven, K., and Feyen, J.: GLUE based assessment on the overall predictions of a MIKE SHE application, Water Resour. Res., 23, 1325–1349, doi:10.1007/s11269-008-9329-6, 2009.

Wang, W., Yang, X., and Yao, T.: Evaluation of ASTER GDEM and SRTM and their suitability in hydraulic modelling of a glacial lake outburst flood in southeast Tibet, Hydrol. Process., 26, 213–225, doi:10.1002/hyp.8127, 2011.

Werner, M. G. F.: Impact of grid size in GIS based flood extent mapping using a 1-D flow model, Phys. Chem. Earth Pt. B, 26, 517–522, 2001.

Wilson, M. D. and Atkinson, P. M.: The use of elevation data in flood inundation modelling: a comparison of ERS interferometric SAR and combined contour and differential GPS data, Intl. J. River Basin Management, 3, 3–20, doi:10.1080/15715124.2005.9635241, 2005.

Yan, K., Di Baldassarre, G., and Solomatine D. P.: Exploring the potential of SRTM topographic data for flood inundation modelling under uncertainty, J. Hydroinf., 15, 849–861, doi:10.2166/hydro.2013.137, 2013.

8

Estimates of global dew collection potential on artificial surfaces

H. Vuollekoski[1], M. Vogt[1,2], V. A. Sinclair[1], J. Duplissy[1], H. Järvinen[1], E.-M. Kyrö[1], R. Makkonen[1], T. Petäjä[1], N. L. Prisle[1], P. Räisänen[3], M. Sipilä[1], J. Ylhäisi[1], and M. Kulmala[1]

[1]University of Helsinki, Department of Physics, Helsinki, Finland
[2]Norwegian Institute for Air Research, Oslo, Norway
[3]Finnish Meteorological Institute, Helsinki, Finland

Correspondence to: H. Vuollekoski (henri.vuollekoski@helsinki.fi)

Abstract. The global potential for collecting usable water from dew on an artificial collector sheet was investigated by utilizing 34 years of meteorological reanalysis data as input to a dew formation model. Continental dew formation was found to be frequent and common, but daily yields were mostly below 0.1 mm. Nevertheless, some water-stressed areas such as parts of the coastal regions of northern Africa and the Arabian Peninsula show potential for large-scale dew harvesting, as the yearly yield may reach up to $100\,\mathrm{L\,m^{-2}}$ for a commonly used polyethylene foil. Statistically significant trends were found in the data, indicating overall changes in dew yields of between $\pm 10\,\%$ over the investigated time period.

1 Introduction

The increasing concern over the diminishing and uneven distribution of fresh water resources affects the daily life and even survival of billions of people. The United Nations Development Programme (2006) estimated that there were already 1.1 billion people in developing countries lacking adequate access to water, a figure that is expected to climb to 3 billion by 2025 due to the increasing population particularly in the most water-stressed parts of the planet.

On the other hand, water exists everywhere in one form or another: ground water, rivers, lakes, seas, glaciers, snow, ice caps, clouds, soil, and as air moisture. In particular, air moisture is present everywhere; even the driest of deserts have some, and warm air can contain more humidity than cold air. The absolute quantities of water by volume of air are of course very small (of the order of grams or some tens of grams per cubic metre), and harvesting it may be expensive or technologically demanding – factors that are rarely met in the areas of most immediate need for sustainable sources of water. Nevertheless, if no other sources of usable water exist nearby, harvesting water from the air might provide an economically sound supply of water for both drinking and agriculture.

Harvesting moisture from the air has two potential pathways: fog and dew. Fog is a highly local phenomenon that occurs, for example, when moist air is cooled by the emission of long-wave radiation or by forced ascent up a mountain slope: the decrease in temperature causes supersaturation and the formation of fog. The droplets may then be harvested by artificial structures resembling tennis nets equipped with rain gutters as has been investigated in many previous studies (e.g. Schemenauer and Cereceda, 1991; Klemm et al., 2012; Fessehaye et al., 2014).

The formation of dew occurs when the temperature of a surface is below the dew point temperature, and water vapour condenses onto the surface. In this study, the surface is assumed to be a macroscopic, artificial structure. Since only a thin layer of air over the surface reaches supersaturation, by volume the formation of dew is a very slow process compared to the formation of fog. Nevertheless, the formation and collection of dew has been studied and has been found to be feasible in several locations around the world (e.g. Nilsson, 1996; Zangvil, 1996; Kidron, 1999; Jacobs et al., 2000; Beysens et al., 2005; Lekouch et al., 2012). Additionally, material design can affect the characteristics of the condensing surface and improve its efficiency for dew collection. For example, the higher the emissivity of the surface, the higher its rate of cooling by radiation. During nights with clear skies, when both sunlight and thermal radiation from clouds are absent, the incoming radiation may be exceeded by the device's own out-going thermal radiation, resulting in a net cooling.

In this global modelling study we focus on the formation of dew onto an artificial surface, and investigate the potential for its collection. This seemingly arbitrary limitation is based on the following facts: (a) the potential for dew formation is almost ubiquitous regardless of orographic features or presence of water in other forms, (b) the formation of dew can be artificially enhanced with relatively minor efforts, (c) the formation of dew is a well-defined mathematical problem suitable for computer modelling at global scales, and (d) we are unaware of any such previous studies.

This paper describes the implementation of a model for dew formation onto an artificial surface, which is upscaled with meteorological input from a long-term reanalysis data set that spans the years 1979–2012. Modelling 34 years of dew formation ensures that the results are statistically robust. Our approach is based on an energy balance model similar to those in e.g. Nilsson (1996), Madeira et al. (2002), Beysens et al. (2005), Jacobs et al. (2008), Richards (2009) and Maestre-Valero et al. (2011), who have demonstrated that their models are able to predict the measured dew yields within reasonable accuracy.

The dew formation model, forced with reanalysis data, provides spatially coarse (80 km) estimates of dew collection yields for given sheet technologies along with the temporal evolution of dew formation. Therefore, the model output allows global maps of dew formation to be produced and areas with potential for large-scale dew collection to be identified. The modelled dew collection estimates can be used as first-order estimates by those who are planning local feasibility studies that include additional factors such as lakes, rivers, and road access. The long time series of our study provides information about the seasonal variation of dew formation as well as long-term trends in dew yield, which could be associated with climate change.

2 Methods

In order to form global estimates of dew collection potential, we combined a computationally efficient dew formation model with historical, global meteorological reanalysis data spanning 34 years. The offline model was run on a computer cluster with 128 cores, which allowed global model runs with different parameterizations to be run in approximately 24 h each.

The program source code, written in Python and Cython, is available at https://github.com/vuolleko/dew_collection/.

2.1 Model description

In implementing the model that describes the formation of dew (represented by mass yield of either liquid water or ice),

Table 1. Some parameters used in the model, unless specified otherwise. The properties of the foil are for common low-density polyethylene with composition according to Nilsson et al. (1994) and radiative properties as found by Clus (2007).

Parameter	Value
Sheet density ρ_c	$920 \, \mathrm{kg \, m^{-3}}$
Sheet thickness δ_c	$0.39 \, \mathrm{mm}$
Sheet specific heat capacity C_c	$2300 \, \mathrm{J \, kg^{-1} \, K^{-1}}$
Sheet IR emissivity e	0.94
Sheet short-wave albedo a	0.84
Time step	$10 \, \mathrm{s}$

we followed the approach presented by Pedro and Gillespie (1982) and Nikolayev et al. (1996), which has been found to agree reasonably well with empirical measurements of dew collection (e.g. Nilsson, 1996; Beysens et al., 2005; Jacobs et al., 2008; Richards, 2009; Maestre-Valero et al., 2011). The algorithm integrates the prognostic equations for the mass and heat balance by turns, thereby describing the temperature of the condenser and the resulting condensation rate onto it. As the model is global and thus incorporates both polar regions, we include the dynamics of water changing phase between liquid and solid. However, for simplicity, here we refer to both phase changes of vapour-to-liquid (condensation) and vapour-to-ice (desublimation) as condensation, and to both liquid and solid phases as water, unless specified otherwise. In our model we consider dew only and the occurrence of precipitation or fog are unaccounted for apart from their potential indirect effects included within the input reanalysis data.

The condenser in our model is a horizontally aligned sheet of some suitable material, such as low-density polyethylene (LDPE) or polymethylmethacrylate (PMMA), and is thermally insulated from the ground at a height of 2 m. Unless specified otherwise, the particular parameter values used in the model (listed in Table 1) match those of the inexpensive LDPE foil used by e.g. the International Organization for Dew Utilization, whose foil composition follows Nilsson et al. (1994).

The heat equation can be written as

$$\frac{dT_c}{dt}(C_c m_c + C_w m_w + C_i m_i) = P_{rad} + P_{cond} + P_{conv} + P_{lat}, \quad (1)$$

where T_c, C_c, and m_c are the condenser's temperature, specific heat capacity and mass, respectively. The condenser's mass is given by $m_c = \rho_c S_c \delta_c$, where ρ_c, S_c and δ_c are its density, surface area (here 1 m^2) and thickness (see Table 1). C_w and m_w are the specific heat capacity and mass of liquid water, representing the cumulative mass of water that has condensed onto the sheet, whereas C_i and m_i are the respective values for ice.

The right-hand side of Eq. (1) describes the powers involved in the heat exchange processes. The radiation term,

Table 2. The data acquired from the ECMWF's ERA-Interim database.

Original parameter	Derived model input
10 metre U wind component	
10 metre V wind component	Wind speed
Forecast surface roughness	
2 metre temperature	Air temperature
2 metre dew point temperature	Dew point
Surface solar radiation downwards	Short-wave radiation in
Surface thermal radiation downwards	Long-wave radiation in

P_{rad}, consists of three parts:

$$P_{rad} = (1 - a)S_c R_{sw} + \varepsilon_c S_c R_{lw} - P_c, \tag{2}$$

where R_{sw} and R_{lw} are the solar and thermal components of the incoming radiation from the input reanalysis data (see Table 2), a is the sheet's albedo and ε_c its emissivity (i.e. the absorbed fraction of radiation) in the infra-red band. Note that the effect of cloudiness is indirectly included via the input radiation terms. The outgoing radiative power, P_c, is given by the Stefan–Boltzmann law,

$$P_c = S_c \varepsilon_c \sigma T_c^4, \tag{3}$$

where σ is the Stefan–Boltzmann constant.

Returning to Eq. (1), the term P_{cond} describes the conductive heat exchange between the condenser surface and the ground. For simplicity, we assume perfect insulation, and the term vanishes.

The convective heat-exchange term, P_{conv}, is given by

$$P_{conv} = S_c h(T_a - T_c), \tag{4}$$

where T_a is the 2 m ambient air temperature and h is the heat transfer coefficient, estimated by a semi-empirical equation (Richards, 2009):

$$h = 5.9 + 4.1u \frac{511 + 294}{511 + T_a} \tag{5}$$

in units $W\,K^{-1}\,m^{-2}$, where u is the prevailing 2 m horizontal wind speed. However, for convenience, the model accepts any parameterization of the heat transfer coefficient (in functional form) as a model input parameter. Please see Sect. 2.3 for more details on the heat transfer coefficient.

The final term in Eq. (1), P_{lat}, represents the latent heat released by the condensation/desublimation of water

$$P_{lat} = \begin{cases} L_{vw} \frac{dm_w}{dt} & \text{if } T_c \geq 0\,°C \\ L_{vi} \frac{dm_i}{dt} & \text{if } T_c < 0\,°C, \end{cases} \tag{6}$$

where L_{vw} and L_{vi} are the specific latent heat of vaporization and desublimation for water, the appropriate one selected

based on whether the temperature of the condenser is above or below the freezing point of water. The algorithm imposes a similar condition for dynamically changing the phase of pre-existing water or ice on the condenser sheet: if liquid water exists (i.e. $m_w > 0$) while $T_c < 0$ and the sheet is losing energy (i.e. the right-hand side of Eq. 1 is negative), instead of solving Eq. (1), the model will keep T_c constant and solve

$$L_{wi} \frac{dm_w}{dt} = P_{rad} + P_{conv} + P_{lat}, \tag{7}$$

where L_{wi} is the latent heat of fusion. The mass of lost (i.e. frozen) water is added to the cumulated mass of ice. A similar equation is solved for m_i in situations when there is ice present on the condenser but the temperature of the condenser is above zero degrees Celsius. Note that Eq. (7) is unrelated to condensation, and only describes the phase transition of already condensed water or ice.

For the rate of condensation (independent of Eq. 7) we can write a mass balance equation

$$\frac{dm}{dt} = \max(0, S_c k(p_{sat}(T_d) - p_c(T_c))), \tag{8}$$

where m represents either m_i or m_w depending on whether $T_c < 0\,°C$ or not, $p_{sat}(T_d)$ is the saturation pressure at the dew point temperature, $p_c(T_c)$ is the vapour pressure over the condenser sheet and k is the mass transfer coefficient, defined through the heat transfer coefficient (Eq. 5)

$$k = \frac{h}{L_{vw}\gamma} = \frac{0.622h}{C_a p}, \tag{9}$$

where γ is the psychrometric constant, p is the atmospheric air pressure and C_a is the specific heat capacity of air. Note that Eq. (8) assumes irreversible condensation, i.e. there is no evaporation or sublimation during daytime even when $T_c > T_a$. This assumption simulates the daily manual collection of the condensed water around sunrise, soon after which the temperature of the sheet often increases above the dew point temperature. In the model we reset the cumulated values for water and ice at local noon, and take the preceding maximum value of $m_w + m_i$ as the representative daily yield.

In our model we approximate the vapour pressure $p_c(T_c)$ in Eq. (8) by the saturation pressure of water at temperature T_c. In reality, the wettability of the surface affects the vapour pressure p_c directly above it: a wetted surface decreases the vapour pressure, and condensation may take place even if $T_c > T_d$ (Beysens, 1995). Beysens et al. (2005) accounted for this effect by including an additional empirical parameter, T_0, such that $p_c(T_c) = p_{sat}(T_c + T_0)$, and found the optimal value of T_0 to be -0.35 K. However, Beysens et al. (2005) used a collector with a different design to that assumed in this study, a more expensive, 5 mm thick PMMA plate, and we were unable to find a reference value for T_0 valid for LDPE. We thus set $T_0 = 0$. This simplification causes a small underestimation of the condensation rate calculated by Eq. (8).

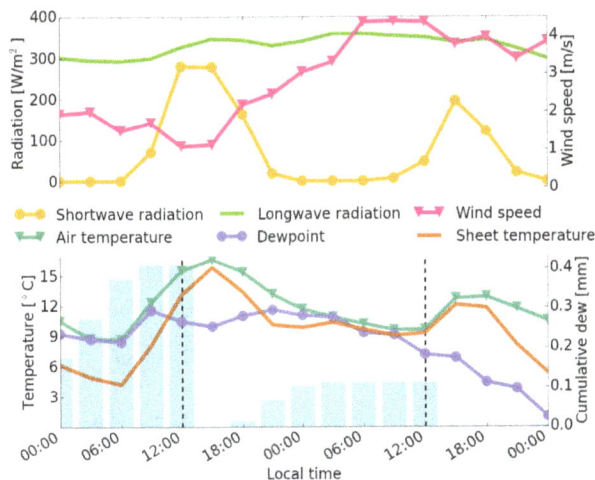

Figure 1. An example of modelled dew formation events on two consecutive days in September 2000 in Helsinki, Finland. The short-wave and long-wave radiation, wind speed, air temperature and dew point are input from the ERA-Interim data set. Note that the cumulated amount of dew (vertical bars) is reset daily at local noon (dashed vertical lines). All data are in 3 h resolution.

The model reads all input data for a given grid point and solves Eqs. (1), (7) and (8) using a fourth-order Runge–Kutta algorithm with a 10 s time step. An example case spanning two consecutive days is presented in Fig. 1, which shows the long- and short-wave radiation components, wind speed, air temperature, dew point temperature as well as the modelled sheet temperature and cumulated dew. During daytime, the incoming short-wave radiation from the sun as well as the atmospheric long-wave radiation act to increase the temperature of the condenser sheet. In contrast, during dark periods, the outgoing thermal radiation exceeds the atmospheric long-wave radiation, the latter of which is greatly influenced by cloudiness: the thermal emission by clouds, especially low clouds, increases the incoming thermal radiation at the surface. As condensation occurs when the temperature of the condenser sheet is below the dew point temperature (Eq. 8), significant dew cumulation can only occur during night-time. The daily collection of dew occurs at noon, depicted by the dashed vertical lines.

2.2 Meteorological input data

The meteorological input data for the dew formation model is obtained from the European Centre for Medium Range Weather Forecasts (ECMWF) Interim Reanalysis (ERA-Interim, Dee et al., 2011). Such reanalysis data sets are produced by combining historical observations from multiple sources with a comprehensive numerical model of the atmosphere using data assimilation systems. As numerical models of the atmosphere are constantly evolving, reanalysis data sets are more appropriate for long-term studies than operational analyses as a fixed numerical model is used. Numer-

ous different global reanalysis data sets are available, for example NASA MERRA (Rienecker et al., 2011), JRA-25 (Onogi et al., 2007) and NCEP-CFSR (Saha et al., 2010), and many inter-comparison studies between the different reanalysis data sets have been conducted (e.g. Lorenz and Kunstmann, 2012; Willett et al., 2013; Simmons et al., 2014). ERA-Interim was selected for this study primarily because it is the only available reanalysis which assimilates two-metre temperature and therefore has a lower two-metre temperature bias than any other available re-analysis (Decker et al., 2012).

ERA-Interim is ECMWF's current global reanalysis data set spanning 1979–present which has a horizontal resolution of 0.75° (approximately 80 km) and 60 levels in the vertical. We use 34 years (1979–2012) of ERA-Interim data and the variables extracted from ERA-Interim to be applied in the dew formation model are listed in Table 2. The data for wind speed, temperature and dew point temperature originate from reanalysis fields valid at 00:00, 06:00, 12:00 and 18:00 UTC, while the data valid at 03:00 and 09:00 UTC (15:00 and 21:00 UTC) are forecast fields based on the reanalysis of 0:00 UTC (12:00 UTC). The radiative parameters are purely forecast fields and are cumulative over the forecast period; in this study we derive a simple average from the difference between adjacent cumulative values to obtain instantaneous values.

The dew formation model requires the wind speed at a height of two metres, whereas only the 10 m wind speed is available in the ERA-Interim reanalysis data set. Therefore, the 2 m wind speed is estimated using the logarithmic wind profile (e.g. Seinfeld and Pandis, 2006) in the positive-definite form

$$u = \frac{\log((2+z_0)/z_0)}{\log((10+z_0)/z_0)} \sqrt{u_{10,x}^2 + u_{10,y}^2}, \quad (10)$$

where z_0 is the forecast surface roughness taken from the ERA-interim reanalysis data set and $u_{10,x}$ and $u_{10,y}$ are the 10 m horizontal wind speed components.

Even by combining the ERA-Interim forecast fields with the analyses fields, the temporal resolution of the meteorological input data is only 3 hours. In contrast, the numerical dew formation model requires meteorological input every time step (10 s). Therefore, the 3-hourly ERA-Interim data is linearly interpolated to 10 s time resolution. This is a disadvantage of using reanalysis data compared to using more frequent observations. However, we believe that this disadvantage is considerably outweighed by the advantages of using reanalysis data – the long time series and the uniform global coverage. Finally, it should be emphasized that in addition to their relatively low resolution, reanalysis data sets have inherent uncertainties and they must not be regarded as exact representations of reality.

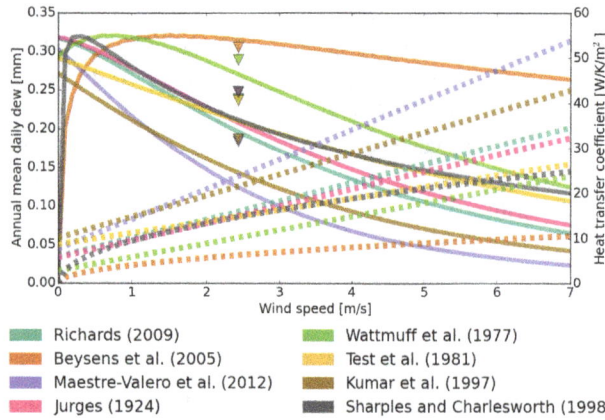

Figure 2. Sensitivity of the model to the heat transfer coefficient. The dashed lines represent heat transfer coefficients for different parameterizations as functions of wind speed. The triangles represent annual mean daily yields of dew using one year of ERA-Interim data for the grid point closest to the Negev Desert, Israel (30.75° N, 34.5° E) in 1992 (here plotted against the annual mean wind speed). The solid lines are the same, but the wind speed has been fixed according to x axis.

2.3 Transfer coefficients

In the model, the heat transfer coefficient determines how effectively the surrounding air heats or cools the condenser surface. During dew formation the surface must be cooler than the air surrounding it, which means that a high heat transfer coefficient impedes dew formation. On the other hand, the mass transfer coefficient is proportional to the heat transfer coefficient, and determines the efficiency of water vapour molecules condensing on the condenser surface. The net effect of a high heat transfer coefficient is therefore ambiguous.

The mentioned transfer phenomena can be divided into free and forced (e.g. wind-driven) convection. In wind-driven atmospheric conditions the heat transfer coefficient is often parameterized as

$$h = a + bu^n, \tag{11}$$

where a, b and n are empirical constants (possibly related to some other parameters). The constant a can be thought to correspond to free convection, although absent in some parameterizations.

Various such parameterizations (with somewhat differing assumptions) for the heat transfer coefficients can be found in the literature; Table 3 lists a few of them. Figure 2 presents these heat transfer coefficients as functions of wind speed (dashed lines). Clearly, the variance is large, especially at larger wind speeds. It should be noted that the authors of these semi-empirical parameterizations have typically assumed a quite narrow range of validity in regard to wind speed. For example, the parameterization by Richards (2009), based on McAdams (1954), is said to be valid for wind speeds $u < 5\,\mathrm{m\,s^{-1}}$. However, 3 h average wind speeds

Table 3. A selection of the various parameterizations for the heat transfer coefficient found in the literature. The first three are studies on dew formation. Here, u and T_a are the horizontal wind speed and air temperature at 2 m height, and D is the characteristic length of the condenser (e.g. 1 m).

Source	Equation
Richards (2009); this study	$h = 5.9 + 4.1u\frac{511+294}{511+T_a}$
Beysens et al. (2005)	$h = 4\sqrt{u/D}$
Maestre-Valero et al. (2011)	$h = 7.6 + 6.6u\frac{511+294}{511+T_a}$
Jürges (1924)	$h = 5.7 + 3.8u$
Watmuff et al. (1977)	$h = 2.8 + 3u$
Test et al. (1981)	$h = 8.55 + 2.56u$
Kumar et al. (1997)	$h = 10.03 + 4.687u$
Sharples and Charlesworth (1998)	$h = 9.4\sqrt{u}$

at 2 m derived from the ERA-Interim data set rarely exceed this value over continental areas.

The mass transfer coefficient is defined through the heat transfer coefficient according to Eq. (9). As noted, the effects of heat and mass transfer are opposite during dew formation.

In order to gain some estimates of the model sensitivity to the transfer coefficients, we performed several series of model runs with different parameterizations. Figure 2 presents the annual mean of the daily yield of dew in 1992 for the grid point closest to the Negev Desert, Israel. For each parameterization, the model was run once with the ERA-Interim data for the year 1992 (triangles). Next, the same model run was repeated so that the wind speed was fixed to one value for the entire year; this was repeated for all wind speeds between 0 and $7\,\mathrm{m\,s^{-1}}$ in $0.1\,\mathrm{m\,s^{-1}}$ resolution (solid lines). Altogether Figure 2 therefore presents $1 + 71$ years of simulations for each parameterization. Clearly, the difference in dew yields between the parameterizations is less pronounced than in the heat transfer coefficients alone. Nevertheless, the difference is significant, and suggests that the choice of transfer coefficients is important for model performance. Note especially the behaviour of parameterizations by Sharples and Charlesworth (1998) and Beysens et al. (2005) for pure forced convection at wind speeds close to zero.

The same test was performed for 10 locations globally (not shown), and the general characteristics are similar to those in Fig. 2, albeit a larger mean wind speed did cause more deviation in some cases.

For the global runs presented in this paper, we chose the parameterization used by Richards (2009), as this heat coefficient is close to the average presented in the literature and is well behaved also at very low wind speeds, see Fig. 2. Additionally, the study was also dedicated to dew, although the condensing surface was an asphalt-shingle roof. The dew study by Maestre-Valero et al. (2011) used the same type of foil as our virtual condenser, albeit inclined at 30°, which

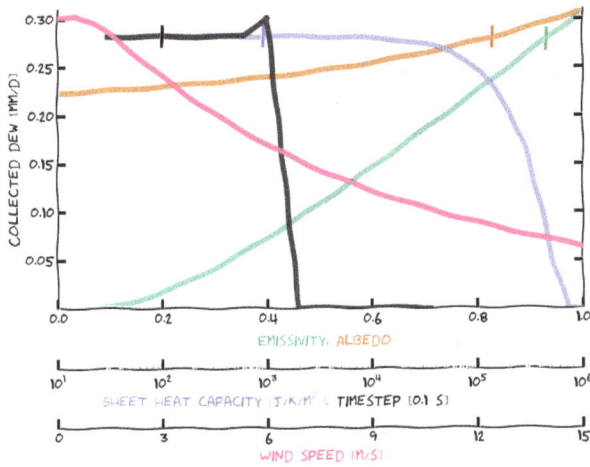

Figure 3. Sensitivity of the model to the emissivity, albedo and heat capacity of the condenser sheet as well as to the wind speed and time step of the model (×10 in figure). The heat capacity is defined here as $C_c\rho_c S_c\delta_c$, i.e. its variation corresponds to varying any of these factors. The input data correspond to Table 1 and the first day of Fig. 1, where applicable. The vertical bars represent these *default* values.

may be the reason for their significantly higher heat transfer coefficient.

For convenience, our model accepts any functional form of the heat transfer coefficient as input to the model, and several are available built-in.

3   Results and discussion

Figure 3 illustrates the sensitivity of the modelled dew yield to changes in the emissivity, albedo, and heat capacity of the sheet as well as to the wind speed and the time step of the model. The dew yield increases almost linearly with the sheet's emissivity, and the emissivity seems to be the most important factor to consider when designing condenser materials (besides economic factors). The albedo of the sheet has a smaller effect as it only affects the sheet's temperature during sunlit hours, when the sheet is anyhow heated convectively by high air temperatures (see Fig. 1). The sheet's heat capacity does not significantly affect the dew yield unless it is either very low or very high (note the logarithmic scale). Interestingly, the issue of heat capacity may have been the key limiting factor in massive ancient dew collection infrastructure (Nikolayev et al., 1996). Note that for the simulated horizontal plane, current technologies already lie close to optima. The model time step was chosen to be 10 s as this keeps the model stable even in the very-high-yield scenario of Fig. 3.

Finally, the effect of wind speed is more complex: decreasing the wind speed reduces the mass flow towards the condenser, whereas increasing the wind speed increases convective heating. It should be noted that the model formulation used in this study assumes a constant supply of atmospheric

moisture defined by the dew point temperature. In a more realistic scenario, the layer of air directly above the surface of the condenser should eventually dry if both vertical mixing and the horizontal wind speed were small, which may become important for very large collectors. On the other hand, the potential for dew collection still exists, and when designing large-scale dew collection, passive air-mixers should be introduced to ensure a supply of moist air. For model sensitivity regarding wind speed, see also Sect. 2.3 and Fig. 2.

The following results originate from a series of global simulations. The model simulations differ only by the parameters of albedo and emissivity that describe the ability of the condenser's sheet to emit and absorb energy by radiation. Recall that the spatial resolution of the meteorological input data is relatively coarse, $0.75° \times 0.75°$ (up to 80 km, depending on latitude), which does limit the model's ability to capture small-scale phenomena such as those caused by local topography. Therefore, this limitation should be considered when interpreting the model results.

Furthermore, Beysens et al. (2005) introduced additional site-specific parameters to the heat and mass transfer coefficients (Eqs. 5, 9) to accommodate for differences in environmental conditions between the condenser surface and the meteorological instruments, as well as a correction in Eq. (8) to account for surface wetting. In our study the difference between the reanalysis data and any real physical location within the area represented by the grid point could arguably be much greater than the differences considered in Beysens et al. (2005), but as we see no means to tailor the model separately for each grid point, we use the theoretical formulation as it is. This assumption will inevitably cause some error in the dew yield estimates, although the large-scale average should be reasonably well predicted.

3.1   Occurrence of dew

First and foremost, it is important to gain insight into how frequently dew forms onto the artificial surface in different areas around the world. Our model results suggest that dew formation is both global and common in continental areas, with surprisingly little seasonal variation in most areas. Figure 4 presents the mean seasonal fraction of days during which the formation of dew onto the collector occurs (i.e. the yield is positive). Apart from very warm and dry deserts, the meteorological conditions on almost all continental areas favour the formation of dew onto the collector.

The lack of dew formation is generally caused by inefficient nocturnal cooling of the surface as a result of high incoming long-wave radiation, which occurs due to a high cloud fraction and high humidity in the atmosphere (although high humidity at surface level favours dew formation).

Perhaps somewhat counter-intuitively, in general the artificial surfaces over oceans do not collect dew as regularly as those over land areas. The lack of oceanic dew formation is probably caused by higher wind speeds and the weaker diur-

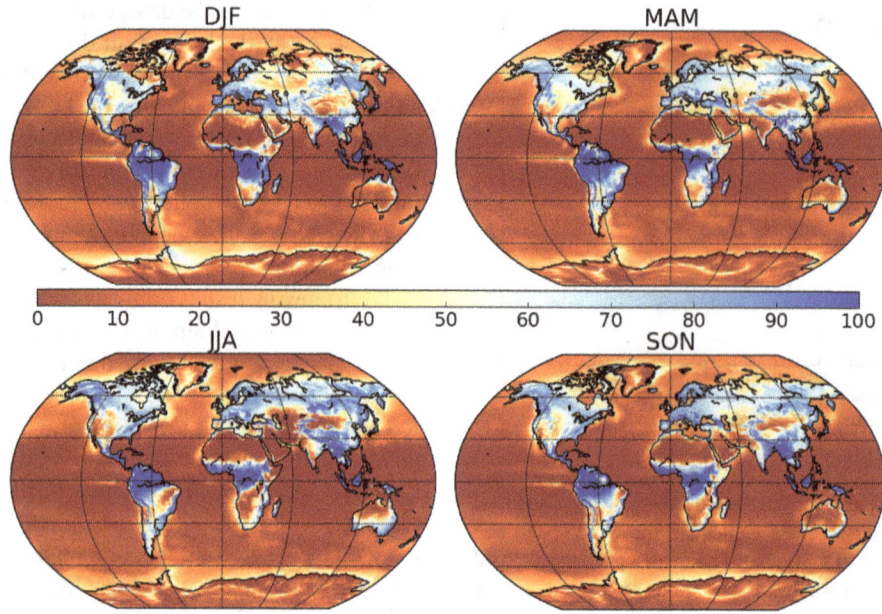

Figure 4. Seasonal occurrence of dew as a fraction of days (%).

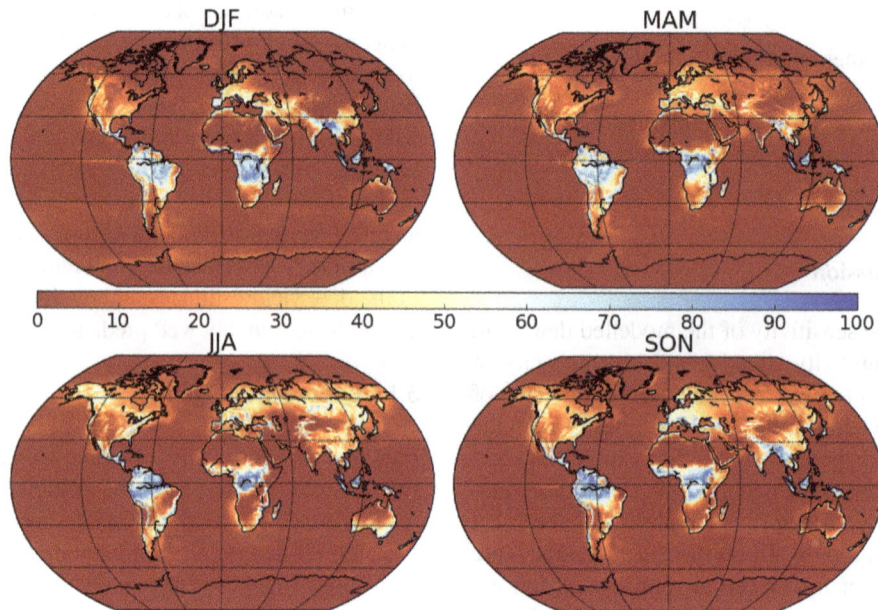

Figure 5. Seasonal occurrence of dew as a fraction of days (%) with a threshold of $0.1 \, \mathrm{mm \, d^{-1}}$.

nal cycle in air temperature, denser average cloud cover (e.g. King et al., 2013) and higher humidity compared to land areas, resulting in amplified long-wave radiation downwards, and therefore weaker cooling.

In most dew events represented by Fig. 4, the cumulated amount of water is insignificant (see Sect. 3.2). Figure 5 shows a similar seasonal occurrence of dew as fraction of days, but only during which more than $0.1 \, \mathrm{mm \, d^{-1}}$ (i.e. $0.1 \, \mathrm{L \, m^{-2} \, d^{-1}}$) can be collected. The contrast between the two figures is notable, as in the latter the seasonal variation

is higher and dew formation occurs regularly in far fewer areas, most of which do not have a water shortage problem. However, in some water-stressed areas, such as the coastal regions of North Africa and the Arabian Peninsula, dew collection may be an alternative source of water worth investigating further.

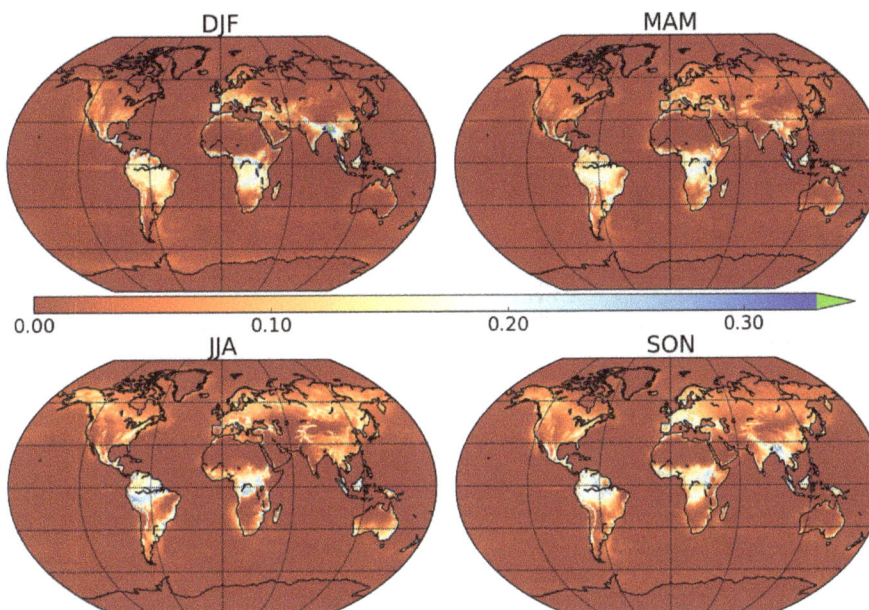

Figure 6. Mean seasonal formation of dew (mm d^{-1}).

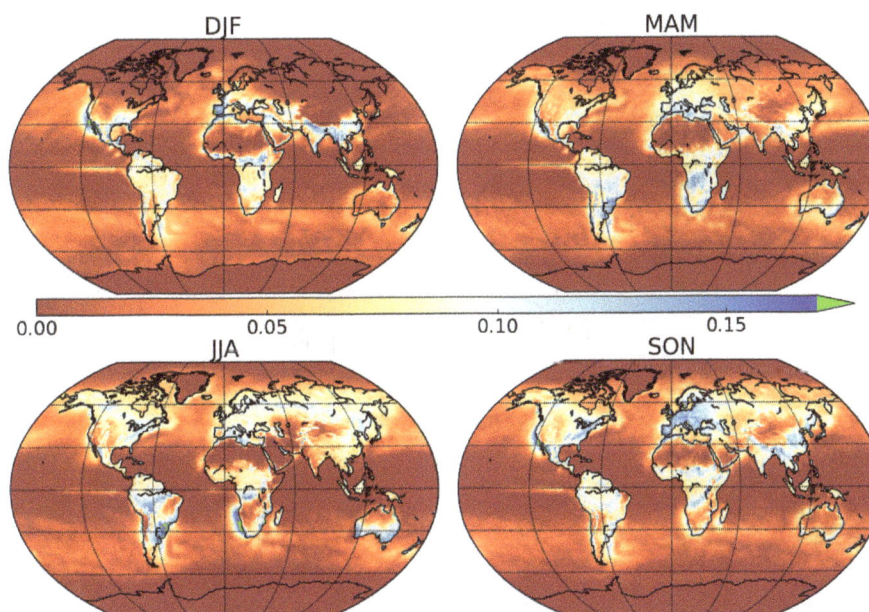

Figure 7. Standard deviation of the seasonal formation of dew (mm d^{-1}).

3.2 Yield of dew

Given the occurrence of dew formation events as presented in Sect. 3.1, we subsequently calculated the mean seasonal values for the actual daily amounts of dew cumulated on the collector sheet. The reported values represent the liquid water-equivalent volumes of the sum of liquid water and ice. For the condenser parameters shown in Table 1, this *dew potential* is presented in Fig. 6. Unsurprisingly, the global distribution of dew potential closely resembles Fig. 5 and indicates

that most areas with the potential to harvest non-negligible quantities of dew are also those with sufficient other sources of water. Note the high seasonal variation especially in equatorial Africa, Southeast Asia and southern Australia.

The standard deviation of the seasonal formation of dew is presented in Fig. 7. The variation is surprisingly zonal compared to Fig. 6. On the other hand, the highest variation is found in regions with the highest dew yields as might be expected. In particular, dew yields in the aforementioned coastal regions of northern Africa and the Arabian Penin-

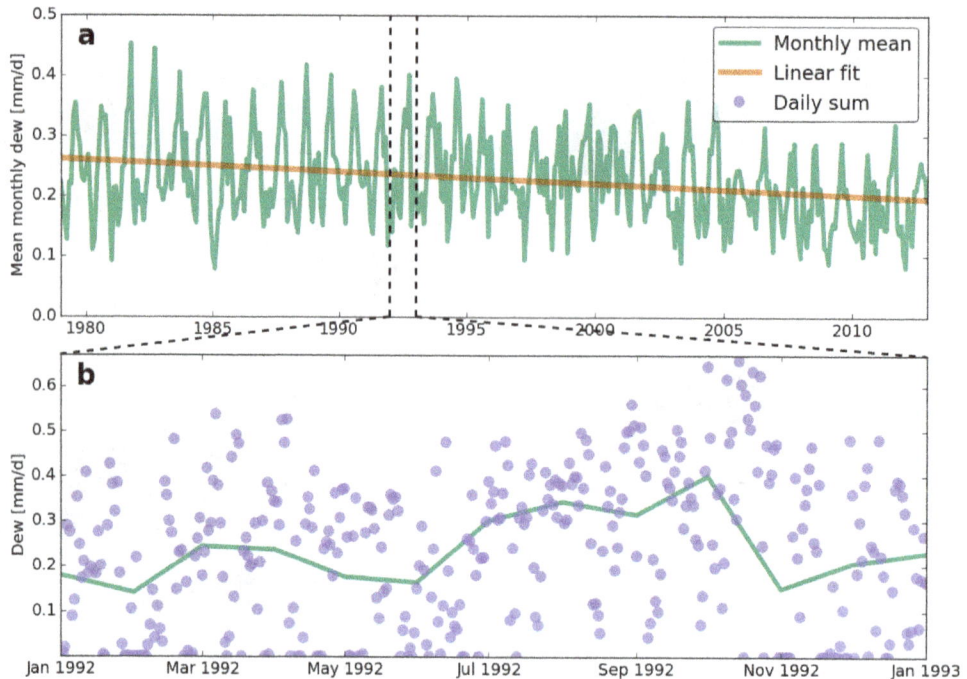

Figure 8. Time series of the modelled dew yield from one grid point, 30.75° N, 34.5° E, located in the Negev desert, Israel: (**a**) the monthly means over the whole data set, as well as a linear fit to the data; (**b**) the monthly means as well as daily values for the year 1992.

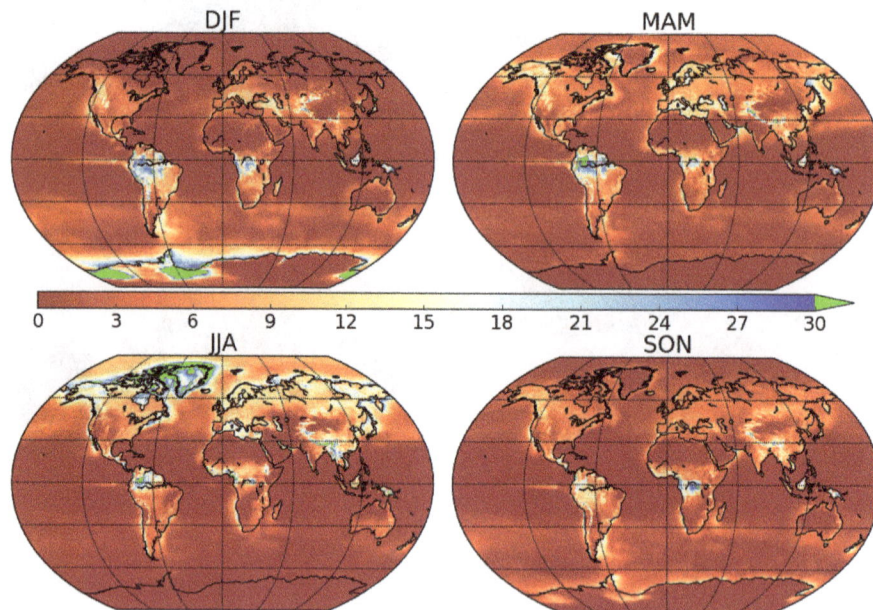

Figure 9. The fractional increase in the seasonal occurrence of dew (%) with a threshold of $0.1\,\mathrm{mm\,d^{-1}}$, when the emissivity of the condenser is increased from 0.94 to 0.999, and the albedo from 0.84 to 0.999.

sula exhibit high standard deviations, suggesting that if large-scale dew collection in these areas was planned, varying dew yields should be expected.

Figure 8 presents a time series of dew yield in the Negev desert, Israel, where natural dew collection has been studied by several authors (e.g. Evenari, 1982; Zangvil, 1996;

Kidron, 1999; Jacobs et al., 2000). The values from our model are significantly higher than most of the reported values in other studies. However, this is expected since the majority of studies report yields of natural dew, which artificial surfaces typically outperform. In any case the coarse resolution of our data, as well as the differences in the collection

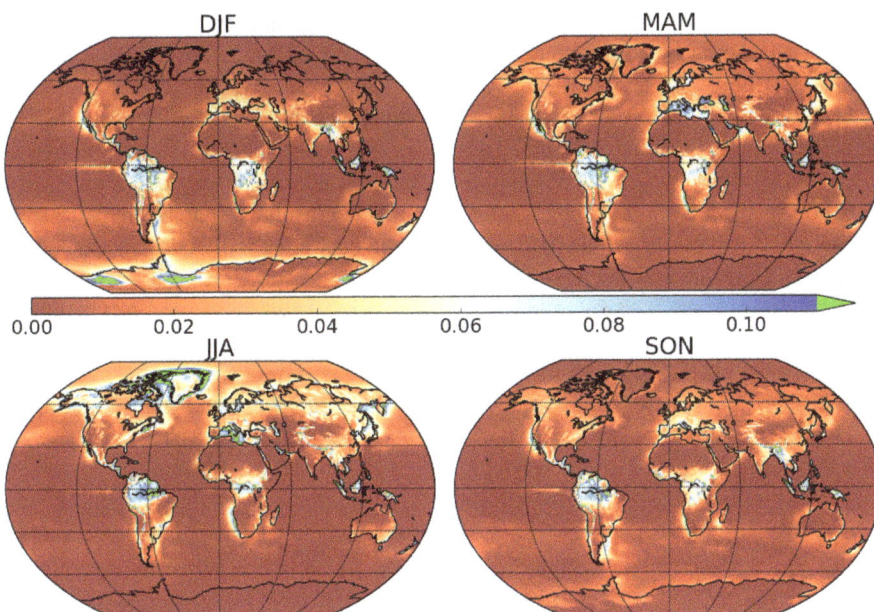

Figure 10. The absolute increase in the mean seasonal formation of dew $(\mathrm{mm\,d^{-1}})$, when the emissivity of the condenser is increased from 0.94 to 0.999, and the albedo from 0.84 to 0.999.

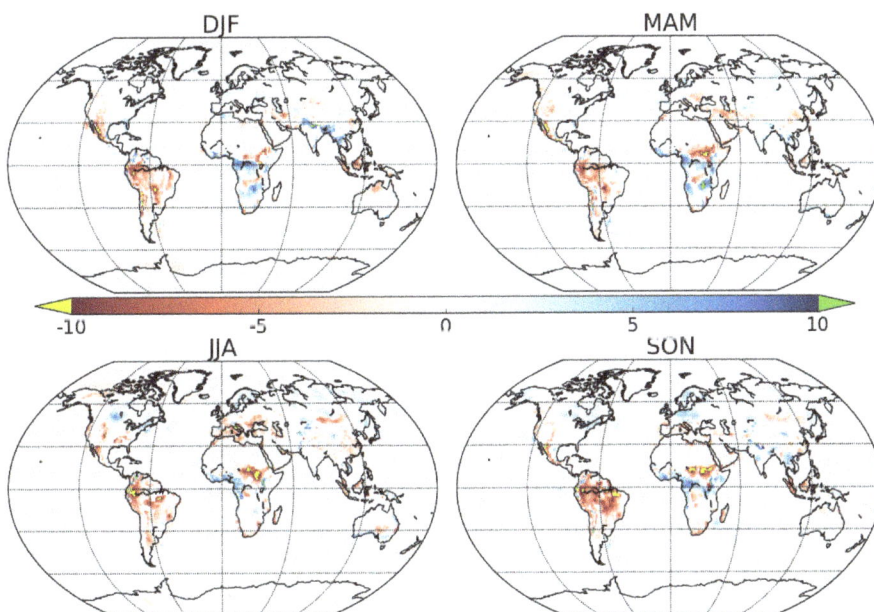

Figure 11. The total change (%) in the mean seasonal formation of dew $(\mathrm{mm\,d^{-1}})$ over the years 1979–2012 as predicted by the Theil–Sen estimator. Only locations with a statistically significant trend ($p < 0.05$) are shown.

methods, make direct comparison with measurements difficult. Note the decreasing trend in the modelled dew yields in Fig. 8.

3.3 Increase of dew

The data presented in Fig. 5 are for a sheet emissivity of 0.94 and albedo of 0.84, both of which can possibly be improved by means of material science. If both the emissivity and the albedo were hypothetically increased to an extreme value of 0.999, the occurrence of dew would increase as presented in Fig. 9. Although this *ideal collector* scenario is exaggerated, these model results suggest that improvements in the emissivity and albedo could have a significant effect on the sheet's ability to condense water, and thus the cost of a high-performance sheet material may be justified. It should be noted that besides increasing the emissivity and the albedo of the sheet, other means of enhancing the condenser's performance exist as well. For example, Beysens et al. (2013) reported an increase in dew yields of up to 400 % for origami-shaped collectors compared to a planar condenser inclined at an angle of $30°$.

In general, the ideal condenser scenario suggests that enhancing the properties of the condenser would increase the occurrence of dew most significantly over the summer-time Northern Hemisphere. In Antarctica and Greenland, the summer-time dew yields increase significantly over the subjective $0.1\,\text{mm}\,\text{d}^{-1}$ limit in the ideal condenser scenario, which results in these regions being highlighted in Fig. 9.

The absolute increase in the mean seasonal formation of dew is presented in Fig. 10, suggesting that the dew yield can be more than doubled in some areas in this extreme scenario. In general, however, the increase in the absolute dew yield is relatively small even in areas where enhancing the condenser's properties significantly increases dew occurrence. This implies that the relative importance of different factors affecting dew formation varies globally, and that radiative cooling is the main limiting factor, for example, in the Mediterranean Sea.

3.4 Trend of dew

With the projected changes in climate and potentially increasing occurrences of drought (Stocker et al., 2013), we investigated the existence of temporal trends in the modelled dew yields. Trends were calculated by applying the Mann–Kendall (i.e. Kendall Tau-b) trend test (e.g. Agresti, 2010) on seasonal means of yearly data. Unsurprisingly, the statistical significance of the trends varies non-uniformly across the globe. Nevertheless, in many regions a statistically significant trend ($p < 0.05$) is found.

Figure 11 presents the overall change in the mean seasonal formation of dew. Only statistically significant ($p < 0.05$) changes are shown, with the trend being equal to the Theil–Sen estimator (Theil, 1950; Sen, 1968). Interestingly, the general trend appears to indicate a decrease in dew potential in most water-stressed areas. The changes appear in large and roughly uniform areas, suggesting that the phenomenon cannot be entirely attributed to noise. A significant decreasing trend is also visible in the case study presented in Fig. 8. In addition, a decreasing trend is also visible in parts of the coastal regions of northern Africa and the Arabian Peninsula, which we identified as regions of high dew collection potential (see previous sections).

4 Conclusions

The global potential for collecting dew on artificial surfaces was investigated by implementing a dew formation model based on solving the heat and mass balance equations. As meteorological input, 34 years of global reanalysis data from ECMWF's ERA-Interim archive was used.

Dew formation was found to be common and frequent, though mostly over land areas where other sources of water exist. Nevertheless, some water-stressed areas, especially parts of the coastal regions of northern Africa and the Arabian Peninsula, might be suitable for economically viable large-scale dew collection, as the yearly yield of dew may reach up to $100\,\text{L}\,\text{m}^{-2}$ for a commonly used LDPE foil. For these locations, more accurate regional modelling and field experiments should be conducted.

The long time series provides some statistical confidence in conducting a trend analysis, and it suggests significant changes in dew yields in some areas up to and exceeding $\pm 10\,\%$ over the investigated time period.

It should be noted that the real-life usefulness of the results presented in this paper depends on several factors not accounted for in this study, such as other sources of water (precipitation, lakes, rivers, desalination of seawater), pipelines, and road access to the location for transportation of water by trucks, as well as financial and technological considerations. Additionally, the uncertainties related to the transfer coefficients, the reanalysis data set and its near-surface application as well as the inherent uncertainties in any global modelling approaches should be acknowledged, and all numbers presented here are rough estimates only.

Acknowledgements. Funding from the Academy of Finland is gratefully acknowledged (Development of cost-effective fog and dew collectors for water management in semiarid and arid regions of developing countries (DF-TRAP), project No. 257382, as well as Centre of Excellence, project No. 272041). We acknowledge CSC – IT Center for Science Ltd for the allocation of computational resources. Technical support and performance tips with large NetCDF files from Russell Rew at Unidata is acknowledged.

Edited by: A. Gelfan

References

Agresti, A.: Analysis of ordinal categorical data, Vol. 656, John Wiley & Sons, New York, 2010.

Beysens, D.: The formation of dew, Atmos. Res., 39, 215–237, 1995.

Beysens, D., Muselli, M., Nikolayev, V., Narhe, R., and Milimouk, I.: Measurement and modelling of dew in island, coastal and alpine areas, Atmos. Res., 73, 1–22, 2005.

Beysens, D., Brogginib, F., Milimouk-Melnytchoukc, I., Ouazzanid, J., and Tixiere, N.: New Architectural Forms to Enhance Dew Collection, Chem. Eng., 34, 79–84, 2013.

Clus, O.: Condenseurs radiatifs de la vapeur d'eau atmosphérique (rosée) comme source alternative d'eau douce, PhD thesis, Université Pascal Paoli, 2007.

Decker, M., Brunke, M. A., Wang, Z., Sakaguchi, K., Zeng, X., and Bosilovich, M. G.: Evaluation of the reanalysis products from

GSFC, NCEP, and ECMWF using flux tower observations, J. Climate, 25, 1916–1944, 2012.

Dee, D. P., Uppala, S. M., Simmons, A. J., Berrisford, P., Poli, P., Kobayashi, S., Andrae, U., Balmaseda, M. A., Balsamo, G., Bauer, P., Bechtold, P., Beljaars, A. C. M., van de Berg, L., Bidlot, J., Bormann, N., Delsol, C., Dragani, R., Fuentes, M., Geer, A. J., Haimberger, L., Healy, S. B., Hersbach, H., Hólm, E. V., Isaksen, L., Kållberg, P., Köhler, M., Matricardi, M., McNally, A. P., Monge-Sanz, B. M., Morcrette, J.-J., Park, B.-K., Peubey, C., de Rosnay, P., Tavolato, C., Thépaut, J.-N., and Vitart, F.: The ERA-Interim reanalysis: configuration and performance of the data assimilation system, Q. J. Roy. Meteorol. Soc., 137, 553–597, 2011.

Evenari, M.: The Negev: the challenge of a desert, Harvard University Press, 1982.

Fessehaye, M., Abdul-Wahab, S. A., Savage, M. J., Kohler, T., Gherezghiher, T., and Hurni, H.: Fog-water collection for community use, Renew. Sustain. Energy Rev., 29, 52–62, 2014.

Jacobs, A., Heusinkveld, B., and Berkowicz, S.: Passive dew collection in a grassland area, the Netherlands, Atmos. Res., 87, 377–385, 2008.

Jacobs, A. F., Heusinkveld, B. G., and Berkowicz, S. M.: Dew measurements along a longitudinal sand dune transect, Negev Desert, Israel, Int. J. Biometeorol., 43, 184–190, 2000.

Jürges, W.: Der Wärmeüubergang an Einer Ebenen Wand, Beihefte zum Gesundheits-Ingenieur, 1, 1227–1249, 1924.

Kidron, G. J.: Altitude dependent dew and fog in the Negev Desert, Israel, Agr. Forest Meteorol., 96, 1–8, 1999.

King, M. D., Platnick, S., Menzel, W. P., Ackerman, S. A., and Hubanks, P. A.: Spatial and Temporal Distribution of Clouds Observed by MODIS Onboard the Terra and Aqua Satellites, Geoscience and Remote Sensing, IEEE Trans.. 51, 3826–3852, 2013.

Klemm, O., Schemenauer, R. S., Lummerich, A., Cereceda, P., Marzol, V., Corell, D., van Heerden, J., Reinhard, D., Gherezghiher, T., Olivier, J., Osses, P., Sarsour, J., Frost, E., Estrela, M. J., Valiente, J. A., and Fessehaye, G. M.: Fog as a fresh-water resource: overview and perspectives, Ambio, 41, 221–234, 2012.

Kumar, S., Sharma, V., Kandpal, T., and Mullick, S.: Wind induced heat losses from outer cover of solar collectors, Renew. Energy, 10, 613–616, 1997.

Lekouch, I., Lekouch, K., Muselli, M., Mongruel, A., Kabbachi, B., and Beysens, D.: Rooftop dew, fog and rain collection in southwest Morocco and predictive dew modeling using neural networks, J. Hydrol., 448, 60–72, 2012.

Lorenz, C. and Kunstmann, H.: The hydrological cycle in three state-of-the-art reanalyses: intercomparison and performance analysis, J. Hydrometeorol., 13, 1397–1420, 2012.

Madeira, A., Kim, K., Taylor, S., and Gleason, M.: A simple cloud-based energy balance model to estimate dew, Agr. Forest Meteorol., 111, 55–63, 2002.

Maestre-Valero, J., Martinez-Alvarez, V., Baille, A., Martín-Górriz, B., and Gallego-Elvira, B.: Comparative analysis of two polyethylene foil materials for dew harvesting in a semi-arid climate, J. Hydrol., 410, 84–91, 2011.

McAdams, W.: Heat Transmission, McGraw-Hill, New York, 1954.

Nikolayev, V., Beysens, D., Gioda, A., Milimouk, I., Katiushin, E., and Morel, J.: Water recovery from dew, J. Hydrol., 182, 19–35, 1996.

Nilsson, T.: Initial experiments on dew collection in Sweden and Tanzania, Solar Energy Mater. Solar Cells, 40, 23–32, 1996.

Nilsson, T., Vargas, W., Niklasson, G., and Granqvist, C.: Condensation of water by radiative cooling, Renew. Energy, 5, 310–317, 1994.

Onogi, K., Tsutsui, J., Koide, H., Sakamoto, M., Kobayashi, S., Hatsushika, H., Matsumoto, T., Yamazaki, N., Kamahori, H., Takahashi, K., Kadokura, S., Wada, K., Kato, K., Oyama, R., Ose, T., Mannoji, N., and Taira, R.: The JRA-25 reanalysis, J. Meteorol. Soc. JPN Ser. II, 85, 369–432, 2007.

Pedro, M. J. and Gillespie, T. J.: Estimating Dew Duration .1. Utilizing Micrometeorological Data, Agr. Meteorol., 25, 283–296, 1982.

Richards, K.: Adaptation of a leaf wetness model to estimate dewfall amount on a roof surface, Agr. Forest Meteorol., 149, 1377–1383, 2009.

Rienecker, M. M., Suarez, M. J., Gelaro, R., Todling, R., Bacmeister, J., Liu, E., Bosilovich, M. G., Schubert, S. D., Takacs, L., Kim, G.-K., Bloom, S., Chen, J., Collins, D., Conaty, A., da Silva, A., Gu, W., Joiner, J., Koster, R. D., Lucchesi, R., Molod, A., Owens, T., Pawson, S., Pegion, P., Redder, C. R., Reichle, R., Robertson, F. R., Ruddick, A. G., Sienkiewicz, M., and Woollen, J.: MERRA: NASA's Modern-Era Retrospective Analysis for Research and Applications, J. Climate, 24, 3624–3648, 2011.

Saha, S., Moorthi, S., Pan, H.-L., Wu, X., Wang, J., Nadiga, S., Tripp, P., Kistler, R., Woollen, J., Behringer, D., Liu, H., Stokes, D., Grumbine, R., Gayno, G., Wang, J., Hou, Y.-T., Chuang, H.-Y., Juang, H.-M. H., Sela, J., Iredell, M., Treadon, R., Kleist, D., Van Delst, P., Keyser, D., Derber, J., Ek, M., Meng, J., Wei, H., Yang, R., Lord, S., Van Den Dool, H., Kumar, A., Wang, W., Long, C., Chelliah, M., Xue, Y., Huang, B., Schemm, J.-K., Ebisuzaki, W., Lin, R., Xie, P., Chen, M., Zhou, S., Higgins, W., Zou, C.-Z., Liu, Q., Chen, Y., Han, Y., Cucurull, L., Reynolds, R. W., Rutledge, G., and Goldberg, M.: The NCEP climate forecast system reanalysis, B. Am. Meteorol. Soc., 91, 1015–1057, 2010.

Schemenauer, R. S. and Cereceda, P.: Fog-Water Collection in Arid Coastal Locations, Ambio, 20, 303–308, 1991.

Seinfeld, J. H. and Pandis, S. N.: Atmospheric Chemistry and Physics, John Wiley & Sons, New York, 2006.

Sen, P. K.: Estimates of the regression coefficient based on Kendall's tau, J. Am. Stat. Assoc., 63, 1379–1389, 1968.

Sharples, S. and Charlesworth, P.: Full-scale measurements of wind-induced convective heat transfer from a roof-mounted flat plate solar collector, Sol. Energy, 62, 69–77, 1998.

Simmons, A. J., Poli, P., Dee, D. P., Berrisford, P., Hersbach, H., Kobayashi, S., and Peubey, C.: Estimating low-frequency variability and trends in atmospheric temperature using ERA-Interim, Q. J. Roy. Meteorol. Soc., 140, 329–353, 2014.

Stocker, T. F., Qin, D., Plattner, G.-K., Tignor, M., Allen, S. K., Boschung, J., Nauels, A., Xia, Y., Bex, V., and Midgley, P. M.: Climate change 2013: The physical science basis, Intergovernmental Panel on Climate Change, Working Group I Contribution to the IPCC Fifth Assessment Report (AR5) (Cambridge Univ Press, New York), 2013.

Test, F., Lessmann, R., and Johary, A.: Heat transfer during wind flow over rectangular bodies in the natural environment, J. Heat Transf., 103, 262–267, 1981.

Theil, H.: A rank-invariant method of linear and polynomial regression analysis, Part 3, Proceedings of Koninalijke Nederlandse Akademie van Weinenschatpen A, 53, 1397–1412, 1950.

United Nations Development Programme: Human Development Report 2006, Beyond scarcity: Power, poverty and the global water crisis, Palgrave Macmillan, New York, 2006.

Watmuff, J., Charters, W., and Proctor, D.: Solar and wind induced external coefficients-solar collectors, Cooperation Mediterraneenne pour l'Energie Solaire, 1, 56, 1977.

Willett, K. M., Dolman, A. J., Hall, B. D., and Thorne, P. W. (Eds.): State of the climate in 2012, Vol. 94, Chap. Global Climate, S7–S46, B. Am. Meteorol. Soc., 2013.

Zangvil, A.: Six years of dew observations in the Negev Desert, Israel, J. Arid Environ., 32, 361–371, 1996.

9

Global trends in extreme precipitation: climate models versus observations

B. Asadieh and N. Y. Krakauer

Civil Engineering Department and NOAA-CREST, The City College of New York,
City University of New York, New York, USA

Correspondence to: B. Asadieh (basadie00@citymail.cuny.edu)

Abstract. Precipitation events are expected to become substantially more intense under global warming, but few global comparisons of observations and climate model simulations are available to constrain predictions of future changes in precipitation extremes. We present a systematic global-scale comparison of changes in historical (1901–2010) annual-maximum daily precipitation between station observations (compiled in HadEX2) and the suite of global climate models contributing to the fifth phase of the Coupled Model Intercomparison Project (CMIP5). We use both parametric and non-parametric methods to quantify the strength of trends in extreme precipitation in observations and models, taking care to sample them spatially and temporally in comparable ways. We find that both observations and models show generally increasing trends in extreme precipitation since 1901, with the largest changes in the deep tropics. Annual-maximum daily precipitation (Rx1day) has increased faster in the observations than in most of the CMIP5 models. On a global scale, the observational annual-maximum daily precipitation has increased by an average of 5.73 mm over the last 110 years, or 8.5 % in relative terms. This corresponds to an increase of 10 % K^{-1} in global warming since 1901, which is larger than the average of climate models, with 8.3 % K^{-1}. The average rate of increase in extreme precipitation per K of warming in both models and observations is higher than the rate of increase in atmospheric water vapor content per K of warming expected from the Clausius–Clapeyron equation. We expect our findings to help inform assessments of precipitation-related hazards such as flooding, droughts and storms.

1 Introduction

Trends in extreme meteorological events have received considerable attention in recent years due to the numerous extreme events such as hurricanes, droughts and floods observed (Easterling et al., 2000). Changes in global climate and alteration of Earth's hydrological cycle (Allen and Ingram, 2002; Held and Soden, 2006; Wentz et al., 2007) have resulted in increased heavy precipitation with consequent increased surface runoff and flooding risk (Trenberth, 1999, 2011), which is likely to continue in the future (Dankers et al., 2013). Anthropogenic climate change is expected to change the distribution, frequency and intensity of precipitation and result in increased intensity and frequency of floods and droughts, with damaging effects on the environment and society (Dankers et al., 2013; Field, 2012; Min et al., 2011; O'Gorman and Schneider, 2009; Solomon et al., 2007; Trenberth, 2011; Trenberth et al., 2003).

As a result of greenhouse gas (GHG) build-up in the atmosphere, global mean near-surface temperature shows an increasing trend since the beginning of the twentieth century (Angeles et al., 2007; Campbell et al., 2011; Singh, 1997; Solomon et al., 2007; Taylor et al., 2007), with greater increases in mean minimum temperature than in mean maximum temperature (Alexander et al., 2006; Peterson, 2002). The Fifth Assessment Report of the Inter-Governmental Panel on Climate Change (IPCC) indicates that, globally, near-surface air temperature has increased by approximately 0.78 °C (0.72 to 0.85) since 1900, with a greater trend slope in recent decades (Stocker et al., 2013).

As a result of global warming, climate models and satellite observations both indicate that atmospheric water vapor content has increased at a rate of approximately 7 % K^{-1} warming (Allen and Ingram, 2002; Held and Soden, 2006; Trenberth et al., 2005; Wentz et al., 2007), as expected from the Clausius–Clapeyron equation under stable relative humidity (Held and Soden, 2006; Pall et al., 2006). Increasing availability of moisture in the atmosphere can be expected to result in increased intensity of extreme precipitation (Allan and Soden, 2008; Allen and Ingram, 2002; O'Gorman and Schneider, 2009; Trenberth, 2011; Trenberth et al., 2003), with a proportionally greater impact than for mean precipitation (Lambert et al., 2008; Pall et al., 2006). An increase in the frequency and intensity of extreme precipitation has already been identified in observations (Alexander et al., 2006; Min et al., 2011; Solomon et al., 2007; Westra et al., 2013) as well as in simulations of climate models (Kharin et al., 2013; Scoccimarro et al., 2013; Toreti et al., 2013). Climate models also indicate that further increases in extreme precipitation would be expected over the next decades (Kharin et al., 2007, 2013; O'Gorman and Schneider, 2009; Pall et al., 2006; Toreti et al., 2013), while in terms of mean precipitation, moist regions become wetter and dry regions drier (Allan and Soden, 2008; Chou and Neelin, 2004; Wentz et al., 2007; Zhang et al., 2007).

Although climate models generally indicate an increase in precipitation and its extremes, the rate of this increase seems to be underestimated (Allan and Soden, 2008; Allen and Ingram, 2002; Min et al., 2011; O'Gorman and Schneider, 2009; Sillmann et al., 2013; Wan et al., 2013; Wentz et al., 2007; Zhang et al., 2007), which implies that future projections of changes in precipitation extremes may also be underpredicted (Allan and Soden, 2008). This underestimation can be a result of differences in scale between climate model grids and observational data (Chen and Knutson, 2008; Sillmann et al., 2013; Toreti et al., 2013; Wan et al., 2013; Zhang et al., 2011) and/or limitations in moist convection or other parameterizations in the models (O'Gorman and Schneider, 2009; Wilcox and Donner, 2007). Assessments of climate models also reveal that the rate of increase in precipitation extremes varies greatly among models, especially in tropical zones (Kharin et al., 2007; O'Gorman and Schneider, 2009), which makes it especially important to compare modeled trends with those identified in observations. However, few global comparisons of observations and climate model simulations are available to constrain predictions of future changes in precipitation extremes. Of the available global-scale studies, some use older versions of climate models or observations and/or use only one or a few climate models (Allan and Soden, 2008; Min et al., 2011; O'Gorman and Schneider, 2009; Wentz et al., 2007; Zhang et al., 2007). Spatial and temporal differences in data coverage between climate models and observations also challenge comparisons.

In this paper, we present a systematic comparison of changes in annual-maximum daily precipitation in weather station observations (compiled in HadEX2) with 15 models from the suite of global climate models contributing to the latest phase of the Coupled Model Intercomparison Project

(CMIP5) (Taylor et al., 2012), as the largest and most recent set of global climate model runs. Both parametric (linear regression) and non-parametric methods – Mann–Kendall (Appendix A1) as well as Sen's slope estimator (Appendix A2) – are utilized to quantify the strength of trends in extreme precipitation in observations and models, taking care to sample them spatially and temporally in comparable ways. We also calculate the rate of change in the defined extreme precipitation index per K of global warming in both observations and models to investigate the relation between global warming and precipitation extremes. Climate models and observation data sets do not provide the same spatial and temporal coverage for precipitation data, leading to some uncertainties in the comparison of the results. In the present study, precipitation data for years/grids of climate models that do not have corresponding observational data are excluded, resulting in a comparable sampling approach for both data sets.

2   Data and methodology

Precipitation data in the Hadley Centre global land-based gridded climate extremes data set (HadEX2) is based on daily observations from about 11 600 precipitation stations gridded on a $2.54° \times 3.75°$ grid from 1901 to 2010 (Donat et al., 2013). Here, gridded HadEX2 annual maximum 1-day precipitation data (Rx1day) are analyzed as the observation data set. The Rx1day extreme precipitation index is defined as the annual-maximum daily precipitation, in which the maximum 1-day precipitation amount is selected for each year. The same index is also obtained for the climate model simulations. Daily precipitation amounts from simulations with 15 models (overall 19 runs) with complete temporal data coverage have been retrieved from the fifth phase of the Coupled Model Intercomparison Project (CMIP5) (Taylor et al., 2012), as the largest and most recent set of global climate model (GCM) runs. The historical data for projections from 1901 to 2005 and the high radiative forcing path scenario (representative concentration pathway, RCP) RCP8.5 (Moss et al., 2010) for projections from 2006 to 2010 are selected. The aforementioned 15 CMIP5 models, provided by the IRI/LDEO Climate Data Library, are BCC-CSM1-1, CMCC-CM, CMCC-CMS, CNRM-CM5, GFDL-CM3, GFDL-ESM2G, HadGEM2-CC, IPSL-CM5A-LR, IPSL-CM5A-MR, IPSL-CM5B-LR, MIROC5 (three runs), MPI-ESM-LR (three runs), MPI-ESM-MR, MRI-CGCM3 and NorESM1-M.

Climate models produce simulated precipitation fields for all years of a specified time interval, covering all coordinates of the globe thoroughly, even the oceans and polar zones. This is completely different from the spatial and temporal coverage of station observation data sets, such as HadEX2, where usually cover only a certain part of the continents, with missing data for a considerable number of years. This differ-

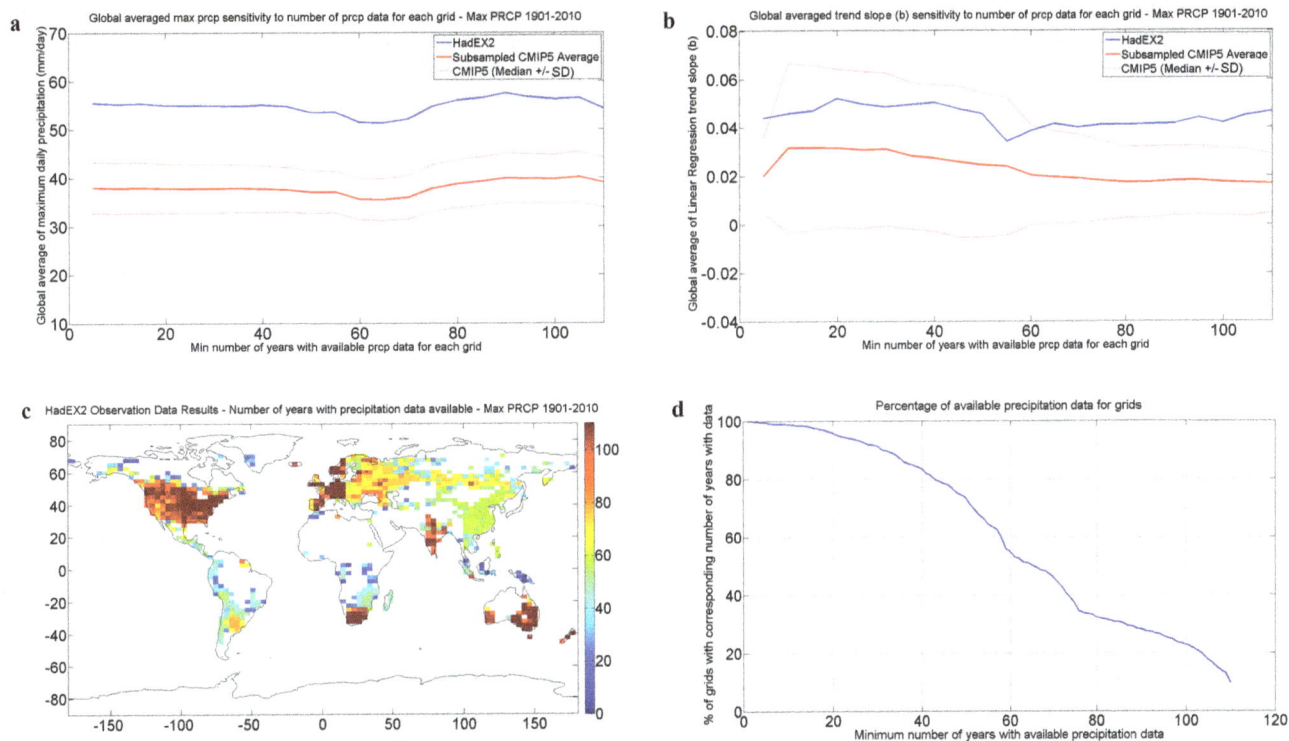

Figure 1. (a) Global averaged extreme precipitation and (b) linear regression trend slope averaged over HadEX2 grid cells with a different minimum number of years with extreme precipitation data available. (c) Map of the number of annual extreme precipitation records in HadEX2 (1901–2010). (d) Minimum number of years with extreme precipitation data available versus the percentage of the grid cells with corresponding coverage.

ence in coverage results in some difficulties in comparison of the two data sets.

As a solution for this issue, a new subsampled data set is created for each of the 19 CMIP5 climate models in which each of the HadEX2 grid cells take the GCM precipitation data of the grid cell in which its geo-referenced coordinates fit. The new data set is created with the same resolution and the same data availability pattern of HadEX2, which means that only data of the grids/years will be assigned to the new data set for which HadEX2 has recorded precipitation data for that year for the corresponding grid cell. The newly created data set is called the subsampled CMIP5 data set.

As stated above, most grid cells in HadEX2 do not have recorded precipitation data for most of the years. A sensitivity analysis of global averaged maximum precipitation and trend slopes to the minimum number of years with precipitation data required for a grid cell to be considered shows that these values do not change drastically (Fig. 1a and b). Selection of only stations with longer records may strengthen the confidence with which trends are quantified, but limits the calculations to smaller spatial coverage of the globe, which is not in line with the scope of this study to evaluate global changes in precipitation. We chose to use the grid cells with at least 30 years of available precipitation data over the last 110 years, which includes more than 90 % of the

766 HadEX2 grid cells that had any Rx1day data (Fig. 1c and d).

Tests for trend detection in time series can be classified as parametric and non-parametric methods. Parametric trend tests require independence and a particular distribution in the data, while non-parametric trend tests require only that the data be independent. The trend slope (b) obtained from the linear regression method, which assumes that the data variability follows a normal distribution, is utilized for trend strength analysis and comparison of the data sets. The relative change in extreme precipitation is defined as the trend slope divided by the average extreme precipitation of the grid cell (b/P). The relative change in extreme precipitation per K of warming is also calculated as an index for the relation between changes in precipitation extremes of each grid cell with global mean near-surface temperature, which indicates the percentage change in extreme precipitation per K global warming. Linear regression is utilized to calculate this parameter, in which global annual mean near-surface temperature obtained from NASA-GISS (Hansen et al., 2010) is selected as the predictor and the natural logarithm of extreme precipitation time series is chosen as the response.

The Z score (Z) obtained from the Mann–Kendall test (Kendall, 1975; Mann, 1945) and the Q median (Q_{med}) from Sen's slope estimator (Sen, 1968) are also applied in order to

Table 1. Statistics of variation of global average extreme precipitation for HadEX2 and the 19 subsampled CMIP5 model runs from 1901 to 2010. The 19 climate model runs give 19 global averages, of which the minimum, maximum, median, mean, and standard deviation are presented.

		Q_{med} (mm day^{-1} yr^{-1})	Z score (–)	Slope of change (b) (mm day^{-1} yr^{-1})	Average of extreme precipitation (P) (mm day^{-1})	Relative change (b/P) (% yr^{-1})	Change per degree warming (% K^{-1})
	Model min.	0.0005	0.0944	0.0023	29.31	0.0118	4.37
	Model max.	0.0648	0.7050	0.1592	48.46	0.3849	28.67
CMIP5 (subsampled)	Model median	0.0218	0.3056	0.0271	37.89	0.0606	7.30
	Model SD	0.0133	0.1555	0.0326	5.08	0.0774	5.16
	Model average	0.0230	0.3330	0.0314	37.85	0.0797	8.43
HadEX2	–	0.0504	0.7242	0.0521	55.03	0.0775	9.99

support the results of linear regression using non-parametric trend detection approaches. It is important to compare the non-parametric trend estimates with those obtained from linear regression, since the extreme precipitation time series need not follow the normal distribution but may instead be better represented by, for example, the generalized extreme value distribution (Katz, 1999; Westra et al., 2013). The trend tests are applied for each grid cell's extreme precipitation time series. The obtained values are averaged globally as well as by continent in order to present the general trend of precipitation extremes in different regions. Continents studied comprise Africa, Asia, Europe, North America, South America and Oceania. The subcontinent of India has results shown separately and is also included in Asia. Results are also averaged by latitude to investigate changes in the tropics versus the northern/southern mid-latitudes.

The statistical significance of the trends, presented in the text as well as the figures, at the 95 % confidence level is based on p values less than 0.05 from the linear regression. The statistical significance of trends estimated from the Mann–Kendall and Sen methods is evaluated differently (Appendix A).

3 Results

Linear regression indicates that 66.2 % of the studied grid cells show a positive trend in annual-maximum daily precipitation during the past 110 years, including 18 % that are statistically significant at the 95 % confidence level. On the other hand, 33.8 % of the studied grids show a negative trend, including only 4 % that are statistically significant at the 95 % confidence level. The results are very similar to those found by Westra et al. (2013) for the same HadEX2 data set (64 %

positive and 36 % negative). Thus, the global record of extreme precipitation shows a meaningful increase over the last century. This increase is expected to continue over the next decades, based on physical arguments and modeling (Kharin et al., 2007, 2013; O'Gorman and Schneider, 2009; Pall et al., 2006; Toreti et al., 2013).

Table 1 presents the statistics of global averaged trend parameters of annual-maximum daily precipitation for HadEX2 and 19 subsampled CMIP5 model runs (from 15 models) from 1901 to 2010. Observation is only one data set; hence, it has one global average for each parameter. The 19 climate model runs give 19 global averages, of which we present the minimum, maximum, median, mean, and standard deviation in Table 1. Figure 2 illustrates the results presented in Table 1 as boxplots of trend parameters and average precipitation for annual-maximum daily precipitation for all 19 subsampled data sets of CMIP5 on global as well as continental scales, showing observations (HadEX2) as blue circles. The boxplots show the minimum, 25th percentile, median, 75th percentile and maximum values obtained from the climate models. As seen in Fig. 2a., the global average of extreme precipitation data shows a higher value than the largest value obtained from the climate models, which indicates that all of the climate models underestimate the annual-maximum daily precipitation. This underestimation can be seen in continental scale averages as well, and is expected given the difference in spatial scale between GCMs and station precipitation gauges.

The mean linear regression slope (b) for HadEX2 observation data globally shows a positive trend of 0.052 mm day^{-1} per year in extreme precipitation over the last 110 years (Table 1). This positive trend is captured by the climate models but is significantly underestimated, since HadEX2 shows a

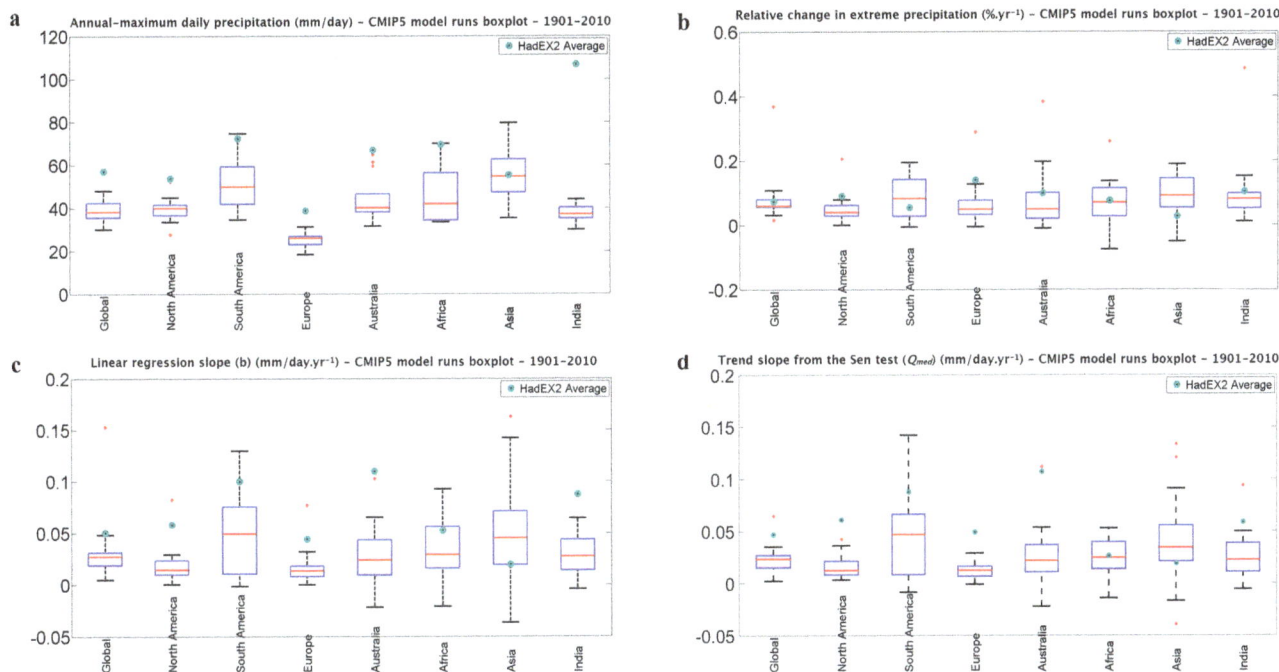

Figure 2. Boxplots of CMIP5 model run averaged results (minimum, 25th percentile, median, 75th percentile and maximum of the 19 model runs) as well as the average of HadEX2 observational data (shown as blue circles) for 1901–2010 extreme precipitation data on global and continental scales – (**a**) annual-maximum daily precipitation (mm day^{-1}), (**b**) relative change in annual-maximum daily precipitation (% yr^{-1}), (**c**) linear regression slope of change in annual-maximum extreme precipitation (mm day^{-1} yr^{-1}), and (**d**) trend slope from the Sen test (Q_{med}) (mm day^{-1} yr^{-1}). The red markers outside the boxes represent model outliers.

greater mean value of b than all but one of the values obtained from CMIP5 models. This underestimation is seen particularly in the continents of America, Europe and Oceania as well as the subcontinent of India. The global average of relative change in precipitation (b/P) for HadEX2 is close to the 75th percentile of the GCMs, which indicates that approximately 75 % of the CMIP5 models have underestimated the relative change in extreme precipitation, but is close to the average value of the CMIP5 models. This substantial difference between the CMIP5 average and median value can be linked to the large and positive skew scatter among the results obtained from the models and the large inter-model standard deviation (Table 1). The observational relative changes in extreme precipitation for North America and Europe are higher than the values obtained from any of the CMIP5 climate models but, for South America, Oceania, Asia and Africa, are lower than the median of the CMIP5 models, suggesting that there may be coherent spatial patterns in the model bias (Fig. 2) analogous to those seen for changes in mean precipitation (Krakauer and Fekete, 2014).

Similar to the linear regression slope (b), Q_{med} from Sen's test shows the direction and magnitude of the trend in a time series, having the advantage of using a non-parametric method for the trend test. The global average of Q_{med} for observations is 0.050 mm day^{-1} per year (Table 1), very close to the average value of b obtained from the linear regression,

which further supports increasing trends in observational annual-maximum daily precipitation. Considering the similar trend magnitudes from parametric and non-parametric methods, similar values for the relative change in annual-maximum daily precipitation are also expected from the two methods. As seen in Fig. 2, the boxplots of the distribution of b and Q_{med} over the climate models show very similar results on global and continental scales (Fig. 2c and d, respectively).

The last column of Table 1 presents a relative change in extreme precipitation per K of global warming (% K^{-1}). On a global scale, the observed annual-maximum daily precipitation has increased by an average of 10 % K^{-1} of global warming since 1901, which is larger than the average of climate models, with 8.3 % K^{-1}. The Clausius–Clapeyron equation under stable relative humidity indicates that atmospheric water vapor content will increase at a rate of approximately 7 % K^{-1} warming (Held and Soden, 2006; Pall et al., 2006). The rates of increase in extreme precipitation per K warming in both models and observations are higher than the rate of increase in atmospheric water vapor content per K warming expected from the Clausius–Clapeyron equation. Observational relative change in extreme precipitation with respect to global warming is also higher than all of the modeled values for North America and Europe, and is higher than the model median for South America, Africa and India, but

a

b

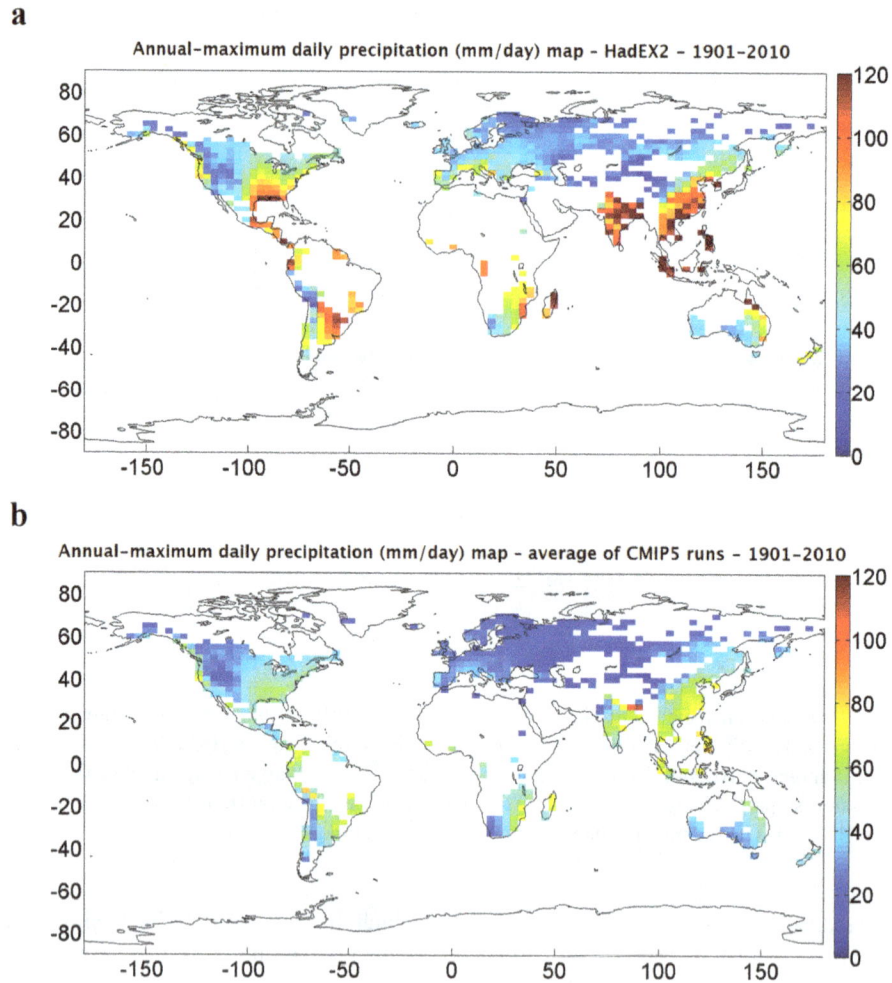

Figure 3. HadEX2 observational data versus CMIP5 averaged results of global extreme precipitation in 1901–2010 – annual-maximum daily precipitation map (mm day^{-1}) for **(a)** HadEX2 and **(b)** the average of CMIP5 model runs.

is lower than the median of the models for Asia and Oceania (Fig. 8a).

Values of the Z score index obtained from the Mann–Kendall method shows the non-parametric confidence level of statistical significance in the identified trends in the data. The expectation might be that observational data would have a lower confidence level in the identified trends due to higher levels of noise in observations compared to climate model simulations. However, Table 1 shows that the global average value of the Z score for HadEX2 is higher than the largest value obtained from the climate models, indicating that the CMIP5 climate models' simulations generally show lower level of confidence in the trends compared to the HadEX2 observations. This interesting finding that the level of internal variability in climate models appears to be too high compared to observations warrants further investigation.

Figure 3 depicts the global maps of the average of annual-maximum daily precipitation (P) for HadEX2 (Fig. 3a) as well as the average of CMIP5 model runs (Fig. 3b). Figure 4

shows the linear regression slope (b) for HadEX2 (Fig. 4a) and the average of CMIP5 model runs (Fig. 4b). Relative change in extreme precipitation (b/P) for HadEX2 as well as the average of CMIP5 model runs are illustrated in Fig. 5a and b, respectively. Stippling in Figs. 4 and 5 means that the grid cell has a significant trend at the 95 % confidence level. In cases of CMIP5 average maps, filled/empty stippling indicates a positive/negative trend on average. While a larger marker size means a larger number of models agreeing on the presented trend, the largest marker size shown indicates only 7 out of 19 model simulations agreeing on the presented trend significance, which also illustrates the discrepancy in the trend significance between the climate models.

Figure 6 shows the average values of extreme precipitation (P), linear regression trend slope (b) and relative change in extreme precipitation (b/P) at each 2.5° latitudinal window (Fig. 6a–c). The figure presents the result of the HadEX2 data set with the average result of CMIP5 data sets as well as their mean ± SD. As seen in Fig. 6a, average extreme precip-

a

b

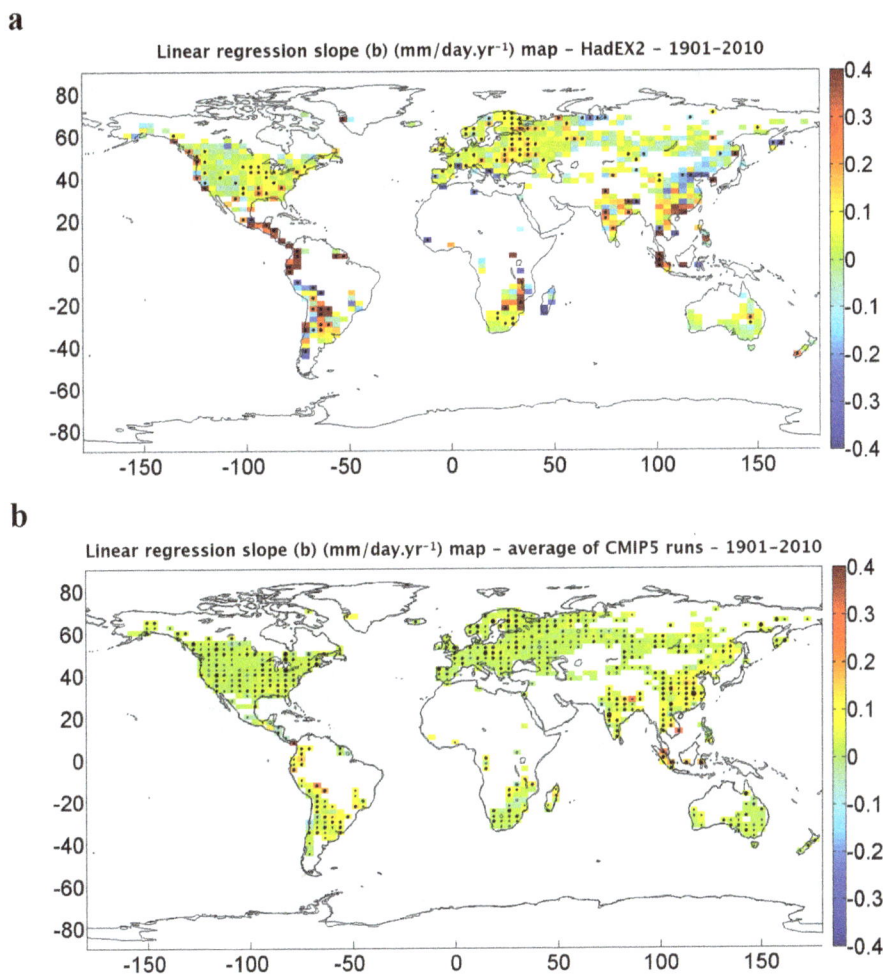

Figure 4. HadEX2 observational data versus CMIP5 averaged results of global extreme precipitation in 1901–2010 – linear regression slope of change in annual-maximum daily precipitation map (mm day^{-1} yr^{-1}) for (**a**) HadEX2 and (**b**) the average of CMIP5 model runs. Stippling indicates significance of the calculated trend at the 95 % confidence level. In cases of CMIP5 average maps, filled/empty stippling indicates a positive/negative trend on average. The larger marker size means a larger number of models agreeing on the presented trend, with the largest one indicating only 7 out of 19 model runs agreeing on the presented trend significance, which also implies the discrepancy in the trend significance between the climate models.

itation observed and simulated in the Northern Hemisphere (NH) is lower than in the Southern Hemisphere (SH), and the underestimation of extreme precipitation by the climate models can also be seen. Figure 6b and c depict the fact that the SH shows larger percentage changes in extreme precipitation than the NH. Tropical zones of the globe show much higher ranges of fluctuations in both observed and simulated extreme precipitation trends compared to mid-latitudes, as well as a larger discrepancy between the observations and simulations (Fig. 6). There is larger uncertainty regarding the results in the tropics, due to fewer numbers of cells with observational data in these regions. The failure of climate models to capture changes in tropical zones has been reported by previous studies as well (Kharin et al., 2007; O'Gorman and Schneider, 2009).

Figure 7 depicts the relative change in extreme precipitation per K of global warming maps for HadEX2 observations (Fig. 7a) and the grid average of CMIP5 model runs (Fig. 7b). Boxplots of CMIP5 model run results as well as HadEX2 observational data (shown as blue circles) for relative change in extreme precipitation per K of global warming on global and continental scales are shown in Fig. 8a. Figure 8b shows the relative change in extreme precipitation per K of global warming at each 2.5° latitudinal window. As seen in Fig. 8b, the Southern Hemisphere shows higher ranges of relative changes in extreme precipitation per K global warming than the Northern Hemisphere. Similar behavior in fluctuations in observational extreme precipitation per K warming can also be seen in Westra et al. (2013) in the HadEX2 data set for 1900–2009, although the aforementioned study

a

Relative change in extreme precipitation (%.yr⁻¹) map – HadEX2 – 1901–2010

b

Relative change in extreme precipitation (%.yr⁻¹) map – average of CMIP5 runs – 1901–2010

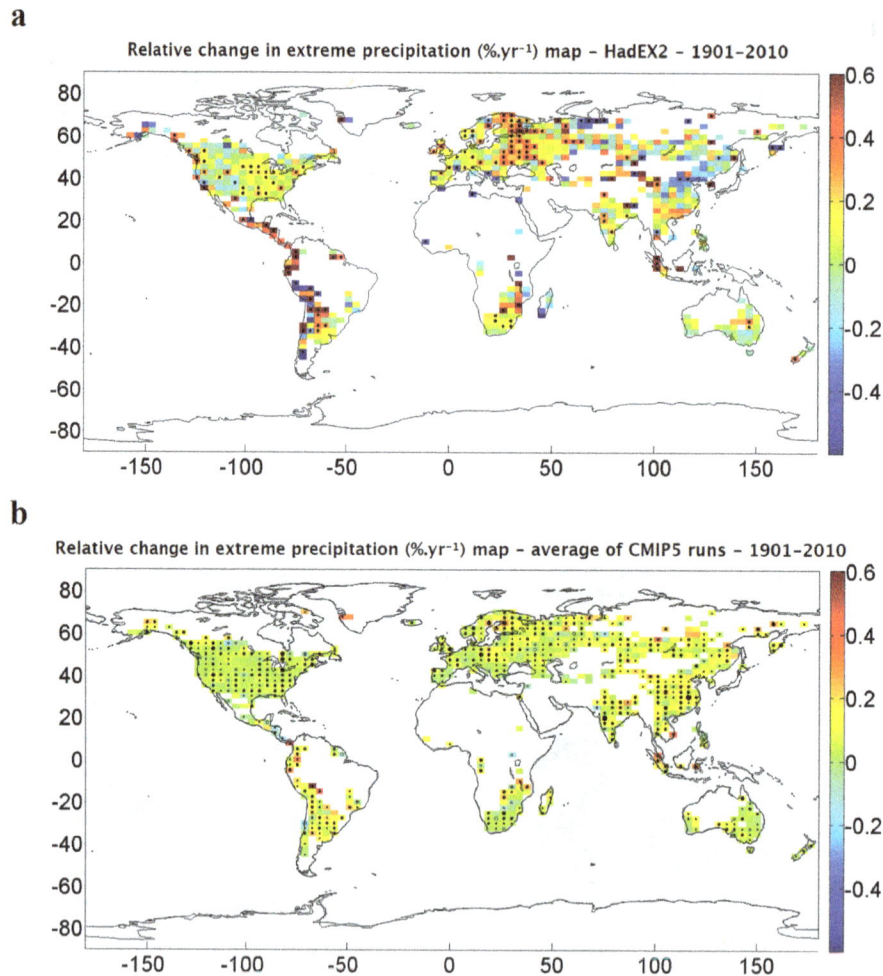

Figure 5. HadEX2 observational data versus CMIP5 averaged results of global extreme precipitation in 1901–2010 – relative change in annual-maximum daily precipitation ($\% \, yr^{-1}$) map for (**a**) HadEX2 and (**b**) the average of CMIP5 model runs. Stippling indicates significance of the calculated trend at the 95 % confidence level. In cases of CMIP5 average maps, filled/empty stippling indicates a positive/negative trend on average. The larger marker size means a larger number of models agreeing on the presented trend, with the largest one indicating only 7 out of 19 model runs agreeing on the presented trend significance.

presents the results as the median of the trends across grid cells instead of the average.

4 Discussion

Results show that both observations and climate models show generally increasing trends in extreme precipitation intensity since 1901. Although the climate models reproduce the direction of observational trends on global and continental scales, the rate of change seems to be underestimated in most models, though the observations fall within the range of inter-model variability for at least the global mean relative change (b/P). Similar discrepancies between observations and climate models have also been reported in earlier studies (Allan and Soden, 2008; Allen and Ingram, 2002; Min et al.,

2011; O'Gorman and Schneider, 2009; Sillmann et al., 2013; Wan et al., 2013; Wentz et al., 2007; Zhang et al., 2007).

The global average of trends from the non-parametric method (Q_{med} from Sen's slope estimator) show similar values to those obtained from the parametric method (b from the linear regression) in observations, confirming the results of the parametric method, which further supports an increasing trend in observational annual-maximum daily precipitation (Table 1 and Fig. 2c and d). Also, the boxplots of b and Q_{med} for climate models are very similar on global and continental scales for different percentiles (Fig. 2c and d, respectively).

Tropical latitudes show higher ranges of fluctuations observed and simulated for extreme precipitation trends compared to mid-latitudes, as well as a larger discrepancy between the observations and simulations (Fig. 6). The high variation of the results for observations as well as models might be due to the small number of data available for those

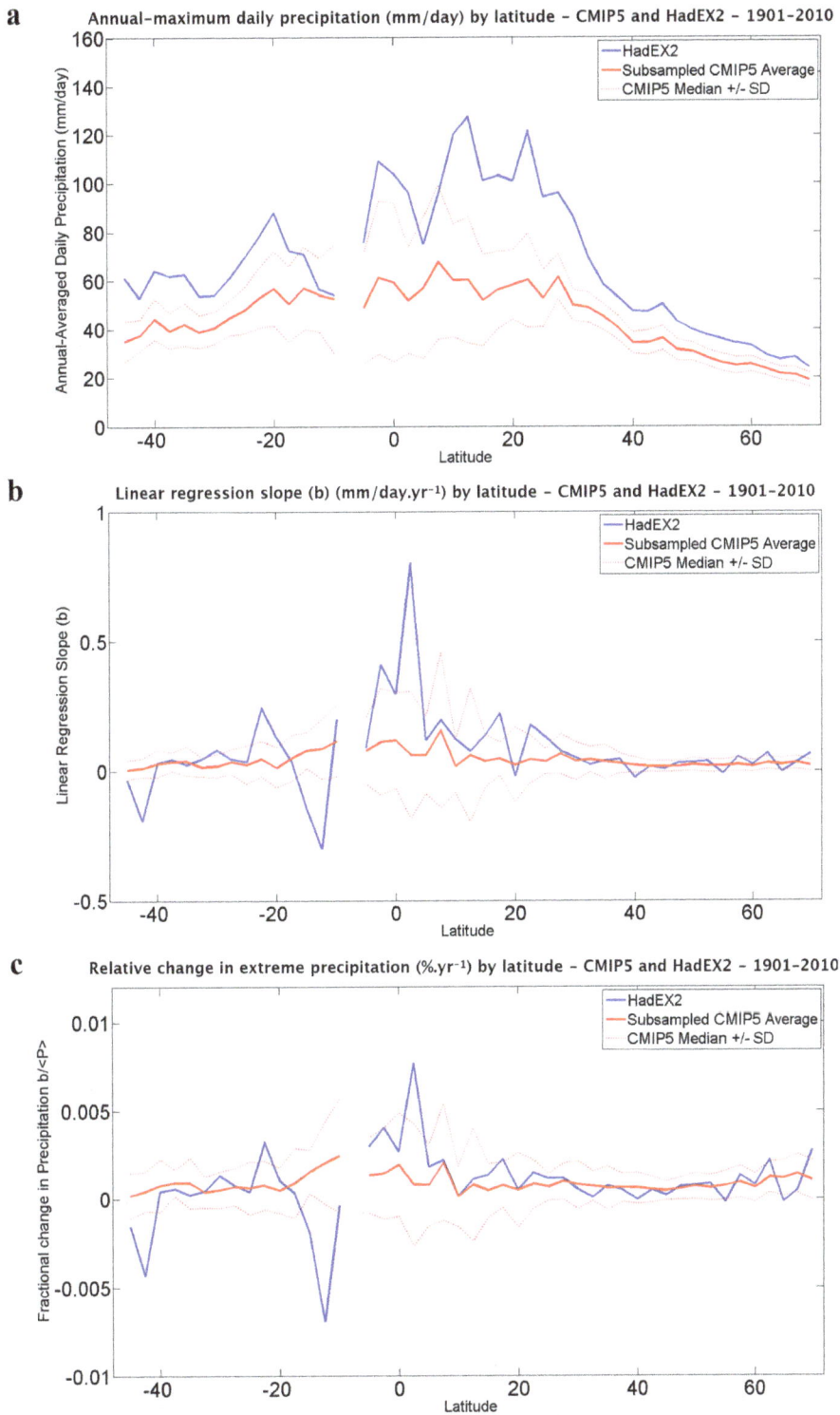

Figure 6. Average parameter value at each 2.5° latitudinal window – (**a**) annual-maximum daily precipitation (mm day^{-1}) for HadEX2 and average CMIP5, (**b**) slope of change in annual-maximum daily extreme precipitation (mm day^{-1} yr^{-1}) for HadEX2 and average CMIP5, and (**c**) relative change in extreme precipitation (% yr^{-1}) for HadEX2 and average CMIP5. Values for the climate models are averages of the 19 runs and the dashed lines are the medians of the models plus/minus the standard deviation of the models. The gap in the tropics indicates the lack of grid cells with more than 30 years of precipitation data available in those zones.

a

Relative Change per global warming (%.K⁻¹) map – HadEX2 – 1901–2010

b

Relative Change per global warming (%.K⁻¹) map – average of CMIP5 runs – 1901–2010

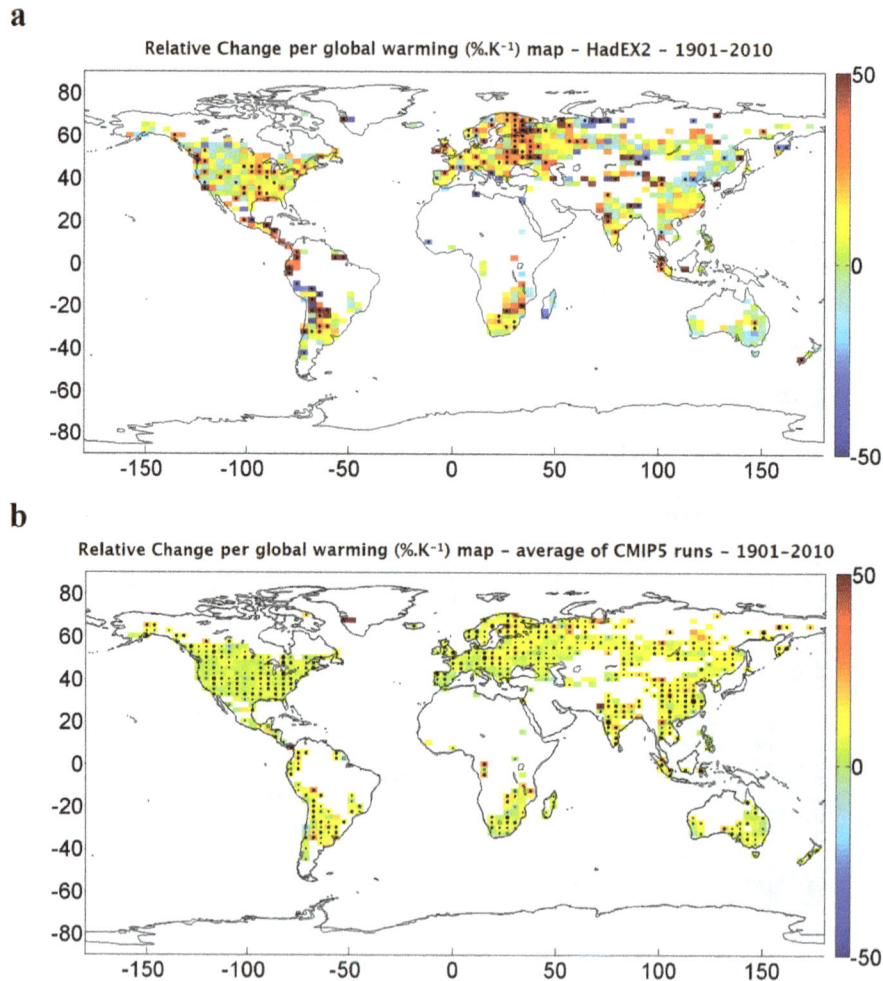

Figure 7. Relative change in extreme precipitation per K of global warming ($\% \, K^{-1}$) maps for 1901–2010 for (**a**) HadEX2 observations and (**b**) the average of CMIP5 model runs.

regions, given that the models are sub-sampled the same way as the available observations. However, the larger discrepancy between observations and models in tropics might also be a result of inaccuracy of the climate models in simulation of tropical climate and of precipitation generated by deep convection, as reported by previous studies (O'Gorman and Schneider, 2009). The continents of North America, Europe and Asia, respectively, contain about 22, 18 and 34 % of total global data grid cells (Fig. 1c). The trend results averaged for the continents of North America and Europe are generally in line with global averaged results. The subcontinent of India generally shows different results from the Asia average, in both observations and models (Figs. 2 and 8a).

The Clausius–Clapeyron equation indicates that atmospheric water vapor content increases at a rate of $7 \, \% \, K^{-1}$ of warming (Held and Soden, 2006; Pall et al., 2006). Although a change in global-mean precipitation with respect to warming does not scale with the Clausius–Clapeyron equation and from energy balance consideration, the rate of increase might

be expected to be around $2 \, \% \, K^{-1}$ (Held and Soden, 2006). The impact of global warming on extreme precipitation is expected to be close to the Clausius–Clapeyron slope (Pall et al., 2006). The results of the present study show that, on average, extreme precipitation since 1901 has increased by $10 \, \% \, K^{-1}$ of global warming in observations and $8.3 \, \% \, K^{-1}$ in climate models over land areas with available station observations (Table 1). North and South America as well as Europe show an even stronger increase in extreme precipitation with respect to global warming (Fig. 8a). These numbers are considerably larger than the $7 \, \% \, K^{-1}$ of the Clausius–Clapeyron equation, which further emphasizes the impact of changes in the Earth's global temperature on precipitation extremes.

As stated earlier, increased availability of moisture in the atmosphere is expected to result in a greater increase in intensity of extreme precipitation than for mean precipitation (Lambert et al., 2008; Pall et al., 2006). Faster change in extreme precipitation than mean precipitation implies a change

a

b

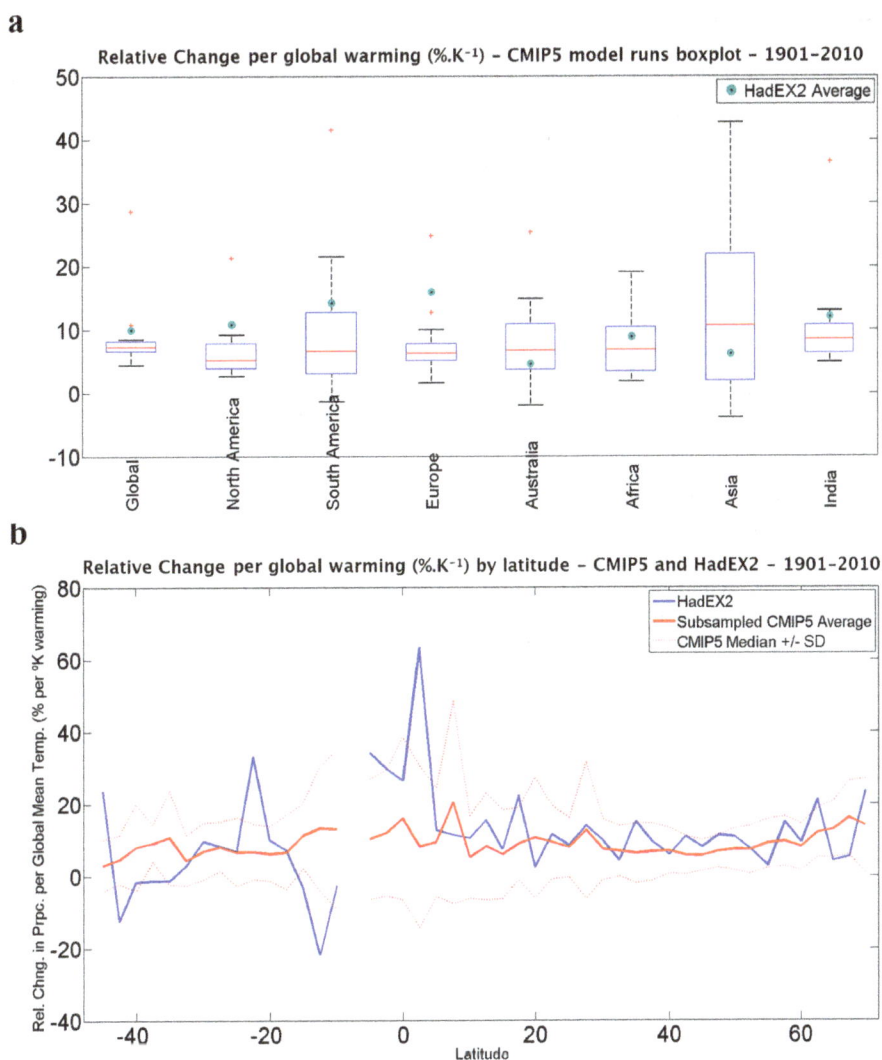

Figure 8. Relative change in extreme precipitation per K of global warming (% K^{-1}) in 1901–2010 – (**a**) boxplots of CMIP5 model run averaged results (minimum, 25th percentile, median, 75th percentile and maximum of the 19 model runs) as well as the average of HadEX2 observational data (shown as blue circles) on global and continental scales and (**b**) average changes at each 2.5° latitudinal window.

in precipitation pattern, where the climate shifts to fewer rainy days and more intense precipitation. This can affect the availability of fresh water resources throughout the year. Such changes in precipitation pattern can affect the capability of reservoirs to capture excessive surface run-off and result in increased flooding events. Failure of the available reservoirs to capture the designed amount of annual surface run-off might also result in a lower total annual amount of water stored in the reservoir, and hence fewer available fresh water resources. The design of newly constructed reservoirs strongly depends on the appropriate prediction of future climate and precipitation extremes, but the available climate models seem to underestimate those for at least some regions. The consequences of changes in both mean and extreme precipitation for water resource system reliability deserve to be investigated further.

5   Conclusions

This study presented a systematic global-scale comparison of changes in historical annual-maximum daily precipitation between the HadEX2 observational records and a CMIP5 ensemble of global climate models. The climate models were spatially and temporally subsampled like the observations, and trends were analyzed for grid cells with at least 30 years of extreme precipitation data over the past 110 years. Both parametric and non-parametric methods were used to quantify the strength of trends in extreme precipitation as well as the confidence level of the identified trends. Results from both parametric and non-parametric tests show that both observations and climate models show generally increasing trends in extreme precipitation since 1901, with larger changes in tropical zones, although annual-maximum daily precipitation has increased faster in the observations than in most of the CMIP5 models. Observations indicate that approximately one-fifth of the global data-covered land area had significant increasing maximum precipitation recorded during the last century. This is more than 4 times larger than the areas with a significant decreasing record, which indicates that the global record of extreme precipitation shows a meaningful increase over the last century. On a global scale, the observational annual-maximum daily precipitation has increased by an average of $5.73 \, \text{mm} \, \text{day}^{-1}$ over the last 110 years, or 8.53% in relative terms. The observational annual-maximum daily precipitation has also increased by an average of $10\% \, \text{K}^{-1}$ of global warming since 1901, which is larger than the average of climate models, with $8.3\% \, \text{K}^{-1}$. The rates of increase in extreme precipitation per K of warming in both models and observations are higher than the rate of increase in atmospheric water vapor content per K of warming expected from the Clausius–Clapeyron equation, which is approximately $7\% \, \text{K}^{-1}$, which highlights the importance of extreme precipitation trends for water resource planning.

Appendix A: Non-parametric trend tests

A1   Mann–Kendall trend test

The MK test is a non-parametric rank-based test (Kendall, 1975; Mann, 1945). The Mann–Kendall test statistic S is calculated as

$$S = \sum_{i=1}^{n-1} \sum_{j=i+1}^{n} \text{sgn}\left(x_j - x_i\right), \tag{A1}$$

where n is the number of data points, x_i and x_j are the data values in time series i and j ($j > i$), respectively, and $\text{sgn}(x_j - x_i)$ is the sign function

$$\text{sgn}\left(x_j - x_i\right) = \begin{cases} +1 & \text{if } x_j - x_i > 0 \\ 0, & \text{if } x_j - x_i = 0 \,. \\ -1 & \text{if } x_j - x_i < 0 \end{cases} \tag{A2}$$

The variance is computed using the equation below:

$$\text{Var}(S) = \frac{n(n-1)(2n+5) - \sum_{i=1}^{m} t_i (t_i - 1)(2t_i + 5)}{18}, \tag{A3}$$

where n is the number of data points, m is the number of tied groups and t_i is the number of ties of extent i. A tied group is a set of sample data having the same value. In cases where the sample size n is greater than 10, the standard normal test statistic Z_S is computed as

$$Z_S = \begin{cases} \frac{S-1}{\sqrt{\text{Var}(S)}}, & \text{if } S > 0 \\ 0, & \text{if } S = 0 \,. \\ \frac{S+1}{\sqrt{\text{Var}(S)}}, & \text{if } S < 0 \end{cases} \tag{A4}$$

The sign of Z_S indicates the trend in the data series, where positive values of Z_S mean an increasing trend, while negative Z_S values show decreasing trends. For the tests at a specific α significance level, if $|Z_S| > Z_{1-\alpha/2}$, the null hypothesis is rejected and the time series has a statistically significant trend.

$Z_{1-\alpha/2}$ is obtained from the standard normal distribution table, where, at the 5 % significance level ($\alpha = 0.05$), the trend is statistically significant if $|Z_S| > 1.96$ and, at the 1 % significance level ($\alpha = 0.01$), the trend is statistically significant if $|Z_S| > 2.576$.

A2   Sen's slope estimator

The non-parametric procedure for estimating the slope of the trend in the sample of N pairs of data was developed by Sen (1968) as

$$Q_i = \frac{x_j - x_k}{j - k} \text{ for } i = 1, \ldots, N, \tag{A5}$$

where x_j and x_k are the data values at times j and k ($j > k$), respectively. N is defined as $\frac{n(n-1)}{2}$, where n is the number of time periods.

If the N values of Q_i are ranked from smallest to largest, the parameter Q_{med} is computed as the median of the Q_i vector. The Q_{med} sign reflects the direction of trend, while its value indicates the magnitude of the trend. To determine whether the median slope is statistically different than zero, the confidence interval of Q_{med} at a specific probability should be computed as follows (Gilbert, 1987; Hollander and Wolfe, 1973):

$$C_\alpha = Z_{1-\alpha/2}\sqrt{\text{Var}(S)}, \tag{A6}$$

where $\text{Var}(S)$ is defined before and $Z_{1-\alpha/2}$ is obtained from the standard normal distribution table. Then, $M_1 = \frac{N - C_\alpha}{2}$ and $M_2 = \frac{N + C_\alpha}{2}$ are computed. The lower and upper limits of the confidence interval, Q_{min} and Q_{max}, are the M_1th largest and the $(M_2 + 1)$th largest of the N ordered slope estimates (Gilbert, 1987). The slope Q_{med} is statistically different than zero if the two limits Q_{min} and Q_{max} have the same sign.

Acknowledgements. The authors gratefully acknowledge support from NOAA under grants NA11SEC4810004 and NA12OAR4310084. All statements made are the views of the authors and not the opinions of the funding agency or the US government.

Edited by: P. Gentine

References

Alexander, L. V., Zhang, X., Peterson, T. C., Caesar, J., Gleason, B., Klein Tank, A. M. G., Haylock, M., Collins, D., Trewin, B., Rahimzadeh, F., Tagipour, A., Rupa Kumar, K., Revadekar, J., Griffiths, G., Vincent, L., Stephenson, D. B., Burn, J., Aguilar, E., Brunet, M., Taylor, M., New, M., Zhai, P., Rusticucci, M., and Vazquez-Aguirre, J. L.: Global observed changes in daily climate extremes of temperature and precipitation, J. Geophys. Res., 111, D05109, doi:10.1029/2005JD006290, 2006.

Allan, R. P. and Soden, B. J.: Atmospheric warming and the amplification of precipitation extremes, Science, 321, 1481–1484, doi:10.1126/science.1160787, 2008.

Allen, M. R. and Ingram, W. J.: Constraints on future changes in climate and the hydrologic cycle, Nature, 419, 224–232, doi:10.1038/nature01092, 2002.

Angeles, M. E., Gonzalez, J. E., Iii, J. E., Hern, L., National, R., and Ridge, O.: Predictions of future climate change in the Caribbean region using global general circulation models, Int. J. Climatol., 27, 555–569, doi:10.1002/joc.1416, 2007.

Campbell, J. D., Taylor, M. A., Stephenson, T. S., Watson, R. A., and Whyte, F. S.: Future climate of the Caribbean from a regional climate model, Int. J. Climatol., 31, 1866–1878, doi:10.1002/joc.2200, 2011.

Chen, C.-T. and Knutson, T.: On the Verification and Comparison of Extreme Rainfall Indices from Climate Models, J. Climate, 21, 1605–1621, doi:10.1175/2007JCLI1494.1, 2008.

Chou, C. and Neelin, J.: Mechanisms of global warming impacts on regional tropical precipitation, J. Climate, 17, 2688–2701, doi:10.1175/1520-0442(2004)017<2688:MOGWIO>2.0.CO;2, 2004.

Dankers, R., Arnell, N. W., Clark, D. B., Falloon, P. D., Fekete, B. M., Gosling, S. N., Heinke, J., Kim, H., Masaki, Y., Satoh, Y., Stacke, T., Wada, Y., and Wisser, D.: First look at changes in flood hazard in the Inter-Sectoral Impact Model Intercomparison Project ensemble, P. Natl. Acad. Sci. USA, 111, 3257–3261, doi:10.1073/pnas.1302078110, 2013.

Donat, M. G., Alexander, L. V., Yang, H., Durre, I., Vose, R., Dunn, R. J. H., Willett, K. M., Aguilar, E., Brunet, M., Caesar, J., Hewitson, B., Jack, C., Klein Tank, A. M. G., Kruger, A. C., Marengo, J., Peterson, T. C., Renom, M., Oria Rojas, C., Rusticucci, M., Salinger, J., Elrayah, A. S., Sekele, S. S., Srivastava, A. K., Trewin, B., Villarroel, C., Vincent, L. A., Zhai, P., Zhang, X., and Kitching, S.: Updated analyses of temperature and precipitation extreme indices since the beginning of the twentieth century: The HadEX2 dataset, J. Geophys. Res.-Atmos., 118, 2098–2118, doi:10.1002/jgrd.50150, 2013.

Easterling, D. R., Evans, J. L., Groisman, P. Y., Karl, T. R., Kunkel, K. E., and Ambenje, P.: Observed Variability and Trends in Extreme Climate Events: A Brief Re-

view, B. Am. Meteorol. Soc., 81, 417–425, doi:10.1175/1520-0477(2000)081<0417:OVATIE>2.3.CO;2, 2000.

Field, C. B.: Managing the risks of extreme events and disasters to advance climate change adaptation: special report of the intergovernmental panel on climate change, Cambridge University Press, Cambridge, UK, 2012.

Gilbert, R. O.: Statistical Methods for Environmental Pollution Monitoring, John Wiley & Sons, New York, USA, 1987.

Hansen, J., Ruedy, R., Sato, M., and Lo, K.: Global surface temperature change, Rev. Geophys., 48, RG4004, doi:10.1029/2010RG000345, 2010.

Held, I. M. and Soden, B. J.: Robust Responses of the Hydrological Cycle to Global Warming, J. Climate, 19, 5686–5699, doi:10.1175/JCLI3990.1, 2006.

Hollander, M. and Wolfe, D. A.: Nonparametric Statistical Methods, J. W. & Sons, New York, USA, 1973.

Katz, R. W.: Extreme value theory for precipitation: sensitivity analysis for climate change, Adv. Water Resour., 23, 133–139, doi:10.1016/S0309-1708(99)00017-2, 1999.

Kendall, M. G.: Rank Correlation Methods, Charless Griffin, London, UK, 1975.

Kharin, V. V., Zwiers, F. W., Zhang, X., and Hegerl, G. C.: Changes in Temperature and Precipitation Extremes in the IPCC Ensemble of Global Coupled Model Simulations, J. Climate, 20, 1419–1444, doi:10.1175/JCLI4066.1, 2007.

Kharin, V. V., Zwiers, F. W., Zhang, X., and Wehner, M.: Changes in temperature and precipitation extremes in the CMIP5 ensemble, Climatic Change, 119, 345–357, doi:10.1007/s10584-013-0705-8, 2013.

Krakauer, N. Y. and Fekete, B. M.: Are climate model simulations useful for forecasting precipitation trends? Hindcast and synthetic-data experiments, Environ. Res. Lett., 9, 024009, doi:10.1088/1748-9326/9/2/024009, 2014.

Lambert, F. H., Stine, A. R., Krakauer, N. Y., and Chiang, J. C. H.: How Much Will Precipitation Increase With Global Warming?, Eos T. Am. Geophys. Un., 89, 193–194, doi:10.1029/2008EO210001, 2008.

Mann, H. B.: Nonparametric tests against trend, Econometrica, 13, 245–259, 1945.

Min, S.-K., Zhang, X., Zwiers, F. W., and Hegerl, G. C.: Human contribution to more-intense precipitation extremes, Nature, 470, 378–81, doi:10.1038/nature09763, 2011.

Moss, R. H., Edmonds, J. A., Hibbard, K. A., Manning, M. R., Rose, S. K., van Vuuren, D. P., Carter, T. R., Emori, S., Kainuma, M., Kram, T., Meehl, G. A., Mitchell, J. F. B., Nakicenovic, N., Riahi, K., Smith, S. J., Stouffer, R. J., Thomson, A. M., Weyant, J. P., and Wilbanks, T. J.: The next generation of scenarios for climate change research and assessment, Nature, 463, 747–756, doi:10.1038/nature08823, 2010.

O'Gorman, P. A. and Schneider, T.: The physical basis for increases in precipitation extremes in simulations of 21st-century climate change, P. Natl. Acad. Sci., 106, 14773–14777, doi:10.1073/pnas.0907610106, 2009.

Pall, P., Allen, M. R., and Stone, D. A.: Testing the Clausius–Clapeyron constraint on changes in extreme precipitation under CO_2 warming, Clim. Dyn., 28, 351–363, doi:10.1007/s00382-006-0180-2, 2006.

Peterson, T. C.: Recent changes in climate extremes in the Caribbean region, J. Geophys. Res., 107, 4601, doi:10.1029/2002JD002251, 2002.

Scoccimarro, E., Gualdi, S., Bellucci, A., Zampieri, M., and Navarra, A.: Heavy Precipitation Events in a Warmer Climate: Results from CMIP5 Models, J. Climate, 26, 7902–7911, doi:10.1175/JCLI-D-12-00850.1, 2013.

Sen, P. K.: Estimates of the regression coefficient based on Kendall's tau, J. Am. Stat. Assoc., 63, 1379–1389, doi:10.1080/01621459.1968.10480934, 1968.

Sillmann, J., Kharin, V. V., Zhang, X., Zwiers, F. W., and Bronaugh, D.: Climate extremes indices in the CMIP5 multimodel ensemble: Part 1. Model evaluation in the present climate, J. Geophys. Res.-Atmos., 118, 1716–1733, doi:10.1002/jgrd.50203, 2013.

Singh, B.: Climate-related global changes in the southern Caribbean: Trinidad and Tobago, Global Planet. Change, 15, 93–111, doi:10.1016/S0921-8181(97)00006-4, 1997.

Solomon, S., Qin, D., Manning, M., Marquis, M., Averyt, K., Tignor, M. M. B., Miller, H. L., and Chen, Z.: Climate Change 2007: The Physical Science Basis, Contribution of Working Group I to the Fourth Assessment Report of the IPCC, Vol. 4, Cambridge University Press, 2007.

Stocker, T. F., Qin, D., Plattner, G.-K., Tignor, M. M. B., Allen, S. K., Boschung, J., Nauels, A., Xia, Y., and Bex, V.: Climate Change 2013: The Physical Science Basis. Working Group I Contribution to the Fifth Assessment Report of the Intergovernmental Panel on Climate Change-Abstract for decision-makers, C/O World Meteorological Organization, Geneva, Switzerland, 2013.

Taylor, K. E., Stouffer, R. J., and Meehl, G. A.: An Overview of CMIP5 and the Experiment Design, B. Am. Meteorol. Soc., 93, 485–498, doi:10.1175/BAMS-D-11-00094.1, 2012.

Taylor, M. A., Centella, A., Charlery, J., Borrajero, I., Bezanilla, A., Campbell, J., and Watson, R.: Glimpses of the future – a briefing from PRECIS Caribbean climate change project, Belmopan, Belize, 2007.

Toreti, A., Naveau, P., Zampieri, M., Schindler, A., Scoccimarro, E., Xoplaki, E., Dijkstra, H. A., Gualdi, S., and Luterbacher, J.: Projections of global changes in precipitation extremes from Coupled Model Intercomparison Project Phase 5 models, Geophys. Res. Lett., 40, 4887–4892, doi:10.1002/grl.50940, 2013.

Trenberth, K. E.: Conceptual Framework for Changes of Extremes of the Hydrological Cycle With Climate Change, Climatic Change, 42, 327–339, doi:10.1023/a:1005488920935, 1999.

Trenberth, K. E.: Changes in precipitation with climate change, Clim. Res., 47, 123–138, doi:10.3354/cr00953, 2011.

Trenberth, K. E., Dai, A., Rasmussen, R. M., and Parsons, D. B.: The Changing Character of Precipitation, B. Am. Meteorol. Soc., 84, 1205–1217, doi:10.1175/BAMS-84-9-1205, 2003.

Trenberth, K. E., Fasullo, J., and Smith, L.: Trends and variability in column-integrated atmospheric water vapor, Clim. Dynam., 24, 741–758, doi:10.1007/s00382-005-0017-4, 2005.

Wan, H., Zhang, X., Zwiers, F. W., and Shiogama, H.: Effect of data coverage on the estimation of mean and variability of precipitation at global and regional scales, J. Geophys. Res.-Atmos., 118, 534–546, doi:10.1002/jgrd.50118, 2013.

Wentz, F. J., Ricciardulli, L., Hilburn, K., and Mears, C.: How much more rain will global warming bring?, Science, 317, 233–235, doi:10.1126/science.1140746, 2007.

Westra, S., Alexander, L. V., and Zwiers, F. W.: Global Increasing Trends in Annual Maximum Daily Precipitation, J. Climate, 26, 3904–3918, doi:10.1175/JCLI-D-12-00502.1, 2013.

Wilcox, E. M. and Donner, L. J.: The Frequency of Extreme Rain Events in Satellite Rain-Rate Estimates and an Atmospheric General Circulation Model, J. Climate, 20, 53–69, doi:10.1175/JCLI3987.1, 2007.

Zhang, X., Zwiers, F. W., Hegerl, G. C., Lambert, F. H., Gillett, N. P., Solomon, S., Stott, P. A., and Nozawa, T.: Detection of human influence on twentieth-century precipitation trends, Nature, 448, 461–465, doi:10.1038/nature06025, 2007.

Zhang, X., Alexander, L., Hegerl, G. C., Jones, P., Tank, A. K., Peterson, T. C., Trewin, B., and Zwiers, F. W.: Indices for monitoring changes in extremes based on daily temperature and precipitation data, Wiley Interdiscip. Rev. Clim. Chang., 2, 851–870, doi:10.1002/wcc.147, 2011.

10

How does bias correction of regional climate model precipitation affect modelled runoff?

J. Teng[1], **N. J. Potter**[1], **F. H. S. Chiew**[1], **L. Zhang**[1], **B. Wang**[1], **J. Vaze**[1], **and J. P. Evans**[2]

[1]CSIRO Land and Water Flagship, Canberra, Australia
[2]Climate Change Research Centre and ARC Centre of Excellence for Climate System Science, University of New South Wales, Sydney, Australia

Correspondence to: J. Teng (jin.teng@csiro.au)

Abstract. Many studies bias correct daily precipitation from climate models to match the observed precipitation statistics, and the bias corrected data are then used for various modelling applications. This paper presents a review of recent methods used to bias correct precipitation from regional climate models (RCMs). The paper then assesses four bias correction methods applied to the weather research and forecasting (WRF) model simulated precipitation, and the follow-on impact on modelled runoff for eight catchments in southeast Australia. Overall, the best results are produced by either quantile mapping or a newly proposed two-state gamma distribution mapping method. However, the differences between the methods are small in the modelling experiments here (and as reported in the literature), mainly due to the substantial corrections required and inconsistent errors over time (non-stationarity). The errors in bias corrected precipitation are typically amplified in modelled runoff. The tested methods cannot overcome limitations of the RCM in simulating precipitation sequence, which affects runoff generation. Results further show that whereas bias correction does not seem to alter change signals in precipitation means, it can introduce additional uncertainty to change signals in high precipitation amounts and, consequently, in runoff. Future climate change impact studies need to take this into account when deciding whether to use raw or bias corrected RCM results. Nevertheless, RCMs will continue to improve and will become increasingly useful for hydrological applications as the bias in RCM simulations reduces.

1 Introduction

Downscaling is a technique commonly used in hydrology when investigating the impact of climate change. It is a way of bridging the gap between low spatial resolution global climate models (GCMs) and the regional-, catchment- or point-scale hydrological models (Fowler et al., 2007). Dynamical downscaling techniques derive regional-scale information by using a high-resolution climate model over a limited area and forcing it with lateral boundary conditions from GCMs or reanalysis products. In brief, it is modelling with a regional climate model, or RCM. With advances in RCMs and the increasing availability of RCM simulations, this type of downscaling is gaining more and more popularity in hydrological impact studies (Dosio et al., 2012; Argüeso et al., 2013; Seaby et al., 2013; Teutschbein and Seibert, 2010; Maraun et al., 2010; Bennett et al., 2012). A drawback, however, is that precipitation simulations from RCMs are "biased": in addition to errors inherited from the driving GCM, there are systematic RCM model errors, due to imperfect conceptualisation and parameterisation, inadequate length and quality of reference data sets, and insufficient spatial resolution (Wilby et al., 2000; Wood et al., 2004; Piani et al., 2010b; Chen et al., 2011a; Christensen et al., 2008; Teutschbein and Seibert, 2010). Various "bias correction" methods have been developed in an attempt to minimise these errors (Boe et al., 2007; Piani et al., 2010a; Johnson and Sharma, 2012; Schmidli et al., 2006; Lenderink et al., 2007).

There have been extensive discussions in the climate change literature on the definition of "bias", and some have recommended limiting its use to refer to the correspondence between a mean forecast and the mean of the observations averaged over a certain area and time (Ehret et al., 2012); others have tried to distinguish model biases from model shortcomings and model errors (Teutschbein and Seibert, 2013). For clarity, in this paper we define bias as the systematic distortion of a statistical outcome from the expected value, and we use "error" or "difference" to refer to the discrepancy between a model output and observations.

Many studies have compared and evaluated different bias correction methods; Table 1 summarises some recent ones and their main conclusions. Most of these studies investigated the impact of bias correction on precipitation and temperature (see column 6, Table 1), yet only Teutschbein and Seibert (2012) and Chen et al. (2013) tested the effect of bias correction on the outputs of hydrological models. Nearly all studies agree that distribution-based bias correction methods (both parametric and non-parametric) give the best performance in terms of reproducing the observed climate, whereas means-based methods, in particular linear scaling (LS), are almost always ranked as the least-skilled bias correction method.

Building on the knowledge gained from previous comparison studies, we have assessed in more detail the best performing bias correction technique – distribution mapping – and compared its performance in several forms against the linear scaling (LS) method as a benchmark (other names of the distribution mapping technique include quantile matching, distribution transformation, probability mapping, and histogram equalisation). Our main interest is to examine the effect of bias correction on modelled runoff. The bias correction methods were applied on modelled precipitation, as it is the most critical and difficult-to-model variable in hydrological studies (Vaze et al., 2011), and evaluated on both precipitation and runoff using a cross-validation method. The raw and bias-corrected precipitation data were used to drive the hydrological models. The key precipitation and runoff characteristics were compared to those of observations to investigate how bias correction affects RCM precipitation, and its follow-on impact on runoff propagating through hydrological models.

Previous studies have shown mixed results in ranking the different types of distribution mapping methods, suggesting that there may be only marginal differences between the methods. For example, some studies have shown that distribution mapping based on theoretical distributions outperforms other bias correction methods (Teutschbein and Seibert, 2013, 2012; Yang et al., 2010). Others have shown that theoretical distribution mapping performs similar to, or only marginally better than, empirical quantile mapping (Berg et al., 2012; Chen et al., 2013). Some studies, on the other hand, show that empirical quantile mapping demonstrates higher skill than theoretical distribution mapping in systematically correcting RCM precipitation (Gudmundsson et al., 2012; Gutjahr and Heinemann, 2013; Li et al., 2010; Lafon, 2013). In view of the discrepancy in the literature, we compared three distribution mapping techniques, each with

increasing degree of dependency on the calibration data, in order to evaluate the methods based on both accuracy and robustness.

Berg et al. (2012) found that 30 years of calibration data are required to produce reasonable accuracy for the estimate of precipitation variance. Due to the difficult parameterisation and expensive computational costs associated with RCMs, this requirement is not easily met in impact studies. In the main modelling experiments, we chose two 8-year-long periods of RCM data (16 years split in half), with significant climatic difference between them, to examine whether the bias correction method derived from one period works for another period, and if not, what causes it to fail. We then validated the generality of our conclusion using two 30-year-long RCM precipitation data sets (60 years split in half).

Chen et al. (2013) concluded that bias correction performance is location dependent and that virtually no bias correction method succeeds in catchments having low coherence between RCM simulated and observed precipitation sequences. We challenged (and confirmed) this conclusion by evaluating the precipitation sequence simulated by the RCM and quantifying the effect of precipitation sequence on modelled runoff.

The impact of bias correction on the change signals (one period vs. another) in both precipitation and runoff was also explored. There were two possible outcomes from this investigation: if bias correction does not alter the change signals in the hydro-climatic projections, then the use of bias correction should be considered either unnecessary or safe to use, depending on the circumstance (Muerth et al., 2013). If it does alter the change signal, bias correction could be reducing or increasing errors in the change signals, either way, it introduces an extra level of uncertainty in the modelling chain.

This paper contributes to the present lively discussion on whether bias correction methods should be applied to global and regional climate model data, a conversation initiated by Christensen et al. (2008), stimulated by Ehret et al. (2012), and continued by more recent studies such as Muerth et al. (2013) and Teutschbein and Seibert (2013).

2 Study area and data

2.1 Study area

The study area was located in the southern Murray–Darling Basin, Australia (Fig. 1). Beginning in the mid-1990s, this area experienced a prolonged drought, a so-called "Millennium Drought", for 10–15 years (Chiew et al., 2010). While the mean annual rainfall over the Millennium Drought was 10–20 % below the long-term mean, in some places the mean annual runoff declined by over 50 %, a reduction unprecedented in historical records (Potter and Chiew, 2011). Eight catchments from the Loddon, Campaspe and Goulburn River

Table 1. Recent studies comparing different RCM bias correction methods.

Study	Study area	Spatial resolution	Validation period	Number of bias correction methods assessed on precipitation	Variable(s) assessed	Statistics evaluated	Metric(s) used	Conclusion
Themeßl et al. (2011)	Domain covering the Alpine region including Austria	Modelled at 10 km resolution. Validated at station scale	11 years of data (1981–1990, 1999) were used in a 11-fold "leave one out" cross-validation	Seven	Precipitation modelled by RCM MM5 forced with the ERA-40 reanalysis data	Median, variability and indicators for extremes	Bias	Quantile mapping shows the best performance, particularly at high quantiles.
Berg et al. (2012)	Domain covering the entirety of Germany and its near surrounding areas	Modelled at 7 km resolution. Validated against 1 km gridded observations	30 years of data (1971–2000). Two realisations were used for calibration and validation respectively	Three	Precipitation, temperature modelled by RCM COSMO-CLM driven by a GCM (ECHAM5-MPIOM)	Mean and variance of temperature and precipitation	Bias	Histogram equalisation (HE) method corrects not only means but also higher moments, but approximations of the transfer function are necessary when applying to new data. About a 30-year-long calibration period is required for a reasonable approximation.
Teutschbein and Seibert (2012, 2013)	Five catchments in Sweden	Modelled at 25 km resolution. Validated at catchment scale (taken from one grid cell)	40 years of data (1961–1990) were used for calibration and validation in the first study and split into warm/cold years and dry/wet years for cross-validation in the second study	Six	Precipitation, temperature modelled by 11 RCMs driven by different GCMs. Streamflow simulated by hydrological model HBV	Mean, standard deviation, 10th and 90th percentile daily temperature during summer and winter. Mean, standard deviation, coefficient of variation, 90th percentile, probability of wet days, and average intensity of wet days during summer and winter. Mean monthly streamflow, spring flood peak, autumn flood peak, total flows, and annual 15 day low flows	Mean absolute error (MAE) on temperature and precipitation CDFs	Distribution mapping performs the best for both climate projections and hydrological impact qualifications. It performs especially well in terms of the simulation of hydrological extremes. It also shows the best transferability to potentially changed climate conditions.
Gudmundsson et al. (2012)	Domain covering Norway and Nordic Arctic	Modelled at 25 km resolution. Validated at station scale	41 years of data (1960–2000) split into 10 subsamples for a 10-fold "leave one out" cross-validation	11	Precipitation modelled by RCM HIRHAM forced with the EAR40 reanalysis data	Precipitation at 0.1, 0.2,... 1.0 percentile	Mean absolute errors (MAEs) at equally spaced probability intervals	Nonparametric methods perform the best in reducing systematic errors, followed by parametric transformations with three or more free parameters, with the distribution derived transformations ranked the lowest.
Lafon et al. (2013)	Seven catchments in Great Britain	Modelled at 25 km resolution. Validated at catchment scale	40 years of data (1961–2000) split into moving window of 10-year subsamples for a 31-fold "leave one out" cross-validation	Four	Precipitation modelled by RCM HadRM3-PPE-UK driven by a GCM (HadCM3)	Mean, standard deviation, coefficient of variation, skewness, kurtosis	Average of the relative differences (ARD)	If both precipitation data sets (modelled and observed) can be approximated by a gamma distribution, the gamma-based quantile mapping method offers the best combination of accuracy and robustness. Otherwise, the nonlinear method is more effective at reducing the bias. The empirical quantile mapping method can be highly accurate, but results are very sensitive to the choice of calibration time period.
Chen et al. (2013)	10 catchments in North America	Modelled at 50 km resolution. Validated at catchment scale	20 years of data (1981–2000) split into odd years and even years for cross-validation	Six	Precipitation modelled by four RCMs (CRCM, HRM3, RCM3 and WRFG) driven by NCEP reanalysis data. Flow discharge simulated using the hydrological model HSAMI	Mean, standard deviation and 95th percentile wet-day precipitation. Mean daily discharge, the mean of 95th percentile spring high flow, and the mean of 5th percentile summer low flow	Absolute relative error (ARE) on precipitation and discharge Nash–Sutcliffe model efficiency coefficient (NSE), root mean square error (RMSE), and transformed root mean square error (TRMSE) for daily discharge	The performance of bias correction is location dependent. The distribution-based methods are consistently better than the mean-based methods for both precipitation projections and hydrological simulations.
Gutjahr and Heinemann (2013)	A German state and its surrounding areas	Modelled at 4.5 km resolution. Validated at station scale	10 years of data (1991–2000), with each year used in a "leave-one-out" cross-validation resulting in 81 combinations	Three distribution-based methods	Precipitation, temperature modelled by RCM COSMO-CLM driven by a GCM (ECHAM5)	Precipitation at 0.1, 0.2,... 1.0 percentile	Mean absolute errors (MAEs) at equally spaced probability intervals	The empirical method outperforms both parametric alternatives.

basins, with areas from 250 to 1033 km^2, were selected for this study. The catchments were mostly unregulated, with continuous climate and streamflow measurements available for 1985–2000, as such the assessment period was chosen. An 8-year period unaffected by the drought (1985–1992) was used as the calibration period, and another 8-year period strongly affected by the drought (1993–2000) was used

as the validation period. Subsequently, they were switched for cross-validation. The observations and RCM simulations are aggregated to each catchment and compared at this level.

2.1.1 Observations

Observed daily precipitation data were derived from 0.05° (∼ 5 km) gridded climate surfaces and averaged over each

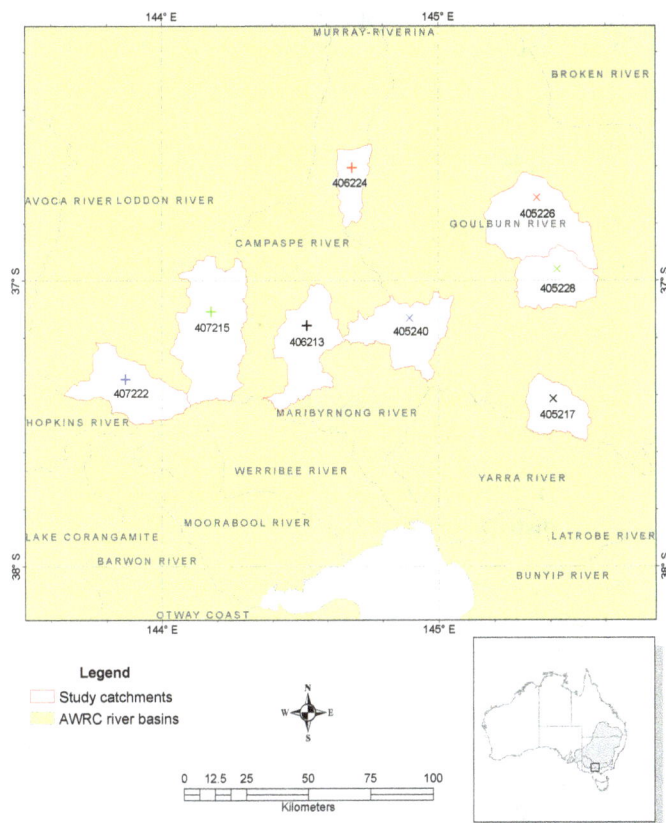

Figure 1. Map showing the eight study catchments (white with red outline, large map), and their locations within the major river basins devised by the Australian Water Resources Council (AWRC). The unique symbols in each catchment identify the catchments and will be used in later figures.

catchment. The source of this data set was the SILO Data Drill (http://www.longpaddock.qld.gov.au/silo) of the Department of Science, Information Technology, Innovation and the Arts, Queensland, Australia (Jeffrey et al., 2001). The SILO gridded climate data sets provide surfaces of daily rainfall and other climate data interpolated from high quality point measurements provided by the Australian Bureau of Meteorology. The daily potential evapotranspiration (PET) sequences used in the hydrological modelling were calculated from SILO climate variables using Morton's wet environment algorithms (Chiew and McMahon, 1991). Measured daily streamflow data were sourced from a previous study (Vaze et al., 2010) and used to calibrate the hydrological models.

2.1.2 RCM data

Most of the analysis in this study was carried out using daily precipitation series for the period 1985–2000, which were simulated by Evans and McCabe (2010) using the weather research and forecasting (WRF) model. Another 60-year-long (1950–2009) WRF precipitation data set (Evans et al.,

2014) was used to validate the conclusion reached by using the shorter data set. For both data sets, WRF was implemented on a 10 km grid using lateral boundary conditions taken from the National Centers for Environmental Prediction (NCEP)/National Center for Atmospheric Research (NCAR) reanalysis data set (Kalnay et al., 1996, see http: //www.cdc.noaa.gov/cdc/reanalysis). The WRF simulations have been found capable of capturing the drought experienced over the study area in another study (Evans and McCabe, 2010). The daily precipitation series for each catchment were aggregated from the WRF simulation by averaging all the grid cells over the catchment.

3 Method

3.1 Bias correction methods

In this study, daily precipitation was the main variable subjected to bias correction. Typically, bias correction methods aim to correct the mean, variance and/or distribution of the modelled precipitation by using a function h:

$$\hat{p}_{\mathrm{obs}} = h(p_{\mathrm{mod}}) \tag{1}$$

so that the transformed precipitation matches the observed data more closely than the modelled precipitation.

3.1.1 Linear scaling (LS)

The simplest choice for h is probably a linear transformation

$$\hat{p}_{\mathrm{obs}} = a\, p_{\mathrm{mod}}, \tag{2}$$

where a is a free parameter that is subject to calibration. This simple form of bias correction is widely used to adjust precipitation from GCMs, RCMs and statistical downscaling methods (Maraun et al., 2010; Teng et al., 2012a). It can efficiently correct the means but does not account for the higher moments. In this study, this method served as the benchmark as LS has been identified in various studies as the least skilful bias correction method (Gudmundsson et al., 2012; Lafon, 2013; Chen et al., 2013; Teutschbein and Seibert, 2012). The LS parameter a was optimised for each season: DJF (December–February), MAM (March–May), JJA (June–August) and SON (September–November) to account for precipitation seasonality. Similarly, seasonal optimisation was also applied for all the other bias correction methods used in this study.

3.1.2 Distribution mapping using the gamma distribution (DMG)

The relation in Eq. (1) can also be modelled so that the distribution of the modelled precipitation matches that of the observations:

$$\hat{p}_{\mathrm{obs}} = F_{\mathrm{obs}}^{-1}(F_{\mathrm{mod}}(p_{\mathrm{mod}})), \tag{3}$$

405240 RCM JJA (1985–1992)

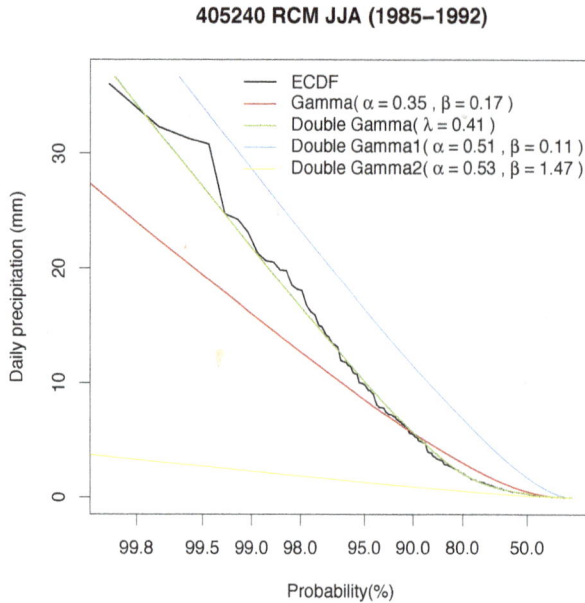

Figure 2. CDF plot comparing a gamma distribution (red) and a double gamma distribution (green), which consists of two gamma distributions (blue and yellow) fitted to the same precipitation data for one study catchment. The empirical distribution is shown in black.

where F_{mod} is the cumulative distribution function (CDF) of P_{mod} and F_{obs}^{-1} is the inverse CDF corresponding to P_{obs}. These CDFs can either be theoretical distributions fitted to the data, or empirical distributions estimated by sorting the data. The gamma distribution with shape parameter α and rate parameter β (Eq. 4) is often used to represent non-zero precipitation amounts (Piani et al., 2010a; Lafon, 2013), as it has the ability to approximate the positively skewed distributions (Yang et al., 2010). The probability density function (PDF) for a gamma random variable is given by:

$$f(p) = \frac{\beta^\alpha p^{\alpha-1} e^{-\beta p}}{\Gamma(\alpha)}, \tag{4}$$

where $\Gamma(\alpha)$ is the gamma function evaluated at α.

When estimating parameters for the gamma distribution, we used the method of maximum-likelihood estimation as it is more accurate (i.e. the standard error of the estimates is lower) compared to the method of moments or least-squares estimation (e.g. Piani et al., 2010a).

Given an occurrence of non-zero precipitation amount $p_i > 0$ for $i = 1, \ldots, n$, the log-likelihood function of the gamma distribution can be written as:

$$l(\alpha, \beta) = \sum_{i=1}^{n} \log f(p_i; \alpha, \beta). \tag{5}$$

The maximum-likelihood estimates for α and β are chosen to maximise this log-likelihood function. To account for dry days, we define the PDF for zero and non-zero precipitation

days $f_0(p)$ as a mixed distribution with an atom of probability at $p = 0$ and a gamma distribution for $p > 0$ so that:

$$f_0(p) = \begin{cases} q_0, & p = 0 \\ \frac{(1-q_0)\beta^\alpha p^{\alpha-1} e^{-\beta p}}{\Gamma(\alpha)}, & p > 0. \end{cases} \tag{6}$$

The maximum likelihood estimate of q_0 depends only on the relative number of zero-precipitation days (n_0):

$$q_0 = n_0/n. \tag{7}$$

The shape and rate parameters α and β are calculated on the non-zero precipitation amounts.

3.1.3 Distribution mapping using a double gamma distribution (DM2G)

Daily precipitation distributions are typically heavily skewed towards high-intensity values. As a result, when fitting a single gamma distribution, the distribution parameters will be dictated by the most frequently occurring values, but may then not accurately represent the extremes. To capture normal precipitation values as well as extremes, different approaches have been tried, but the most common is to divide the precipitation distribution into segments and fit separate distributions to each segment (Yang et al., 2010; Grillakis et al., 2013; Gutjahr and Heinemann, 2013; Smith et al., 2014). Instead of introducing arbitrary cut-offs, we propose what can be interpreted as a two-state distribution. It is a mix of two gamma distributions which can model non-zero precipitation amounts:

$$f(p) = \lambda \frac{\beta_1^{\alpha_1} p^{\alpha_1-1} e^{-\beta_1 p}}{\Gamma(\alpha_1)} + (1-\lambda) \frac{\beta_2^{\alpha_2} p^{\alpha_2-1} e^{-\beta_2 p}}{\Gamma(\alpha_2)} \tag{8}$$

with $0 < \lambda < 1$. The parameter λ is the relative occurrence of the states, and, fitted correctly, the two gamma distributions represent rainfall occurring in high and low rainfall states. The advantage of this approach compared to segmenting the distribution is that all parameters can be estimated simultaneously using maximum-likelihood estimation. Thus, six parameters – q_0, α_1, β_1, α_2, β_2 and λ – were estimated from observations and from the RCM output for the calibration period; they were then used to correct the RCM output for the validation period.

The Kolmogorov–Smirnov (KS) test (Chakravarti and Laha, 1967) performed on both observations and RCM simulations confirmed that the double gamma distribution gives better fittings compared to the gamma distribution (a table of KS test results is provided as the Supplement). Figure 2 compares the empirical distribution, gamma and double gamma distribution for one catchment. A significant improvement in fit is achieved by the double gamma distribution compared to the gamma distribution, especially at the high end.

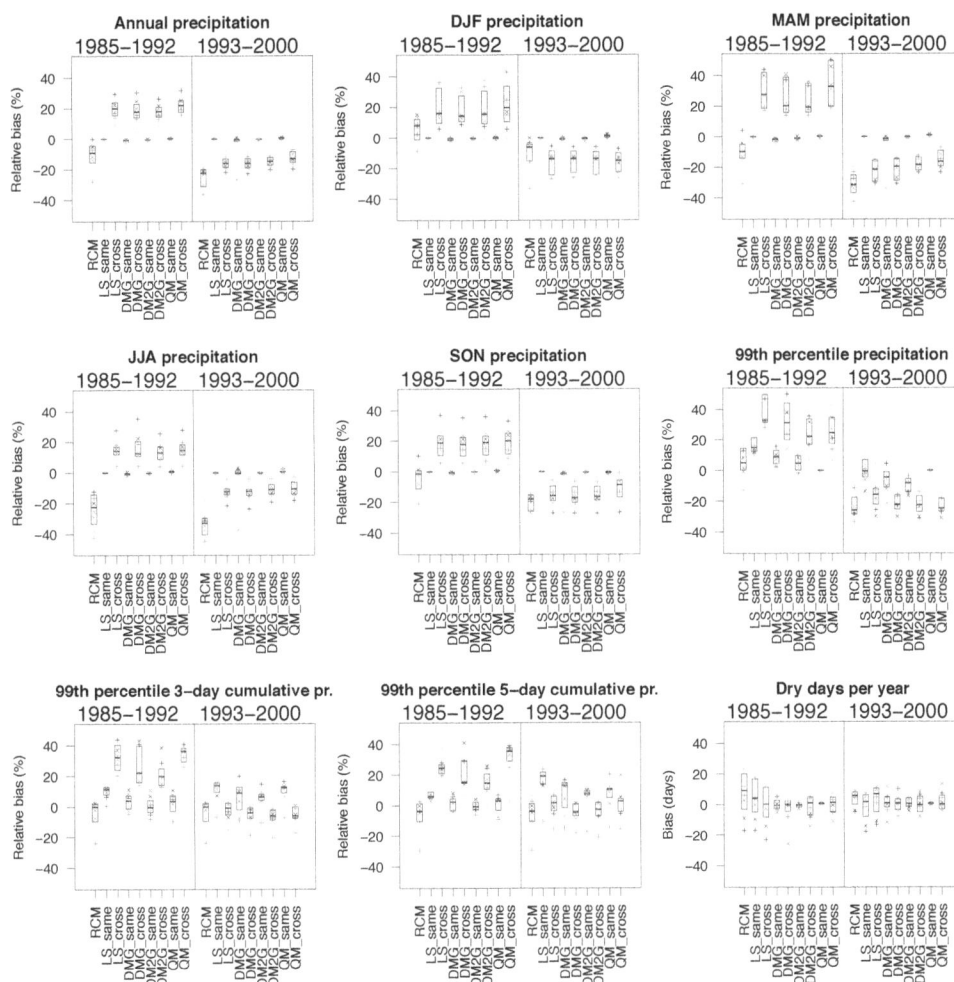

Figure 3. Relative bias of precipitation characteristics, expressed as percentage differences relative to observations between raw RCM and bias corrected RCM precipitation. Each panel displays a different characteristic (title at top of panel) and the percentages were calculated after applying four different bias correction methods (key at bottom) to eight catchments over two periods (1985–1992, left; 1993–2000, right). The bias correction methods were LS, DMG, DM2G and QM (see text), the "_same" suffix denotes calibration and the "_cross" refers to cross-validation. Boxes indicate interquartile range; markers indicate the numbers from each catchment (markers are constrained to the edges of the plotting area if the values exceed the range of plotting); the symbol for each catchment can be found in Fig. 1.

3.1.4 Empirical quantile mapping (QM)

Apart from using theoretical distributions, the empirical CDF is also commonly used to solve Eq. (3) (Themeßl et al., 2011; Gudmundsson et al., 2012; Boe et al., 2007; Bennett et al., 2014). Here the empirical CDFs of observed and modelled precipitation were estimated using empirical percentiles. Values in between the percentiles were approximated using linear interpolation. In cases where new RCM values (such as from the validation period) were larger than the calibration values used to estimate the empirical CDF, a linear regression fit on the last five data points was used to extrapolate beyond the range of observations and allow for possible "new extremes".

3.2 Hydrological modelling (HM)

Two lumped conceptual daily rainfall–runoff models – GR4J (Perrin et al., 2003) and Sacramento (Burnash et al., 1973) – were used to model runoff. The model versions were very similar to those described in foregoing references and in Vaze et al. (2010). Both models have interconnected soil moisture stores and algorithms that mimic the hydrological processes of water moving into and out of soil moisture stores. The choice of models did not have a large effect on the conclusions of this study because the errors associated with hydrological models are relatively small compared to errors in GCM/downscaling (Teng et al., 2012b; Chen et al., 2011b). In this study, 4 and 14 parameters were calibrated for GR4J and Sacramento respectively. The models were calibrated against observations for the two periods separately, with the

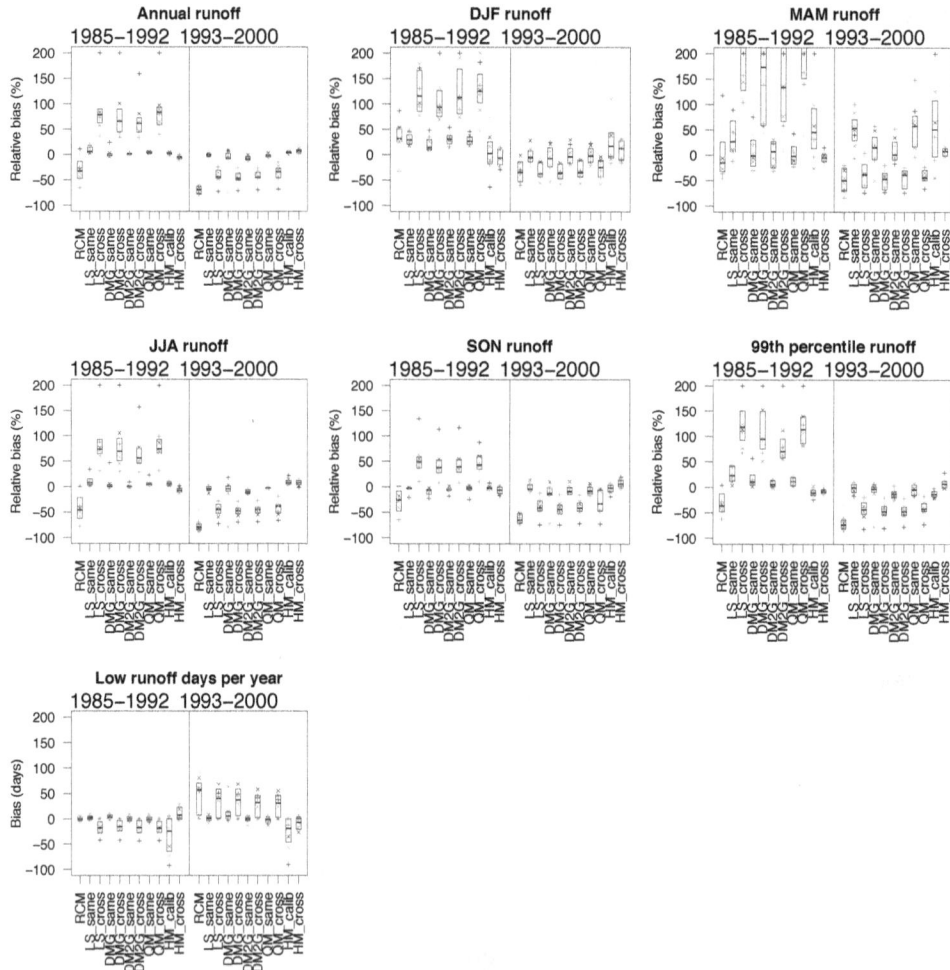

Figure 4. Relative bias in runoff characteristics derived from precipitation-driven hydrological model GR4J. Values are percentage differences, relative to runoff modelled from observed precipitation, when GR4J was driven by raw RCM precipitation and bias corrected RCM precipitation. Same layout as Fig. 3, with additional HM_calib and HM_cross, which represent calibration errors and cross-validation errors from GR4J alone.

model parameters optimised to maximise the NSE-bias objective function; this function is a weighted combination of the Nash–Sutcliffe efficiency (Nash and Sutcliffe, 1970) and a logarithmic function of bias in the modelled mean annual streamflow (Viney et al., 2009). The models were run at a daily time step. To estimate the impact of bias correction on runoff, the models were driven by WRF precipitation before and after bias correction using the optimised parameters derived from the calibrations described above. The same PET data set calculated using observed climate variables was used throughout the hydrological modelling. By keeping PET the same, the possible impact of PET and the correlation between RCM precipitation and PET was not considered in this study to isolate the impact of precipitation.

3.3 Evaluating performance

Comparison of the bias correction methods was based on a split-sample cross-validation approach. The 16 years of data (1985–2000) were split into two periods of 8 years each (1985–1992 and 1993–2000). The bias correction methods were trained using one period and tested against the same period ("same") as well as against the other period ("cross"), and vice versa. Similarly, the hydrological models were calibrated using one period and the parameters were used in the "cross" experiments, treating the validation period as though there were no other information except climate from RCM, like a future period. Compared to studies that have used "odd-year/even-year" or "leave-one-out" validation methods, the design of this experiment puts the bias correction methods to a stricter test (contrasting between wet and dry) so that the impact of different climatic conditions can be clearly identified.

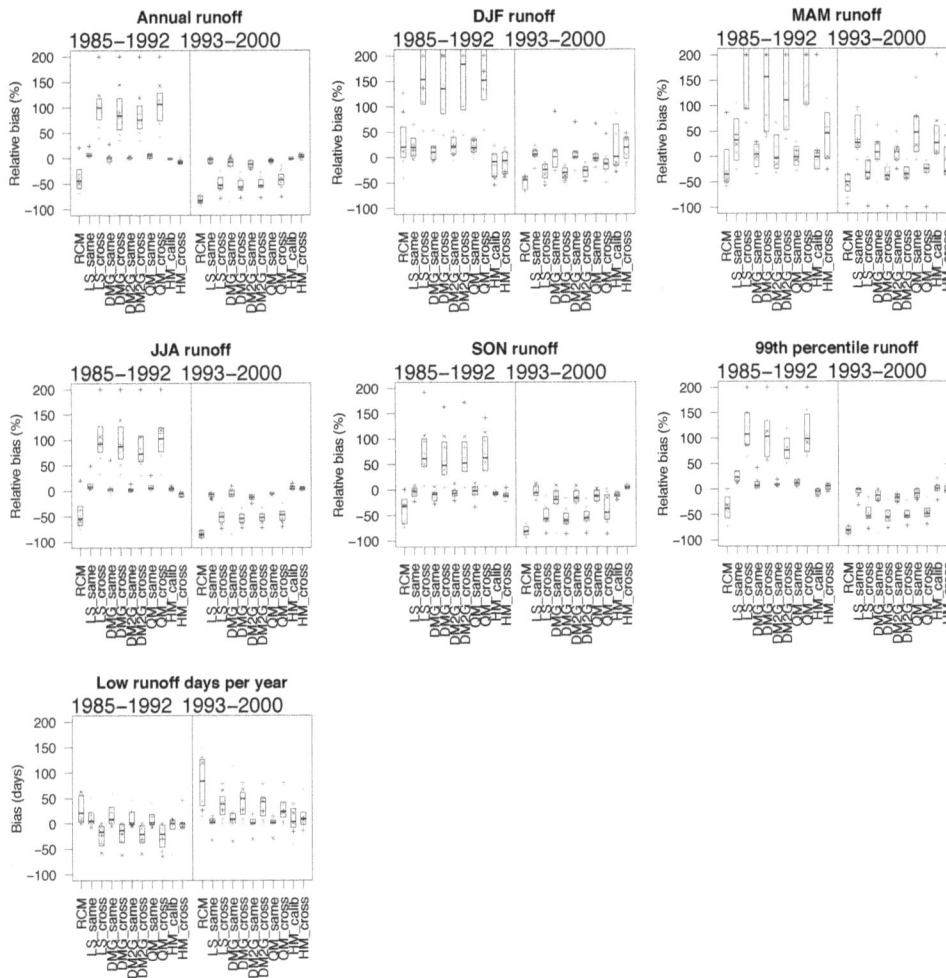

Figure 5. As for Fig. 4, but using results from hydrological model Sacramento.

To gauge the impact of bias correction methods on precipitation, we compared the RCM precipitation before and after bias correction with the observations using salient metrics: annual and seasonal means, 99th percentile precipitation as an indicator of high precipitation events, number of dry days (daily precipitation less than 0.1 mm) per year as an indicator of low precipitation, and 99th percentile of 3- and 5-day cumulative precipitation as indicators of runoff-generating events. The runoff modelled using RCM precipitation before and after bias correction was also evaluated against key runoff characteristics: annual and seasonal means, 99th percentile runoff as an indicator of high-flow events, and number of low-flow days (daily runoff less than 0.01 mm) as an indicator of low-flow conditions. We also looked at the effect of bias correction methods on change signals by comparing the relative difference in precipitation and runoff between the two periods derived from various methods.

4   Results

Figure 3 shows the percentage difference in raw RCM and bias corrected RCM precipitation relative to observations for annual and seasonal means, 99th percentile precipitation, 99th percentile 3- and 5-day cumulative precipitation, and the difference in number of dry days per year. Generally, the raw RCM precipitation exhibits negative errors in annual and seasonal means, with the median errors in raw RCM annual means being −9.1 and −22.5 % for the two periods respectively. There are larger (further away from zero) errors in the drier period (1993–2000). The 99th percentile precipitation is mostly overestimated in one period (1985–1992) and underestimated in the other. The raw RCM performs quite well in reproducing 99th percentile 3- and 5-day cumulative precipitation but slightly overestimates the number of low precipitation (< 0.1 mm) days.

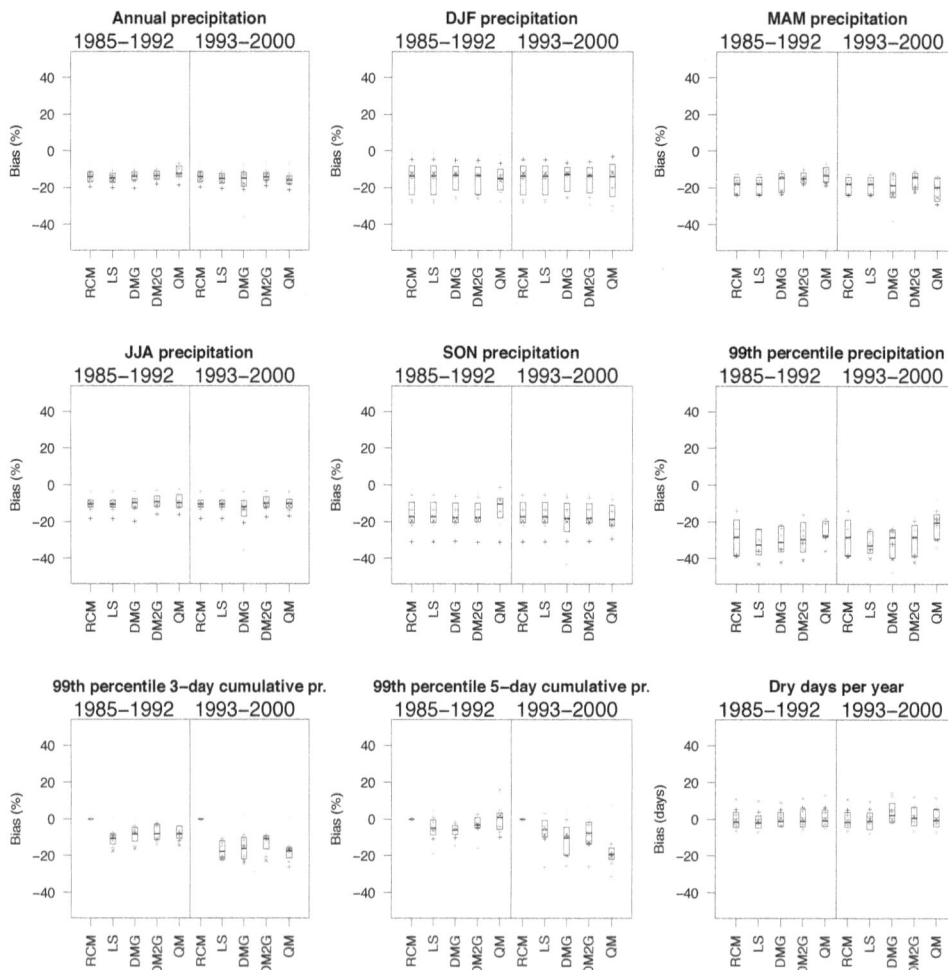

Figure 6. Differences between RCM simulations and observations in change signals in precipitation characteristics between periods 1985–1992 and 1993–2000. The left and right panels indicate the validation periods in each case.

4.1   Impact on precipitation

The calibration results in Fig. 3 (denoted by "_same") show that, as expected, all the bias correction methods are able to match the annual and seasonal means of precipitation when validating on the same period as the calibration period (see LS_same, DMG_same, DM2G_same and QM_same in the boxplots of annual and seasonal means in Fig. 3). For instance, LS perfectly corrects the median errors in annual means for the two periods (0 and 0 %), followed by DM2G (−0.3 and −0.2 %), QM (0.4 and 0.6 %) and DMG (−0.7 and −0.7 %). Only the distribution mapping methods (DMG, DM2G and QM) are able to reduce the errors in the high- and low-precipitation characteristics; QM in particular performs exceptionally well in reproducing 99th percentile precipitation and number of dry days per year. LS is not only unable to reduce the errors in high- and low-precipitation characteristics, but also increases the errors in some cases, as seen in the 99th percentile precipitation for 1985–1992 (period I, left panels) and number of dry days per year for 1993–2000 (pe-

riod II, right panels). This is consistent with the findings from previous studies (Chen et al., 2013; Teutschbein and Seibert, 2012).

By contrast, the cross-validation results (denoted by "_cross") seem to depend on the period, with most of the bias correction methods reducing the raw RCM errors (closer to zero) in period II but increasing the raw RCM errors (away from zero) in period I. The exception is DJF mean precipitation, where all the bias correction methods increase the errors in both periods. Although DM2G performs better compared to other bias correction methods for nearly all precipitation characteristics (shown by lower median errors given by DM2G_cross in Fig. 3), the difference between the bias correction methods is small compared to the overall large overestimation in period I and underestimation in period II. The cause of this "period dependency" is discussed in Sect. 5.1. It is notable that the errors in DJF and MAM mean precipitation are generally larger than in other seasons; the reason is that the amounts of precipitation in DJF and MAM are smaller in these catchments since they are dominated by winter precipi-

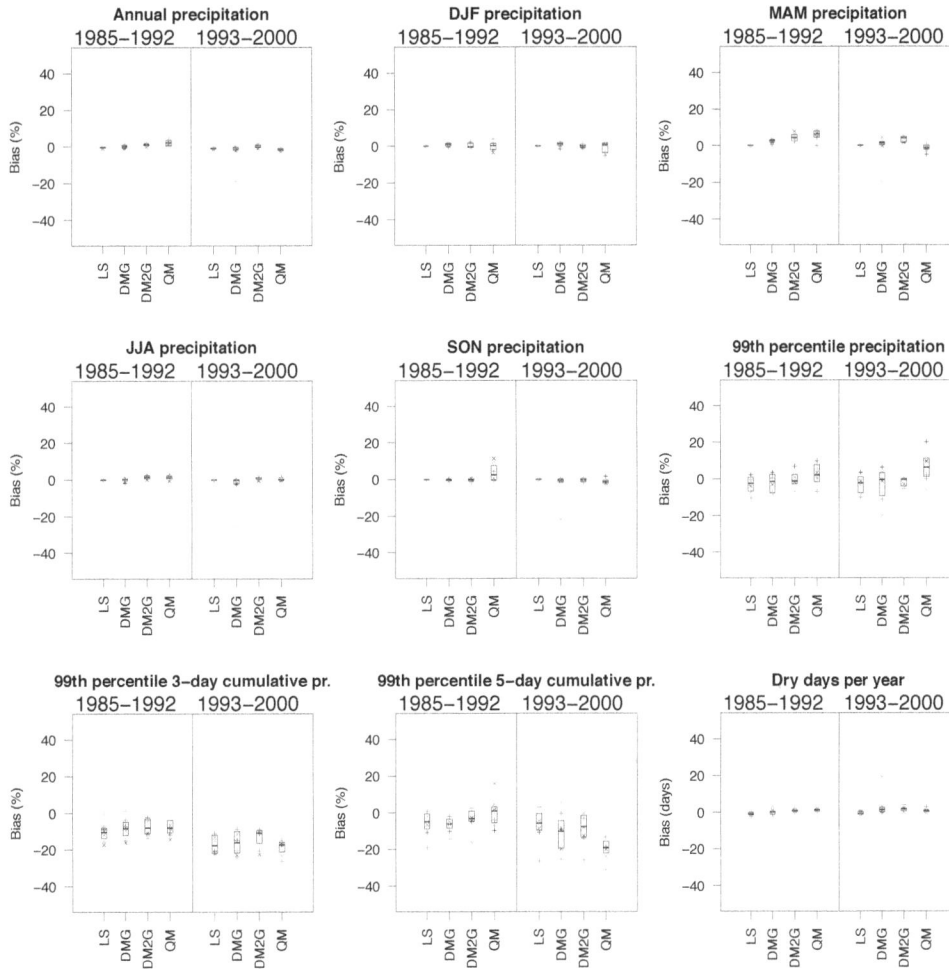

Figure 7. Differences between bias-corrected RCM and raw RCM simulations in change signals in precipitation characteristics between the periods 1985–1992 and 1993–2000. The left and right panels indicate the validation periods in each case.

tation, with more than 60 % of the precipitation coming from JJA and SON.

4.2 Impact on runoff

Figure 4 presents the relative differences in runoff characteristics simulated by GR4J using raw RCM and bias corrected precipitation when compared to those modelled using observed precipitation. The layout is similar to Fig. 3 except, for perspective, two boxes are added to each panel to show the conventional hydrological model errors: "HM_calib", which represents the calibration error (when runoff from GR4J driven by observed precipitation is compared with observed streamflow); and "HM_cross", which represents the cross-validation error (when GR4J runoff driven by observed precipitation and using parameters calibrated to the same period is compared with those using parameters calibrated to a different period).

The errors in runoff show a similar pattern to those for precipitation, but are much larger. They are also consider-

ably larger than the hydrological model errors. For instance, the median errors in mean annual runoff simulated using raw RCM precipitation increase to −33.1 % (period I) and −69.5 % (period II). The calibration results show that LS is no longer able to correct the errors in annual and seasonal mean runoff to zero due to errors in high-percentile precipitation (see the 99th percentile precipitation plot in Fig. 3) and, consequently, in high runoff. QM does not perform very well in correcting the high- and low-runoff characteristics as it was able to do for the high- and low-precipitation characteristics which may relate to its weakness (as shown in Fig. 3) in reproducing 3-day and 5-day cumulative precipitation. These results highlight the importance of precipitation sequence in runoff production, as discussed in Sect. 5.3.

The cross-validation results show that, after bias correction, the median errors in period I are increased to 62–84 % by various bias correction methods. While the median errors in period II are decreased, an error of −34 to −48 % is still considered large compared to the conventional hydrological

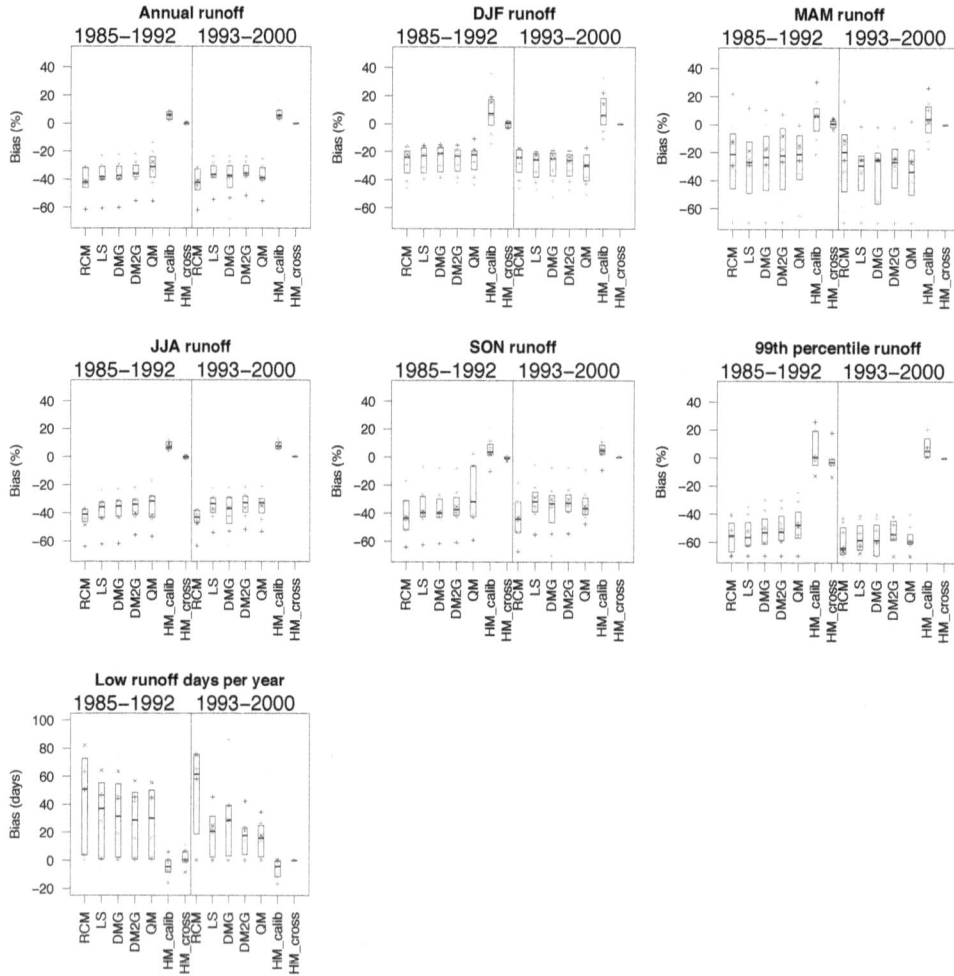

Figure 8. As for Fig. 6, but showing runoff characteristics modelled by GR4J.

model error of less than 10 %, as shown by HM_calib and HM_cross.

Figure 5 shows the same results as for Fig. 4 but using the Sacramento hydrological model. Similar observations can be made from this figure but with a larger range of the errors, probably because the Sacramento model errors (HM_calib and HM_cross) are larger in some seasons. Sacramento does better in reproducing observed low flows in period I but slightly worse than GR4J in reproducing high flows. In general, the bias correction affects two hydrological models similarly, although the magnitude of impact can be different. As the focus of this study is on the impact of bias correction method, only results from GR4J are presented and discussed in the following sections.

4.3 Impact on change signals

Figure 6 presents the differences in precipitation change signals when comparing raw RCM simulation and bias corrected RCM simulations to observations. Here, the "change" (ΔP) is defined as the relative difference of various charac-

teristics between period II and period I (Eq. 9):

$$\Delta P = \frac{P_{II} - P_{I}}{P_{I}} \cdot 100\%. \tag{9}$$

The baseline in Fig. 6 is the change derived from observations (ΔP_{obs}); the "difference" is between the baseline and the change derived from the raw RCM ($\Delta P_{RCM} - \Delta P_{obs}$) and bias-corrected RCM simulations ($\Delta P_{BC} - \Delta P_{obs}$). The ΔP_{BC} values used to plot the left panel of each plot in Fig. 6 were derived assuming period I is the validation period and period II the calibration period:

$$\Delta P_{BC}^{I} = \frac{P_{II\_same} - P_{I\_cross}}{P_{I\_cross}} \cdot 100\%. \tag{10}$$

Similarly, the ΔP_{BC} values used to plot the right panel were derived assuming period II is the validation period and period I the calibration period:

$$\Delta P_{BC}^{II} = \frac{P_{II\_cross} - P_{I\_same}}{P_{I\_same}} \cdot 100\%. \tag{11}$$

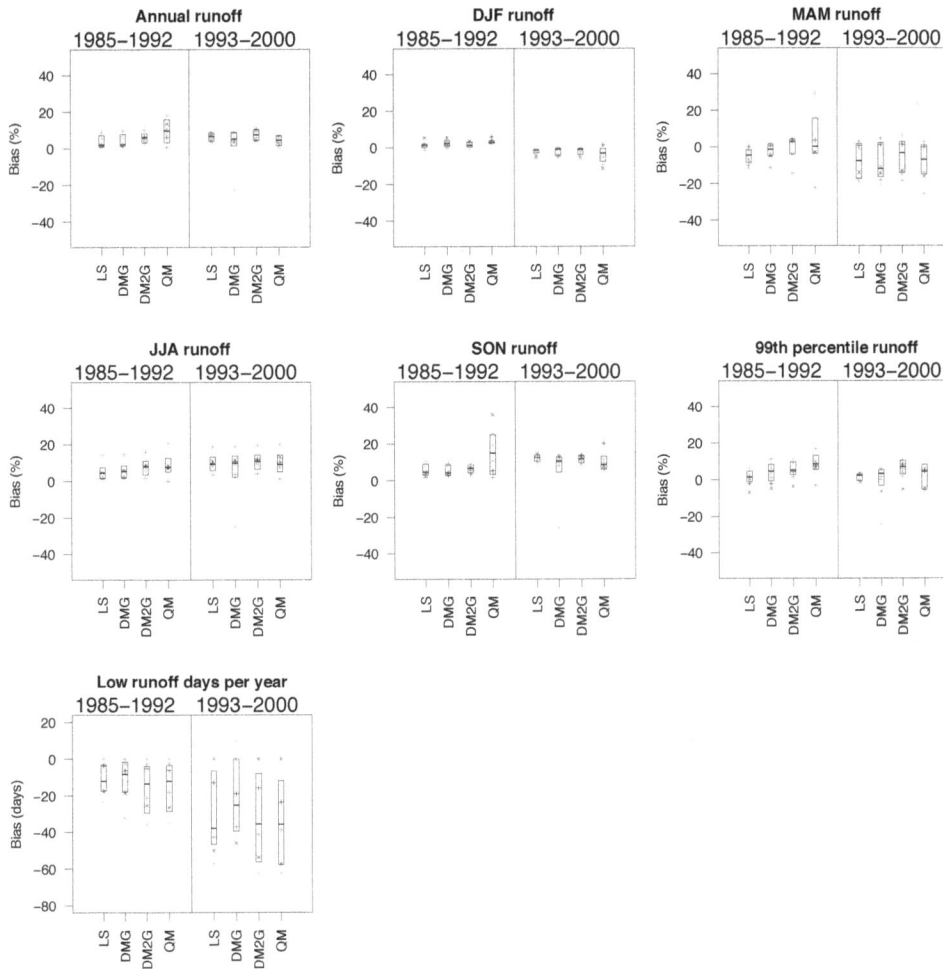

Figure 9. As for Fig. 7, but showing runoff characteristics modelled by GR4J.

For the majority of precipitation characteristics, all bias correction methods seem to produce a similar range and median of differences as given by the raw RCM, except for 3- and 5-day cumulative precipitation, where the raw RCM does better than the bias-corrected simulations. To take a closer look, we altered the baseline from change in observations (ΔP_{obs}) to change in raw RCM (ΔP_{RCM}), and the results ($\Delta P_{\text{BC}} - \Delta P_{\text{RCM}}$) are presented in Fig. 7. While the bias correction methods do not seem to affect changes in precipitation means, they do modify changes in high precipitation characteristics as shown in Fig. 7 as a large range of differences given by LS, DMG, DM2G and QM in 99th percentile precipitation, and 99th percentile 3- and 5-day cumulative precipitation plots.

The follow-on effects on runoff can be seen in Figs. 8 and 9 which show differences in runoff changes (substitute P with Q in Eqs. 9–11) corresponding to Figs. 6 and 7. The differences in runoff changes are much larger compared to those in precipitation changes. The bias correction methods affect change signals in every runoff characteristic (Fig. 9), especially high flows. This finding is consistent with Hagemann

et al. (2011), Cloke et al. (2013) and Gutjahr and Heinemann (2013), who showed that bias correction can alter climate change signals, a result slightly different from that of Muerth et al. (2013) who concluded that the impact of bias correction on change signals in flow is weak (except for the timing of the spring flood peak).

5 Discussion

5.1 Non-stationarity of the RCM bias

As shown in Figs. 3–5, the cross-validation results are period-dependent. When the errors in the calibration period are larger than, or in a different direction to, the errors in the validation period, all the bias correction methods overcorrect the errors in the validation period. When the errors in the calibration period are smaller than, and in the same direction as, the errors in the validation period, all the bias correction methods can reduce errors somewhat even though the under-correction can still be substantial. This is mainly

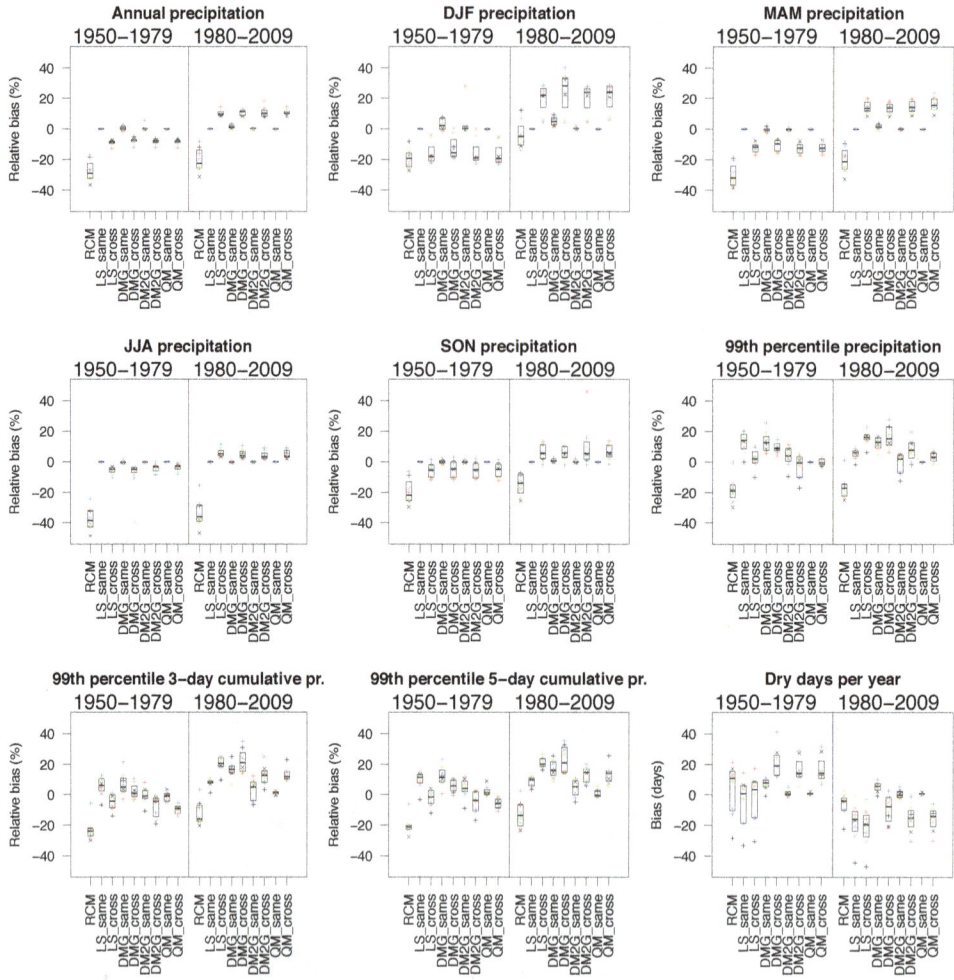

Figure 10. As for Fig. 3, but showing results from the long-term (two 30-year-long) experiments.

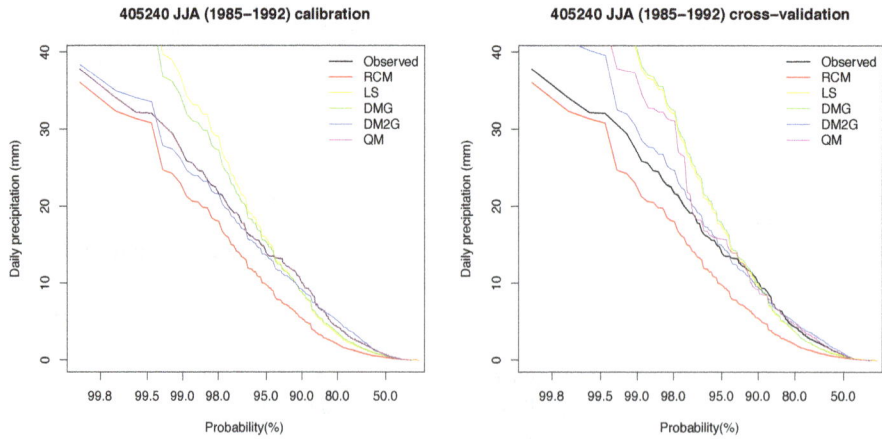

Figure 11. Comparison of CDFs derived from observed, raw RCM and bias-corrected RCM daily precipitation data for one study catchment. The left plot shows calibration results (note that the observed precipitation is completely hidden by QM in this plot) and the right plot shows cross-validation results.

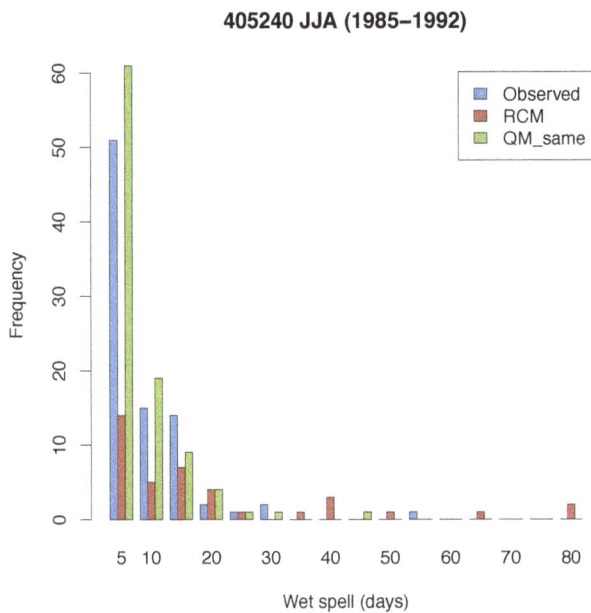

Figure 12. Histograms of consecutive wet days for observed, raw RCM and QM bias-corrected RCM precipitation for one study catchment.

due to the inconsistent errors over time. The large magnitude of errors to be corrected amplifies the differences in the bias correction relationships and results in clear under-correction in one period and over-correction in another.

The differences in errors from the two periods may be a result of insufficient length of data to achieve robust calibration (Berg et al., 2012), or it could be due to the non-stationarity of RCM bias. It is difficult to assess the non-stationarity of biases because time series long enough to achieve robust calibration and validation are rare (Maraun, 2012), and the definition of "long enough" varies for dry and wet regions. However, the probability of bias non-stationarity is high (Ehret et al., 2012). Thus, the results shown here serve as a good indicator for what could happen if bias were to vary over time.

Using a longer record is likely to improve the outcome because it better represents the complete variability, and has less likelihood of calibration and validation periods being very different. To test this, we repeated the same analysis on precipitation using a 60-year-long RCM simulation split in half – 30 years for calibration and 30 years for cross-validation. The results (Fig. 10) show improved cross-validation performance across bias correction methods and across characteristics. Nevertheless, the under-correction in the first period (1950–1979), and the over-correction in the second (1980–2009) are still apparent in most of the characteristics. Note that the runoff experiments cannot be repeated using the longer data set due to limited streamflow data, but it is reasonable to assume that this tendency will have a larger manifestation in modelled runoff.

The results suggest that non-stationarity of the RCM bias is one of the main obstacles preventing bias correction from achieving good outcomes, which makes the choice of bias correction method a secondary issue. When applying bias correction to a future period (as in most climate change impact studies), it is better to calibrate using a long data set (30 years or more), or at least a data period that best reflects the future (e.g. calibrate over a dry period and apply to a dry future RCM simulation, and vice versa). As the bias correction relationship is unlikely to be the same for two periods, the more different the periods are (different means, extremes, low-frequency variability, etc.), and the larger the magnitude of bias to be corrected, the smaller the chance of getting satisfactory results from bias correction. These problems have implications on the application of bias correction to climate model outputs in hydrological impact studies and related sectors (even more so at extremes like floods). Projections derived from bias corrected climate input should therefore be interpreted cautiously and/or combined with other approaches (Cloke et al., 2013; Smith et al., 2014).

5.2 Performance of the bias correction methods

Figure 11 shows a selected example comparing daily CDF of the four bias correction methods (LS, DMG, DM2G and QM) for calibration and cross-validation experiments. The LS performs poorly in both calibration and cross-validation as it under-estimates small and medium rainfall values (< 95th percentile) and over-estimates the very high rainfall values (> 95th percentile). The DMG performs significantly better than the LS because it attempts to correct the distribution rather than simply scaling the data with one factor. The DM2G performs better than DMG for its better representation of distribution, especially at the high end (as shown in Fig. 2). By definition, the QM will always give perfect results in calibration but the over-fitting can lead to poorer performance in cross-validation, particularly when the errors in the two periods are very different.

In general, the best results are produced by either QM or DM2G in this study. The non-parametric QM fits every part of the entire distribution and performs the best when the errors in the two periods are similar. When the errors are different in the two periods, the DM2G is likely to be more robust (theoretical distribution with six parameters) and has less chance of over-fitting like QM is liable to do. Nevertheless, the difference between the three distribution mapping methods is very small in our modelling experiments (and as reported in the literature) because of the large corrections required which are then amplified by the inconsistent errors in different periods, as discussed in Sect. 5.1.

5.3 Importance of precipitation sequence

The errors in the bias corrected precipitation are significantly amplified in modelled runoff. The choice of hydrological

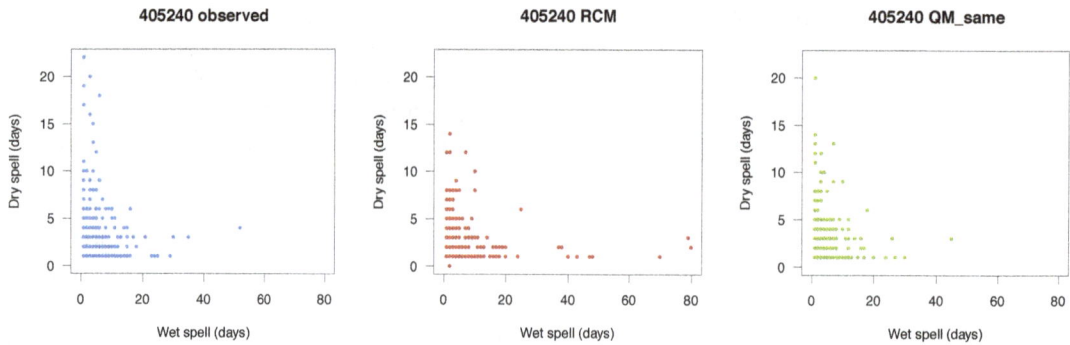

Figure 13. Scatter plots showing the length of each dry spell on y axes and the length of the following wet spell on x axes for observed, raw RCM and QM bias-corrected RCM precipitation for an example catchment.

models does not have big impact because of the relatively small errors associated with hydrological models. Although the RCM precipitation can be bias corrected to practically match the observed precipitation means and high precipitation amounts (see calibration results in Fig. 3), there can still be considerable errors in the modelled runoff (see calibration results in Fig. 4).

Apart from precipitation intensity, other aspects of precipitation can also affect runoff. Precipitation sequence is one of them as runoff generation is driven by high precipitation events that last over several days, and preceding events influence runoff by changing soil moisture content. The importance of precipitation sequence can be quantified in our modelling experiments by analysing calibration results for the QM. As shown in the calibration plot (left) in Fig. 11, QM corrects the RCM daily precipitation to perfectly match the observed daily precipitation distribution; therefore the errors in the modelled runoff should mainly reflect the differences in the precipitation sequences between RCM and observations. To examine whether this is the case, we compared the wet spell histograms of observations, RCM and QM corrected RCM precipitation. Figure 12 shows the results for one of the study catchments. Compared to observations, the raw RCM simulation shows a lack of short events and an excess of very long events. This is the widely reported "drizzle effect" – RCMs simulate too many low-intensity precipitation events and too few high-intensity precipitation events (Gutowski et al., 2003). Although the QM is able to break long events into many shorter ones by reducing the "drizzles" with intensity below probability q_0 (Eq. 7) to zero, as shown by the increased number of short events, there are still differences in the wet spell frequencies. These differences are due to the lack of short wet spells followed by long dry spells, as seen in the scatter plots in Fig. 13, which show length of dry spells on the y axes and the following wet spells on the x axes.

The runoff errors in the QM calibration results, for the eight catchments in the two calibration periods, range from -17 to 24% (median of 5%) for the 99th percentile runoff

and -7 to 7% (median of 1%) for mean annual runoff ("QM_same" in Fig. 4). These errors are significant considering that the precipitation distribution perfectly matches that of the observations. But they are relatively small compared to the errors in runoff from bias correction for cross-validation periods.

These results show that the bias correction methods tested here are unable to overcome the discrepancy in the precipitation sequence. It is important for the RCMs to better simulate the number and length of storms, and the dry periods that intersperse them. After all, the ultimate approach to reduce errors in models is to improve the models themselves. This will require better process descriptions and implementations, higher spatiotemporal resolution, and perhaps using multi-model/multi-physics ensembles, as seen in recent developments in this area (Ji et al., 2014; Evans et al., 2012; Flaounas et al., 2011).

6 Conclusions

This paper reviewed recent studies comparing various bias correction methods as applied to RCM simulated precipitation. The distribution mapping techniques were selected to remove errors (relative to observations) in daily precipitation series simulated by the weather research and forecasting (WRF) model for eight catchments in southeast Australia. The performance of three different techniques – DMG, DM2G, QM – and a linear scaling method (LS) as a benchmark, was evaluated with the focus on the follow-on impact on runoff modelling.

The results confirm the relatively higher skill of the distribution-based methods, compared to the linear scaling method, in correcting key precipitation characteristics. The best results are produced by either QM or DM2G. The non-parametric QM fits every part of the entire distribution and performs the best when the errors in the calibration and validation periods are similar. When errors in the two periods are different, DM2G can be more robust as it has a smaller number of parameters and so there is less chance of over-fitting.

However, the difference between the distribution mapping methods tested here is small because of the large corrections required and the inconsistent errors in the calibration and validation periods (non-stationarity).

The errors in bias corrected precipitation lead to amplified errors in modelled runoff. The bias correction methods tested here cannot overcome the limitations of the RCM in simulating all precipitation features that influence runoff, in particular, daily precipitation sequence. The errors in modelled runoff are strongly influenced by the inconsistent RCM errors over time, although this can be partially overcome by using a long calibration data set.

Results further show that whereas bias correction does not seem to affect the change signals in precipitation means, it can introduce extra uncertainty to the change signals in high precipitation amounts, and consequently, in runoff. Future climate change impact studies need to take this into account when deciding whether to use raw or bias corrected RCM results.

These problems associated with bias correction in general have implications on its application to climate model outputs in hydrological impact studies. Projections derived from bias corrected climate input should therefore be interpreted with caution and/or combined with other approaches. Nevertheless, the bias in RCM simulations will continue to reduce as RCM accuracy improves and RCMs will become increasingly useful for hydrological studies.

Acknowledgements. This work was carried out in the CSIRO Land and Water Flagship and was funded by the Victorian Climate Initiative (VicCI). J. P. Evans was funded through the Australian Research Council as part of Discovery Project DP0772665 and Future Fellowship FT110100576. We thank James Bennett and Steve Charles for comments on an earlier version of this manuscript, as well as the anonymous HESS reviewers for their constructive feedback.

Edited by: H. Cloke

References

Argüeso, D., Evans, J. P., and Fita, L.: Precipitation bias correction of very high resolution regional climate models, Hydrol. Earth Syst. Sci., 17, 4379–4388, doi:10.5194/hess-17-4379-2013, 2013.

Bennett, J. C., Ling, F. L. N., Post, D. A., Grose, M. R., Corney, S. P., Graham, B., Holz, G. K., Katzfey, J. J., and Bindoff, N. L.: High-resolution projections of surface water availability for Tasmania, Australia, Hydrol. Earth Syst. Sci., 16, 1287–1303, doi:10.5194/hess-16-1287-2012, 2012.

Bennett, J. C., Grose, M. R., Corney, S. P., White, C. J., Holz, G. K., Katzfey, J. J., Post, D. A., and Bindoff, N. L.: Performance of an empirical bias-correction of a high-resolution climate data set, Int. J. Climatol., 34, 2189–2204, doi:10.1002/joc.3830, 2014.

Berg, P., Feldmann, H., and Panitz, H. J.: Bias correction of high resolution regional climate model data, J. Hydrol., 448–449, 80–92, doi:10.1016/j.jhydrol.2012.04.026, 2012.

Boe, J., Terray, L., Habets, F., and Martin, E.: Statistical and dynamical downscaling of the Seine basin climate for hydro-meteorological studies, Int. J. Climatol., 27, 1643–1655, doi:10.1002/joc.1602, 2007.

Burnash, R. J. C., Ferral, R. L., and McGuire, R. A.: A Generalised Streamflow Simulation System – Conceptual Modelling for Digital Computers, Technical Report, Joint Federal and State River Forecast Center, Sacramento, 204 pp., 1973.

Chakravarti, I. M. and Laha, R. G.: Handbook of Methods of Applied Statistics, John Wiley & Sons, New York, 460 pp., 1967.

Chen, J., Brissette, F. P., and Leconte, R.: Uncertainty of downscaling method in quantifying the impact of climate change on hydrology, J. Hydrol., 401, 190–202, doi:10.1016/j.jhydrol.2011.02.020, 2011a.

Chen, J., Brissette, F. P., Poulin, A., and Leconte, R.: Overall uncertainty study of the hydrological impacts of climate change for a Canadian watershed, Water Resour. Res., 47, W12509, doi:10.1029/2011wr010602, 2011b.

Chen, J., Brissette, F. P., Chaumont, D., and Braun, M.: Finding appropriate bias correction methods in downscaling precipitation for hydrologic impact studies over North America, Water Resour. Res., 49, 4187–4205, doi:10.1002/wrcr.20331, 2013.

Chiew, F. H. S. and McMahon, T. A.: The applicability of morton and penman evapotranspiration estimates in rainfall–runoff modeling, Water Resour. Bull., 27, 611–620, doi:10.1111/j.1752-1688.1991.tb01462.x, 1991.

Chiew, F. H. S., Young, W., Cai, W., and Teng, J.: Current drought and future hydroclimate projections in southeast Australia and implications for water resources management, Stoch. Env. Res. Risk A., 25, 601–612, doi:10.1007/s00477-010-0424-x, 2010.

Christensen, J. H., Boberg, F., Christensen, O. B., and Lucas-Picher, P.: On the need for bias correction of regional climate change projections of temperature and precipitation, Geophys. Res. Lett., 35, L20709, doi:10.1029/2008gl035694, 2008.

Cloke, H. L., Wetterhall, F., He, Y., Freer, J. E., and Pappenberger, F.: Modelling climate impact on floods with ensemble climate projections, Q. J. Roy. Meteor. Soc., 139, 282–297, doi:10.1002/qj.1998, 2013.

Dosio, A., Paruolo, P., and Rojas, R.: Bias correction of the ENSEMBLES high resolution climate change projections for use by impact models: analysis of the climate change signal, J. Geophys. Res., 117, D17110, doi:10.1029/2012jd017968, 2012.

Ehret, U., Zehe, E., Wulfmeyer, V., Warrach-Sagi, K., and Liebert, J.: HESS Opinions "Should we apply bias correction to global and regional climate model data?", Hydrol. Earth Syst. Sci., 16, 3391–3404, doi:10.5194/hess-16-3391-2012, 2012.

Evans, J. P. and McCabe, M. F.: Regional climate simulation over Australia's Murray–Darling basin: a multitemporal assessment, J. Geophys. Res., 115, D14114, doi:10.1029/2010jd013816, 2010.

Evans, J. P., Ekström, M., and Ji, F.: Evaluating the performance of a WRF physics ensemble over South-East Australia, Clim. Dynam., 39, 1241–1258, doi:10.1007/s00382-011-1244-5, 2012.

Evans, J. P., Ji, F., Lee, C., Smith, P., Argüeso, D., and Fita, L.: Design of a regional climate modelling projection ensemble experiment – NARCliM, Geosci. Model Dev., 7, 621–629, doi:10.5194/gmd-7-621-2014, 2014.

Flaounas, E., Bastin, S., and Janicot, S.: Regional climate modelling of the 2006 West African monsoon: sensitivity to convection and planetary boundary layer parameterisation using WRF, Clim. Dynam., 36, 1083–1105, doi:10.1007/s00382-010-0785-3, 2011.

Fowler, H. J., Blenkinsop, S., and Tebaldi, C.: Linking climate change modelling to impacts studies: recent advances in downscaling techniques for hydrological modelling, Int. J. Climatol., 27, 1547–1578, doi:10.1002/joc.1556, 2007.

Grillakis, M. G., Koutroulis, A. G., and Tsanis, I. K.: Multisegment statistical bias correction of daily GCM precipitation output, J. Geophys. Res., 118, 3150–3162, doi:10.1002/jgrd.50323, 2013.

Gudmundsson, L., Bremnes, J. B., Haugen, J. E., and Engen-Skaugen, T.: Technical Note: Downscaling RCM precipitation to the station scale using statistical transformations – a comparison of methods, Hydrol. Earth Syst. Sci., 16, 3383–3390, doi:10.5194/hess-16-3383-2012, 2012.

Gutjahr, O. and Heinemann, G.: Comparing precipitation bias correction methods for high-resolution regional climate simulations using COSMO-CLM, Theor. Appl. Climatol., 114, 511–529, doi:10.1007/s00704-013-0834-z, 2013.

Gutowski, W. J., Decker, S. G., Donavon, R. A., Pan, Z. T., Arritt, R. W., and Takle, E. S.: Temporal-spatial scales of observed and simulated precipitation in central U.S. climate, J. Climate, 16, 3841–3847, doi:10.1175/1520-0442(2003)016<3841:tsooas>2.0.co;2, 2003.

Hagemann, S., Chen, C., Haerter, J. O., Heinke, J., Gerten, D., and Piani, C.: Impact of a statistical bias correction on the projected hydrological changes obtained from three GCMs and two hydrology models, J. Hydrometeorol., 12, 556–578, doi:10.1175/2011jhm1336.1, 2011.

Jeffrey, S. J., Carter, J. O., Moodie, K. B., and Beswick, A. R.: Using spatial interpolation to construct a comprehensive archive of Australian climate data, Environ. Modell. Softw., 16, 309–330, doi:10.1016/S1364-8152(01)00008-1, 2001.

Ji, F., Ekström, M., Evans, J., and Teng, J.: Evaluating rainfall patterns using physics scheme ensembles from a regional atmospheric model, Theor. Appl. Climatol., 115, 297–304, doi:10.1007/s00704-013-0904-2, 2014.

Johnson, F. and Sharma, A.: A nesting model for bias correction of variability at multiple time scales in general circulation model precipitation simulations, Water Resour. Res., 48, W01504, doi:10.1029/2011wr010464, 2012.

Kalnay, E., Kanamitsu, M., Kistler, R., Collins, W., Deaven, D., Gandin, L., Iredell, M., Saha, S., White, G., Woollen, J., Zhu, Y., Chelliah, M., Ebisuzaki, W., Higgins, W., Janowiak, J., Mo, K. C., Ropelewski, C., Wang, J., Leetmaa, A., Reynolds, R., Jenne, R., and Joseph, D.: The NCEP/NCAR 40-year reanalysis project, B. Am. Meteorol. Soc., 77, 437–471, doi:10.1175/1520-0477(1996)077<0437:tnyrp>2.0.co;2, 1996.

Lafon, T.: Bias correction of daily precipitation simulated by a regional climate model: a comparison of methods, Int. J. Climatol., 33, 1367–1381, doi:10.1002/joc.3518, 2013.

Lenderink, G., Buishand, A., and van Deursen, W.: Estimates of future discharges of the river Rhine using two scenario methodologies: direct versus delta approach, Hydrol. Earth Syst. Sci., 11, 1145–1159, doi:10.5194/hess-11-1145-2007, 2007.

Li, H., Sheffield, J., and Wood, E. F.: Bias correction of monthly precipitation and temperature fields from Intergovernmental Panel on Climate Change AR4 models using equidistant quantile matching, J. Geophys. Res., 115, D10101, doi:10.1029/2009jd012882, 2010.

Maraun, D.: Nonstationarities of regional climate model biases in European seasonal mean temperature and precipitation sums, Geophys. Res. Lett., 39, L06706, doi:10.1029/2012gl051210, 2012.

Maraun, D., Wetterhall, F., Ireson, A. M., Chandler, R. E., Kendon, E. J., Widmann, M., Brienen, S., Rust, H. W., Sauter, T., Themeßl, M., Venema, V. K. C., Chun, K. P., Goodess, C. M., Jones, R. G., Onof, C., Vrac, M., and Thiele-Eich, I.: Precipitation downscaling under climate change: recent developments to bridge the gap between dynamical models and the end user, Rev. Geophys., 48, RG3003, doi:10.1029/2009rg000314, 2010.

Muerth, M. J., Gauvin St-Denis, B., Ricard, S., Velázquez, J. A., Schmid, J., Minville, M., Caya, D., Chaumont, D., Ludwig, R., and Turcotte, R.: On the need for bias correction in regional climate scenarios to assess climate change impacts on river runoff, Hydrol. Earth Syst. Sci., 17, 1189–1204, doi:10.5194/hess-17-1189-2013, 2013.

Nash, J. E. and Sutcliffe, J. V.: River flow forecasting through conceptual models part I – A discussion of principles, J. Hydrol., 10, 282–290, doi:10.1016/0022-1694(70)90255-6, 1970.

Perrin, C., Michel, C., and Andreassian, V.: Improvement of a parsimonious model for streamflow simulation, J. Hydrol., 279, 275–289, doi:10.1016/s0022-1694(03)00225-7, 2003.

Piani, C., Haerter, J. O., and Coppola, E.: Statistical bias correction for daily precipitation in regional climate models over Europe, Theor. Appl. Climatol., 99, 187–192, doi:10.1007/s00704-009-0134-9, 2010a.

Piani, C., Weedon, G. P., Best, M., Gomes, S. M., Viterbo, P., Hagemann, S., and Haerter, J. O.: Statistical bias correction of global simulated daily precipitation and temperature for the application of hydrological models, J. Hydrol., 395, 199–215, doi:10.1016/j.jhydrol.2010.10.024, 2010b.

Potter, N. J. and Chiew, F. H. S.: An investigation into changes in climate characteristics causing the recent very low runoff in the southern Murray–Darling Basin using rainfall–runoff models, Water Resour. Res., 47, W00G10, doi:10.1029/2010wr010333, 2011.

Schmidli, J., Frei, C., and Vidale, P. L.: Downscaling from GCM precipitation: a benchmark for dynamical and statistical downscaling methods, Int. J. Climatol., 26, 679–689, doi:10.1002/joc.1287, 2006.

Seaby, L. P., Refsgaard, J. C., Sonnenborg, T. O., Stisen, S., Christensen, J. H., and Jensen, K. H.: Assessment of robustness and significance of climate change signals for an ensemble of distribution-based scaled climate projections, J. Hydrol., 486, 479–493, doi:10.1016/j.jhydrol.2013.02.015, 2013.

Smith, A., Freer, J., Bates, P., and Sampson, C.: Comparing ensemble projections of flooding against flood estimation by continuous simulation, J. Hydrol., 511, 205–219, doi:10.1016/j.jhydrol.2014.01.045, 2014.

Teng, J., Chiew, F. H. S., Timbal, B., Wang, Y., Vaze, J., and Wang, B.: Assessment of an analogue downscaling method for modelling climate change impacts on runoff, J. Hydrol., 472–473, 111–125, doi:10.1016/j.jhydrol.2012.09.024, 2012a.

Teng, J., Vaze, J., Chiew, F. H. S., Wang, B., and Perraud, J.-M.: Estimating the relative uncertainties sourced from GCMs and hydrological models in modeling climate change impact on runoff, J. Hydrometeorol., 13, 122–139, doi:10.1175/jhm-d-11-058.1, 2012b.

Teutschbein, C. and Seibert, J.: Regional climate models for hydrological impact studies at the catchment scale: a review of recent modeling strategies, Geography Compass, 4, 834–860, doi:10.1111/j.1749-8198.2010.00357.x, 2010.

Teutschbein, C. and Seibert, J.: Bias correction of regional climate model simulations for hydrological climate-change impact studies: review and evaluation of different methods, J. Hydrol., 456–457, 12–29, doi:10.1016/j.jhydrol.2012.05.052, 2012.

Teutschbein, C. and Seibert, J.: Is bias correction of regional climate model (RCM) simulations possible for non-stationary conditions?, Hydrol. Earth Syst. Sci., 17, 5061–5077, doi:10.5194/hess-17-5061-2013, 2013.

Themeßl, M. J., Gobiet, A., and Leuprecht, A.: Empirical-statistical downscaling and error correction of daily precipitation from regional climate models, Int. J. Climatol., 31, 1530–1544, doi:10.1002/joc.2168, 2011.

Vaze, J., Chiew, F. H. S., Perraud, J.-M., Viney, N., Post, D. A., Teng, J., Wang, B., Lerat, J., and Goswami, M.: Rainfall–runoff modelling across southeast Australia: data sets, models and results, Aust. J. Water Resour., 14, 101–116, 2010.

Vaze, J., Post, D. A., Chiew, F. H. S., Perraud, J.-M., Viney, N., and Teng, J.: Conceptual rainfall-runoff model performance with different spatial rainfall inputs, J. Hydrometeorol., 12, 1100–1112, doi:10.1175/2011JHM1340.1, 2011.

Viney, N. R., Perraud, J.-M., Vaze, J., Chiew, F. H. S., Post, D. A., and Yang, A.: The usefulness of bias constraints in model calibration for regionalisation to ungauged catchments, 18th World IMACS Congress and MODSIM09 International Congress on Modelling and Simulation, Cairns, Australia, 13–17 July 2009, 3421–3427, 2009.

Wilby, R. L., Hay, L. E., Gutowski, W. J., Arritt, R. W., Takle, E. S., Pan, Z., Leavesley, G. H., and Clark, M. P.: Hydrological responses to dynamically and statistically downscaled climate model output, Geophys. Res. Lett., 27, 1199–1202, doi:10.1029/1999gl006078, 2000.

Wood, A. W., Leung, L. R., Sridhar, V., and Lettenmaier, D. P.: Hydrologic implications of dynamical and statistical approaches to downscaling climate model outputs, Clim. Change, 62, 189–216, doi:10.1023/B:CLIM.0000013685.99609.9e, 2004.

Yang, W., Andreasson, J., Graham, L. P., Olsson, J., Rosberg, J., and Wetterhall, F.: Distribution-based scaling to improve usability of regional climate model projections for hydrological climate change impacts studies, Hydrol. Res., 41, 211–229, doi:10.2166/nh.2010.004, 2010.

11

Evolving flood patterns in a Mediterranean region (1301–2012) and climatic factors – the case of Catalonia

A. Barrera-Escoda[1] **and M. C. Llasat**[2]

[1]Climate Change Unit, Meteorological Service of Catalonia, Barcelona, Spain
[2]Meteorological Hazard Analysis Team (GAMA), Department of Astronomy and Meteorology, University of Barcelona, Barcelona, Spain

Correspondence to: A. Barrera-Escoda (tbarrera@meteo.cat)

Abstract. Data on flood occurrence and flood impacts for the last seven centuries in the northeastern Iberian Peninsula have been analysed in order to characterise long-term trends, anomalous periods and their relationship with different climatic factors such as precipitation, general circulation and solar activity. Catastrophic floods (those that produce complete or partial destruction of infrastructure close to the river, and major damages in the overflowed area, including some zones away from the channels) do not present a statistically significant trend, whereas extraordinary floods (the channel is overflowed and some punctual severe damages can be produced in the infrastructures placed in the rivercourse or near it, but usually damages are slight) have seen a significant rise, especially from 1850 on, and were responsible for the total increase in flooding in the region. This rise can be mainly attributed to small coastal catchments, which have experienced a marked increase in developed land and population, resulting in changes in land use and greater vulnerability. Changes in precipitation alone cannot explain the variation in flood patterns, although a certain increase was shown in late summer–early autumn, when extraordinary floods are most frequently recorded. The relationship between the North Atlantic circulation and floods is not as strong, due to the important role of mesoscale factors in heavy precipitation in the northwest of the Mediterranean region. However, it can explain the variance to some extent, mainly in relation to the catastrophic floods experienced during the autumn. Solar activity has some impact on changes in catastrophic floods, with cycles related to the quasi-biennial oscillation (QBO) and the Gleissberg solar cycle. In addition, anomalous periods of high flood frequency in autumn generally occurred during periods of increased solar activity. The physical influence of the latter in general circulation patterns, the high troposphere and the stratosphere, has been analysed in order to ascertain its role in causing floods.

1 Introduction

Floods are the natural hazard with the largest socio-economic impact in the world, and they are responsible for the highest number of deaths and the most damage caused by natural hazards worldwide (Munich Re, 2006; IPCC, 2012). This problem is exacerbated by the acceleration of the hydrological cycle and other extreme events, which are considered a consequence of climate change. Some studies show that water extremes (floods and droughts), which currently have a return period of 100 years, may recur every 10 to 50 years by 2070 in some regions in Europe (Lehner et al., 2006). Precipitation intensity is also projected to increase in some regions, leading to more flooding (Dankers and Feyen, 2009). However, keeping the previous IPCC assessment report in mind (IPCC, 2013), these results are not representative enough for a full understanding of the impact of climatic change on future floods, due to the complexity of the different factors involved in causing floods and the impact they have, as well as significant questions about future changes to precipitation.

With regard to present trends in precipitation extremes and floods, the latest IPCC report on extremes (IPCC, 2012) states that there is "limited to medium" evidence available to assess climate-driven changes to the magnitude and frequency of floods on a regional scale, with "low agreement" evidence and "low confidence" on a global scale for the signs of said changes. In the case of the Iberian Peninsula (IP), there are some controversial results, as for the rest of the Mediterranean area, with regard to signs of changes in precipitation. This is due to the diverse periods considered by different authors, the regions in question, and the varying approaches taken as the methodology applied (Barrera and Llasat, 2004; Llasat and Quintas, 2004; González-Hidalgo et al., 2009; Pryor et al., 2009; Turco and Llasat, 2011). Some of them point to a decrease in the intensity of daily precipitation and an increase in the number of days with light rain (García et al., 2007; Rodrigo and Trigo, 2007; Rodrigo, 2010; López-Moreno et al., 2010) for much of the Iberian Peninsula. On the contrary, other studies show that less-intensive rainy days are more frequent and that intensive precipitation episodes have increased along the Mediterranean coast (Alpert et al., 2002; Goodess and Jones, 2002), but that these results cannot be generalised (Altava-Ortiz et al., 2010). These findings are consistent with the "low-agreement" findings of the IPCC (2012), which is notable in the Mediterranean region, where mesoscale mechanisms and convective precipitation play a major role in the temporal and spatial distribution of precipitation.

Apart from the low significance and agreement in precipitation trends, the complexity associated with flooding requires further analysis of the relationship between changes to floods and rainfall and the evolution of climatic and hydrologic parameters (Di Baldassarre et al., 2009; Heine and Pinter, 2012; Remo et al., 2012; Hall et al., 2014). Most of these authors stress the importance of changes to land use (Naef et al., 2002), or possible changes in runoff coefficients (Sivapalan et al., 2005; Mouri et al., 2011). In short, and as indicated by Mertz et al. (2012), changes to the river itself, to the basin and to the atmosphere should be considered to be possible physical causes of the changes to the flood regime. The problem is exacerbated when flood series are built from proxy data (i.e. flood impacts), where case factors such as vulnerability, exposure and perception play a more important role. For instance, the trends observed in changes to floods in the northwest of the Mediterranean region could be related principally to changes in vulnerability and land use, and also to changes in perception and exposed assets (Barrera et al., 2006; Barredo et al., 2012; Llasat et al., 2013).

Historical flood evidence is mainly based on the impact descriptions and, consequently, it refers to the floods as a holistic risk, it being difficult to separate the "natural" causes from the rest. The flood chronologies that can be constructed from instrumental records and flow series for Europe do not usually extend further back than the nineteenth century (the twentieth century for Spain). Flood historical records can arrive until the fourteenth century, except for those in Italy dating from the Roman Empire. Besides this, information density in the past is heterogeneous, not only due to the lack of records (i.e. Macdonald, 2014), but also due to the relative youth of the science that encompasses histori-

cal climatology with the modern understanding of climate dynamics, meteorology and hydrology (Glaser, 1996; Camuffo and Enzi, 1996; Brázdil et al., 1999; Lang and Cœur, 2002). The major documentary historical sources containing climatic information and details of its effects are local and state government records, religious collections, private collections, notaries' archives and taxation records (Barriendos et al., 2003; Brázdil et al., 2014). Whenever possible, the historical flood classification should be based on discharge estimates, with a sensitivity analysis to assess the specific errors of the hydraulic model for the conversion of historical flood levels into discharge (Brázdil et al., 2006; Herget et al., 2014). On the contrary, in order to have the longest possible flood series, a scale of event magnitude can be proposed using the effects of the floods on the river channel system and surrounding areas. This is the approach more commonly used (Llasat et al., 2005; Barriendos et al., 2014; Retsö, 2014). In this sense, the objective of the FLOOD-CHANGE project is to improve, on a European scale, the construction of long historical flood records in order to build a flood-change model (http://floodchange.hydro.tuwien.ac.at/deciphering-river-flood-change/). We would like to direct the reader to the papers published in this special issue to find more details about historical flood data and their analysis (Kiss et al., 2014).

In this context, analysis of long-term homogeneous flood series and the corresponding causes is necessary in order to have a better understanding of how they evolve. That is the aim of this paper. Subsequently, based on early works on the evolution of floods in the northeast of the IP from the fourteenth century until 2002, the historical flood database of Catalonia (north-eastern Spain) has been updated until 2012 to review and identify any significant long-term trends or anomalies, and to analyse the potential relationship with climatic and non-climatic factors. This database, together with data on precipitation and information on solar activity and general circulation, is presented in Sect. 2. The methodology on building flood index series and carrying out statistical analysis on the same is explained in Sect. 3. Finally, the main results obtained on flood evolution and the potential relationship between the same and different climatic factors is presented and discussed in Sect. 4.

2 Area of study and data

Catalonia (north-eastern Spain) is characterised by a complex topography that has a considerable impact on the region's climatology and atmospheric circulation patterns: there are two mountain ranges with average heights of around 500 m a.s.l. (littoral zone) and 1500 m a.s.l. (prelittoral zone), located parallel to the coastline, and the Pyrenees, with summits above 3000 m a.s.l. (Fig. 1). These orographic factors, together with the influence of the Mediterranean Sea and the associated Mediterranean air mass (Jansà,

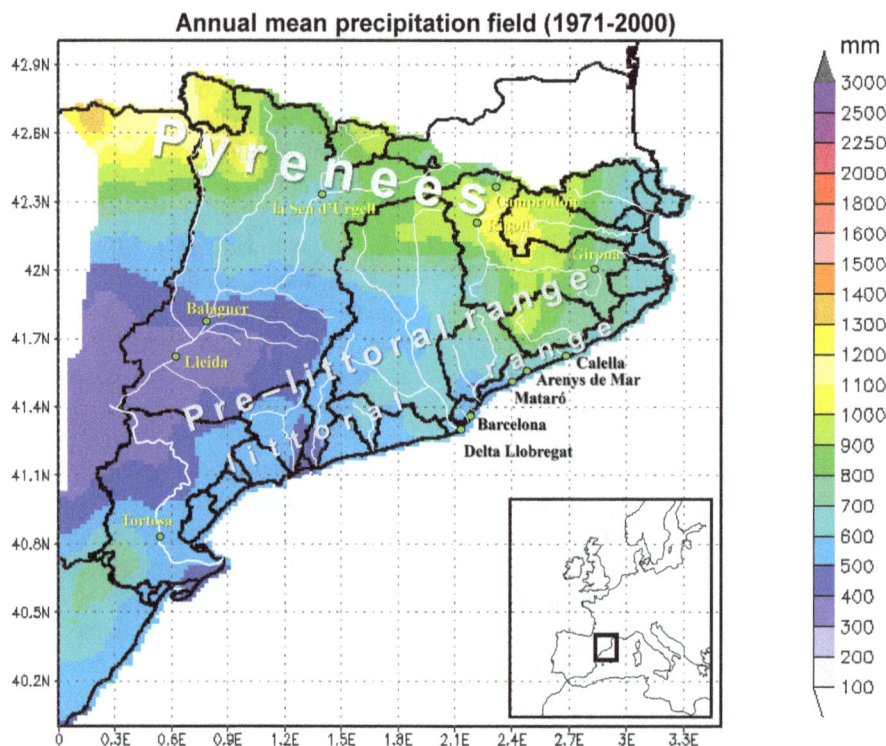

Figure 1. Annual mean precipitation field for Catalonia (1971–2000) at 5 km resolution and computed from a high-density network of observations (adapted from Altava-Ortiz, 2010). The three main mountain ranges of Catalonia are shown over the map. The Catalan river basins and their related main water courses are also displayed. Finally, the location of the analysed flood chronologies is also shown.

1997), as well as the Atlantic influence on the northwestern side of the region, produce high climatic and meteorological contrasts between the different areas. Subsequently, precipitation is characterised by significant spatial and temporal variability. Annual mean rainfall varies from 400 mm in the south to 1300 mm in the north, while extreme daily values can surpass 300 mm, mainly in areas located near the coast or in the Pyrenees.

Heavy rainfall is usually due to convective precipitation normally caused by mesoscale systems and multicellular structures that occur in the late summer, autumn and spring. While isolated thunderstorms associated with convective events are a typical feature of summer weather, convection embedded in stratiform precipitation associated with slight convective events is more frequent during late autumn and winter (Llasat, 2001; Rigo and Llasat, 2004; Barnolas et al., 2010). The annual cycle of convective rainfall shows that maximum levels fall between May and November, with the highest rainfall in August at 64 % of convective precipitation (Llasat et al., 2007).

As a result, this region frequently experiences floods. More than 40 % of municipalities in Catalonia have a high or very high flood risk, according to INUNCAT, a civil protection plan covering flood risks in Catalonia (DGPC, 2012). Generally, flash floods have an impact on coastal torrential basins and cause some level of damage. They are associated with highly convective and locally concentrated heavy rainfall, where accumulated precipitation does not surpass 100 mm. On certain occasions, heavy rainfall produced by multicellular or mesoscale systems can be more extensive in terms of duration and area covered, producing more than 200 mm in less than 3 h or surpassing 400 mm in 24 h. The coastal fringe, where the majority of the population is concentrated, is the most flood-prone area due to the presence of numerous small torrential catchments, delta plains and river or stream mouths (Barnolas and Llasat, 2007). In these populated areas, the relationship between human activity and flooding (in terms of the environment, land-use changes, vulnerability, etc.) is very complex. In addition, some of the catchments are characterised by a non-permanent flooding regime, which means that it is not possible to gauge data on discharge flows.

In this paper we have used flood chronologies for the Ebro, Segre, Ter and Llobregat rivers, in the Maresme and Barcelona regions (with non-permanent flow), which are representative of the key climatic and hydrological regions within Catalonia (Fig. 1). Data measured from the fourteenth century until 2002 were provided within the framework of the SPHERE project (EVG1-CT-1999-00010, Barriendos et al., 2003; Llasat et al., 2005). This database has been updated to cover the temporal period until the year 2012, following the PhD by Barrera-Escoda (2008) and research carried out

Table 1. Main characteristics of the updated flood chronologies: basins, locations, sub-basin surface at the location, temporal coverage and the number of extraordinary (EXT), catastrophic (CAT) and total (TOT = EXT + CAT) floods.

Basin	Location	Surface (km^2)	Period	EXT	CAT	TOT
Segre	La Seu d'Urgell	1233	1451–2012	16	19	35
	Balaguer	7796	1616–2012	9	14	23
	Lleida	11 389	1301–2012	24	27	51
	Total	12 879				
Ebro	Tortosa	82 763	1351–2012	34	15	49
	Total	83 093				
Llobregat	Martorell	4561	1301–2012	98	25	123
	Total	4957				
Ter	Camprodon	280	1616–2012	6	4	10
	Ripoll	738	1576–2012	10	7	17
	Girona	1802	1301–2012	112	22	134
	Total	3010				
Maresme	Calella	10	1671–2012	31	15	46
	Arenys de Mar	28	1666–2012	59	35	94
	Mataró	22	1739–2012	78	39	117
	Total	342				
Pla de Barcelona	Barcelona	100	1351–2012	157	43	200

by the authors until the present day. Flood data have been obtained from documentary sources (from the fourteenth to the twentieth century), and newspapers and technical reports (from the end of the nineteenth century). Flood data basically consist of the recorded date for the flood and details of the damage caused. The locations of all of the flood series are shown in Fig. 1, while the main characteristics of each are shown in Table 1.

Instrumental data are only available for the last two centuries. The monthly (from 1786 onwards) and daily (from 1854 onwards) rainfall series for Barcelona (Barriendos et al., 1997; Barrera-Escoda, 2008) have been updated until the year 2012. Taking into account its location and climate, Barcelona is a good representative of precipitation behaviour along the Catalan coastal region, despite the fact that some flash floods that occurred in the city can be due to local rainfall only that can miss the city or rainfall in Barcelona missing other catchments. This series is the longest set of instrumental data available for rainfall in the western Mediterranean. A representative annual mean average areal precipitation series for the north-eastern Iberian Peninsula (NEIP), with records for over 100 years, has also been used to complete the analysis. This areal precipitation series has been computed from all available monthly precipitation series within the NEIP with continuous temporal records of more than 90 years (Barrera and Llasat, 2004).

A seasonal reconstruction (1500–2000) for the North Atlantic Oscillation (NAO) developed by Luterbacher et al. (2002) was used in order to analyse the relationship be-

tween floods and general circulation. The NAO is a large-scale seesaw in atmospheric mass between the subtropical high and the polar low. It is also the dominant mode of winter climate variability in the North Atlantic region, ranging from central North America to Europe and covering much of northern Asia, although it has an impact on every season. Therefore, it is a good indicator of general circulation in Europe (Hurrell et al., 2003).

Finally, solar activity is taken into account using annually resolved [10]Be measurements for the past 600 years (1389–1994) taken from the North Greenland Ice Sheet Project (NGRIP) 1997 S2 ice core (Berggren et al., 2009). These data are used to study the possible relationship with flood frequency. It is currently the only available series for annually resolved solar activity and covers almost the same period as the flood data.

3 Methodology

The criterion for flood classification (Barriendos et al., 2003; Llasat et al., 2005) is as follows:

- Ordinary or small floods (ORD) do not cause rivers to overflow their banks, cause some damage if activities are being carried out in or near the river at the time, and cause minor damage to hydraulic installations.

- Extraordinary or intermediate floods (EXT) cause the overtopping of riverbanks, inconveniences in the daily life of the local population, and damage to structures near the river or torrent, with possible partial destruction.

- Catastrophic or large floods (CAT) cause the overtopping of river banks and lead to serious damage to or destruction of hydraulic installations, infrastructures, paths and roadways, buildings, livestock, crops, and so on.

This classification allows us to compare historic floods and those that were documented with instrumental records. It also matches similar criteria or methodologies used in other European countries such as in the studies by Sturm et al. (2001), Glaser et al. (2010) or Petrucci et al. (2012). This kind of classification refers to the flood as a risk, including all the factors that could be involved in the produced impact (hazard, vulnerability, exposure, emergency management, etc.). Consequently, the change in any one of these factors may affect the evolution of risk and impact.

For each type of flood and location, a flood frequency index has been compiled to show the annual scale. Ordinary floods have not been considered due to overall heterogeneity (most of them were not recorded throughout history). Each flood series is normalised by taking into account the annual mean value and standard deviation of flood occurrence for the 1901–2000 period as follows:

$$x = (n - m)/s, \tag{1}$$

where n is the yearly flood occurrence, m is the annual mean value, and s is the standard deviation. This procedure has also been applied in Barriendos et al. (2003), where the homogeneity of the series was analysed following the methodology proposed by Lang et al. (1999). On the other hand, this normalisation is necessary in order to cope with different data series and to construct a geographically representative series. Finally, in order to show the changes in the flood indices clearly, all the values are smoothed by low-pass Gaussian filters of 11 and 31 years (Llasat et al., 2005; Barrera et al., 2006). A representative flood index for each category and for Catalonia as a whole has been developed by averaging out the normalised flood series.

Temporal trends are calculated using the flood index series (not smoothed) by means of a linear regression testing its significance level following a Monte Carlo method (Lizevey and Chen, 1983). This technique consists of the following steps: (1) calculation of the linear trend of the original series by the linear fitting of data (minimum squares or linear regression). (2) Generation of 10 000 random permutations of the original series. (3) Calculation of the linear trends for each 10 000 generated series. (4) Calculation of the 97.5 and 2.5 percentiles for the 10 000 calculated linear trends. (5) Finally, if the first linear trend calculated was higher (lower) than the 97.5 (2.5) percentile for its positive (negative) value, then the obtained trend would be significant

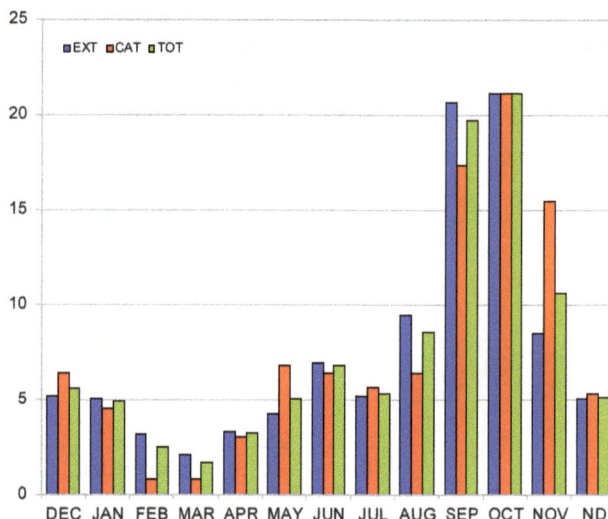

Figure 2. Temporal distribution of historical floods in Catalonia (1301–2012). Extraordinary (EXT), catastrophic (CAT) and total (TOT = EXT + CAT) floods. ND indicates floods with an unknown exact date.

at 95 %. Temporal correlations are calculated using Pearson's linear coefficient, and are applied to the raw data without being smoothed. Spectral analysis is carried out by means of Tukey's power spectrum with a confidence level of 95 %, computed using unsmoothed data. The anomalous periods are those with a high frequency in flood occurrence, which are estimated from the mean value plus the standard deviation of a temporal series. In this work, they have been obtained only for catastrophic floods which are most related to climatic factors. The catastrophic flood series have been smoothed using a low-pass Gaussian filter of 31 years like in other studies (i.e. Llasat et al., 2005; Glaser et al., 2010), and the threshold to consider an anomalous period is defined by flood indexes greater than or equal to 0.1 (mean value + standard deviation ~ 0.1).

4   Results and discussions

4.1   Flood variability

Seasonal flood distribution shows that autumn is the season with the greatest number of floods (54 %), followed by summer (21 %), with the highest number in October (21 %), followed by September (20 %) (Fig. 2). Catastrophic floods are mainly concentrated between September and November, while August also records a high frequency of extraordinary floods (23 % of catastrophic floods are recorded during the summer). These summer events are usually associated with coastal flash floods produced in short, torrential watercourses, which cause extraordinary damage (Llasat et al., 2013). Barcelona is a good example, with a flood percentage of 29 % during the summer.

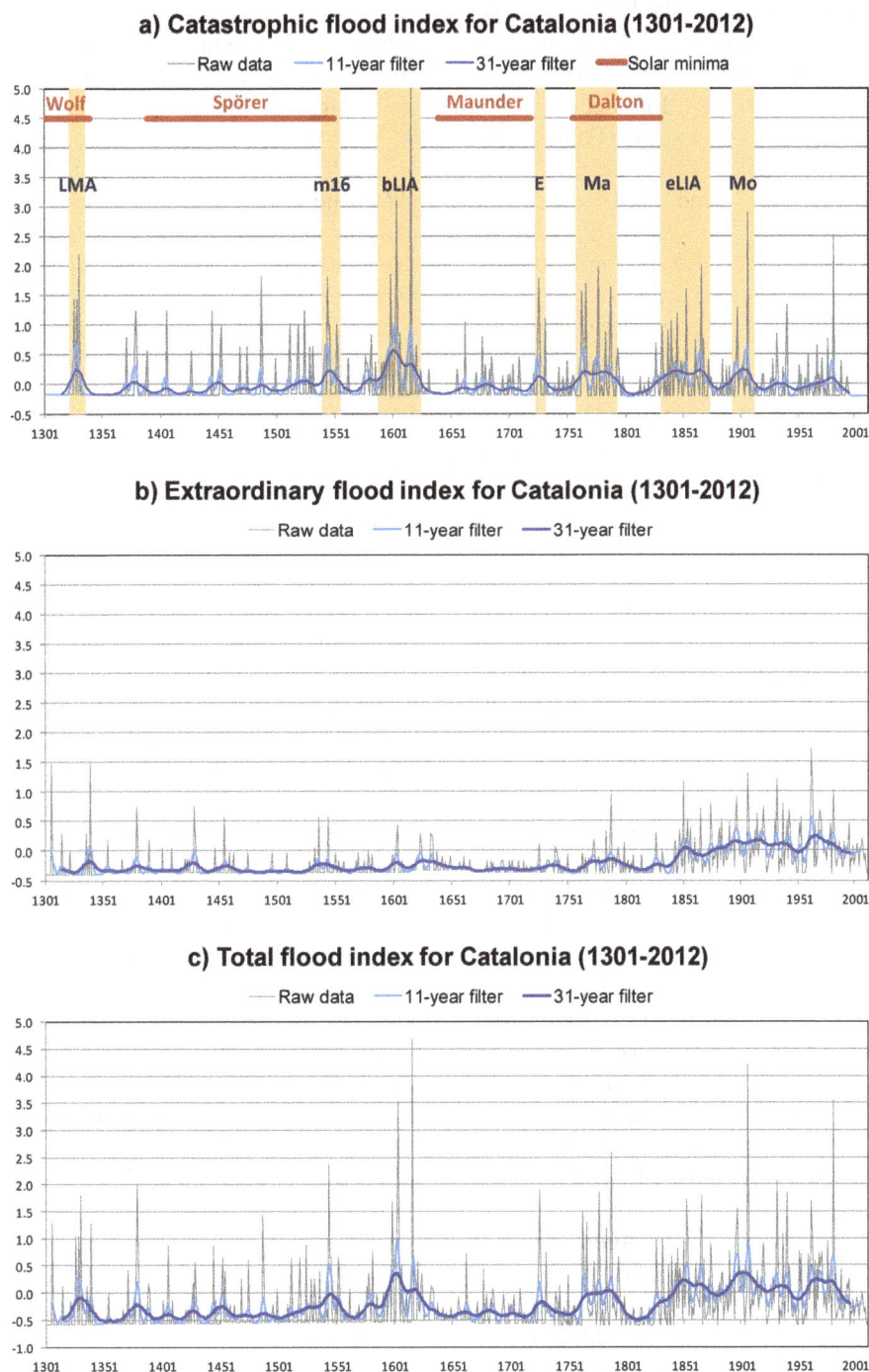

Figure 3. Temporal evolution of catastrophic (**a**), extraordinary (**b**) and total (**c**) flood index series for Catalonia (1301–2012). Data smoothed by low-pass Gaussian filters of 31 and 11 years are also displayed. Anomalous periods of high catastrophic flood frequency and periods of solar minimum (following Usoskin et al., 2007) are highlighted in (**a**): LMA (late Middle Age oscillation), m16 (mid-sixteenth century osc.), bLIA (beginning of LIA osc.), E ("Enlightened" osc.), Ma (Maldà osc.), eLIA (end of LIA oscillation) and Mo ("Modernist" osc.).

The temporal evolution of the annual catastrophic flood index for Catalonia (Fig. 3a) shows a greater inter-annual variability up until the nineteenth century than for the last century alone ($\sigma = 0.26$ vs. $\sigma = 0.22$). This evolution also shows different periods of high and low catastrophic flood frequency.

The periods with the lowest frequency of catastrophic floods are the mid-fourteenth and mid-seventeenth centuries, the beginning of the eighteenth century and the end of the twentieth century and beginning of the twenty-first century. On the other hand, seven different anomalous periods of high flood frequency are highlighted in Fig. 3a: (1) 1325–1334 (late Middle Age oscillation), (2) 1541–1552 (mid-sixteenth century oscillation, Brázdil et al., 1999), (3) 1591–1623 (beginning of Little Ice Age oscillation, LIA, Jones et al., 2001), (4) 1725–1729 (what we propose calling the "Enlightened" oscillation), (5) 1761–1790 (Maldà oscillation, Barriendos and Llasat, 2003), (6) 1833–1871 (end of LIA oscillation, Llasat et al., 2005), and finally (7) 1895–1910 (what we propose calling the "Modernist" oscillation). Similar anomalous periods were found in other flood series for central Europe (Glaser and Stangl, 2004), southern France (Lang et al., 2000; Lang and Cœur, 2002) and northern Italy (Camuffo and Enzi, 1996), especially the third, fifth and sixth oscillations. Most recently, Glaser et al. (2010) have identified four common anomalous periods of high flood frequency for central and eastern Europe: 1540–1600, 1640–1700, 1730–1790 and 1790–1840, which match to a certain extent with the second, third, fourth, fifth and sixth oscillations found in Catalonia. However, most catastrophic floods produced in central and eastern Europe are due to thawing in spring after anomalous high accumulation of ice jams combined with abundant rainfalls. This is not the usual case for Catalonia, but periods with a high frequency of catastrophic floods in both zones are a result of climate anomalies affecting the whole continent.

Trend analysis of temporal evolution for flood indices in Catalonia shows that catastrophic floods do not present a statistically significant trend, whereas extraordinary floods have seen a significant increase, especially from 1850 to the last decades of the twentieth century (Fig. 3b). Extraordinary floods are responsible for the total increase in flooding in Catalonia (Fig. 3c). Due to the diversity of the catchments studied in this paper, we have separated them into inland basins (the Ebro, Segre and Ter catchments) and coastal basins (the Llobregat estuary, and the Barcelona and Maresme regions) for a more in-depth analysis. This means that Fig. 4 shows how small and torrential coastal basins have seen a significant increase in flood frequency ($+0.11/100$ yr), which is 2 times or more the magnitude of those in inland basins. Thus, coastal basins are mainly responsible for the increase in flood frequency, especially for extraordinary floods ($+0.08/100$ yr). In particular, the detailed analysis of urban floods in the city of Barcelona from 1351 onwards shows a significant trend ($+0.26/100$ yr) that is mainly due to the increase in extraordinary floods in the summer ($+0.13/100$ yr). This trend could be due to the strong flood occurrence increase in the middle of the nineteenth century (Fig. 5), which could either be related to the end of the LIA (in the case of France; see Lang et al., 2002) and a possible corresponding increase in convective precipitation, or the notable urban changes in the city. The expansion of the city to the river flanks, but especially the demolition of the walls that frequently acted as flood protection barriers, increased the flood vulnerability and exposure in the new and old city during a period of increasing frequency of high rainfall events (Llasat et al., 2005; Barrera et al., 2006). However, the construction

of the drainage network and the coverage of the wadis in the late nineteenth century and early twentieth century again decreased the vulnerability (Martín-Pascual, 2009).

The large reservoirs built to generate electricity and supply water for agriculture in the main drainage basins in Catalonia during the second half of the twentieth century (between the 1960s and 1970s) have probably lessened the flood hazards in the inland basins, producing less catastrophic floods and a slight increase in extraordinary ones, as is shown in Fig. 4. Elements such as further mitigation measures, changes in land use, exposed assets and climatic factors should also be considered. A larger population living in flood-prone areas and exposed assets in coastal regions could be one of the key factors responsible for the rise in extraordinary floods in the region.

Spectral analysis applied to the annual catastrophic flood index series for Catalonia shows two main significant periodicities (Fig. 6) of 71 and 2.6 years. The first could be related to the Gleissberg solar cycle (~ 70–100 yr) and the second to the quasi-biennial oscillation (QBO, ~ 28–29 months ~ 2.33–2.42 yr), which is present in almost all temporal series involving climatological and meteorological variables (Baldwin et al., 2001). Two less significant periodicities of 4.2 and 2.2 years are also shown, which could also be associated with the QBO.

4.2 Floods versus rainfall

Autumn precipitation contributes less to annual precipitation than autumn floods contribute to the annual total, but it is nonetheless the rainiest season in most of Catalonia. In Barcelona, for example, autumn precipitation represents 37 % of annual precipitation, followed by spring (25 %), for the period 1786–2012. Summer is the driest season, contributing just 18 % of total annual precipitation in Barcelona. However, summer precipitation is mainly convective (nearly 65 % in August; see Llasat, 2001), and is usually associated with thunderstorms or localised heavy rain that produces flash floods in coastal water streams.

On an annual scale, the temporal evolution of precipitation anomalies in Barcelona since 1786 and the same variable for areal precipitation in the NEIP do not show any significant long-term trends (Fig. 7). These same findings are shown for changes to annual maximum daily precipitation in Barcelona (Barrera et al., 2006), and for the number of days that exceed different daily precipitation thresholds (20, 30, 50 and 100 mm in one day; Fig. 8). Subsequently, when considering this common non-significant trend, we can state that annual extreme precipitation has not increased. These results can be extrapolated to the entire Catalan coastal region, although the precipitation series is smaller (Llasat et al., 2009). Similarly, Turco and Llasat (2011) analysed the evolution of extreme precipitation through the ETCCDI (Expert Team on Climate Change Detection and Indices; Zhang et al., 2011) in Catalonia from 1951 to 2003, and did not find any signif-

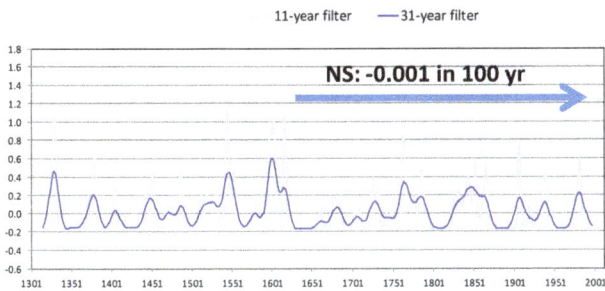

a) Catastrophic flood index for Inland basins of Catalonia (1301-2012)

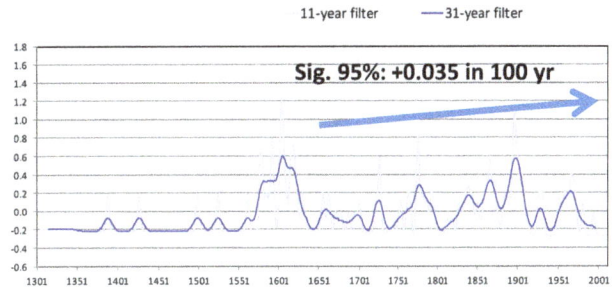

a) Catastrophic flood index for Coastal basins of Catalonia (1301-2012)

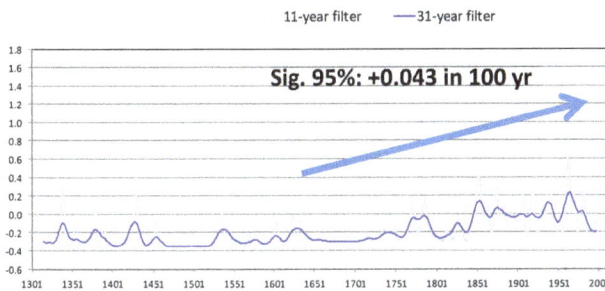

b) Extraordinary flood index for Inland basins of Catalonia (1301-2012)

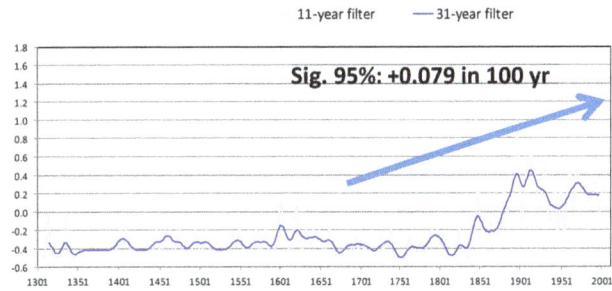

b) Extraordinary flood index for Coastal basins of Catalonia (1301-2012)

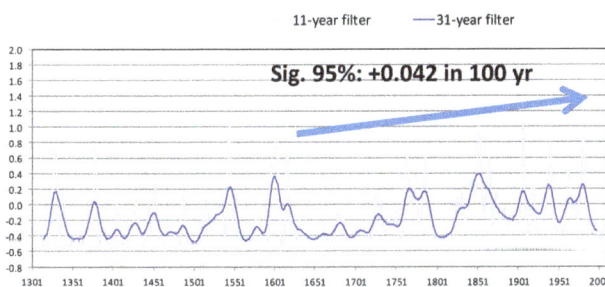

c) Total flood index for Inland basins of Catalonia (1301-2012)

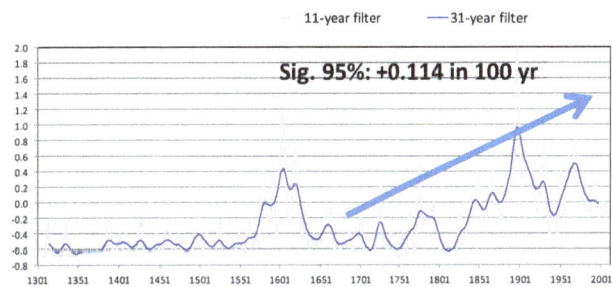

c) Total flood index for Coastal basins of Catalonia (1301-2012)

Figure 4. Temporal evolution (1301–2012) of catastrophic (**a**), extraordinary (**b**) and total (**c**) flood index series for inland (Ebro, Segre and Ter basins; left panels) and coastal basins (Llobregat mouth, Barcelona County and Maresme basins; right panels). The result of applying a trend analysis for each series is also shown.

icant increase in total precipitation, in the highest precipitation amount in a 5 day period (RX5DAY), in the mean precipitation amount on a wet day (SDII), or in the proportion of heavy rainfall (R95p). Working on a seasonal scale and focusing on the city of Barcelona, some trends are shown if the confidence level is reduced to 90 %. In this case, a positive trend of $+0.12\,\mathrm{mm\,yr^{-1}}$ can be found for summer precipitation in the city, which corresponds to a certain extent to the increase in extraordinary floods in the summer (Fig. 5). These results match the increase in precipitation in August in the city, with a significant value of $+2.73\,\mathrm{mm\,yr^{-1}}$ for the 1850–1984 period, as shown by Altava-Ortiz et al. (2010).

The correlation between annual precipitation and the annual index for different types of floods (1786–2012) is very low, and changes over time. The 31 year moving correla-

tions between annual precipitation and catastrophic floods in Barcelona (Fig. 7a) show a maximum value ($r = +0.59$) for 1860–1890 related to the end of the LIA (characterised by very wet years and without flood protection measures). From 1957 on, there is null correlation, maybe as a result of different hydraulic works developed within the city to diminish flood risk and lessen climatic variability (Barrera et al., 2006; Martín-Pascual, 2009).

In Barcelona, the temporal correlation (Fig. 8) between total annual floods and the number of days exceeding thresholds of 20, 30, 50 and $100\,\mathrm{mm\,day^{-1}}$ is relatively low for the 1854–2012 period, which shows the most significant correlation for the number of days exceeding 50 mm ($+0.24$). The correlation between the previous thresholds and the catastrophic flood index also shows the same pattern. Barrera et

a) Catastrophic flood index for Barcelona (1351-2012)

b) Extraordinary flood index for Barcelona (1351-2012)

c) Total flood index for Barcelona (1351-2012)

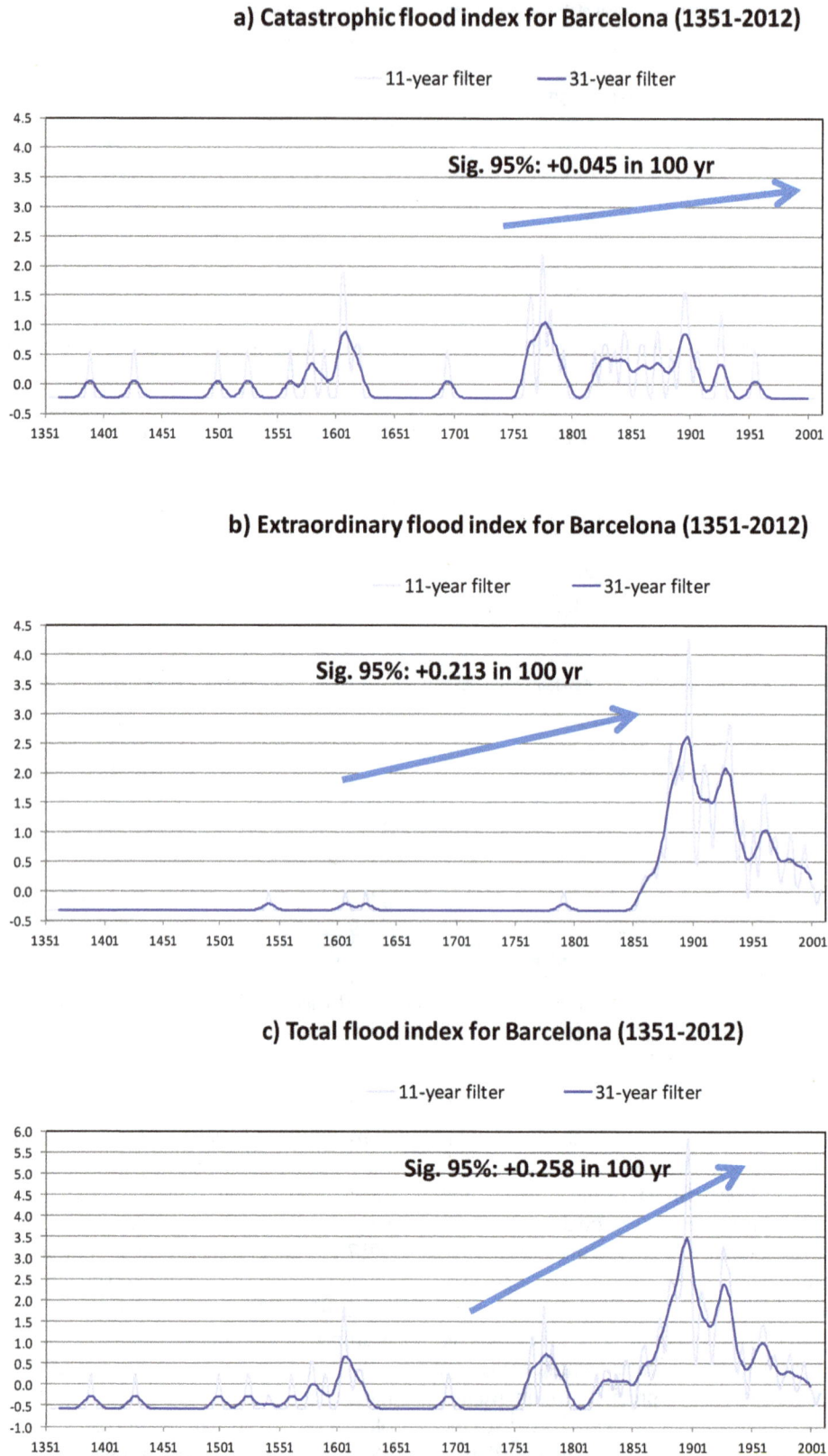

Figure 5. Temporal evolution of catastrophic (**a**), extraordinary (**b**) and total (**c**) urban flood index series for Barcelona (1351–2012). Data are smoothed by low-pass Gaussian filters of 31 and 11 years. The result of applying a trend analysis for each series is also shown.

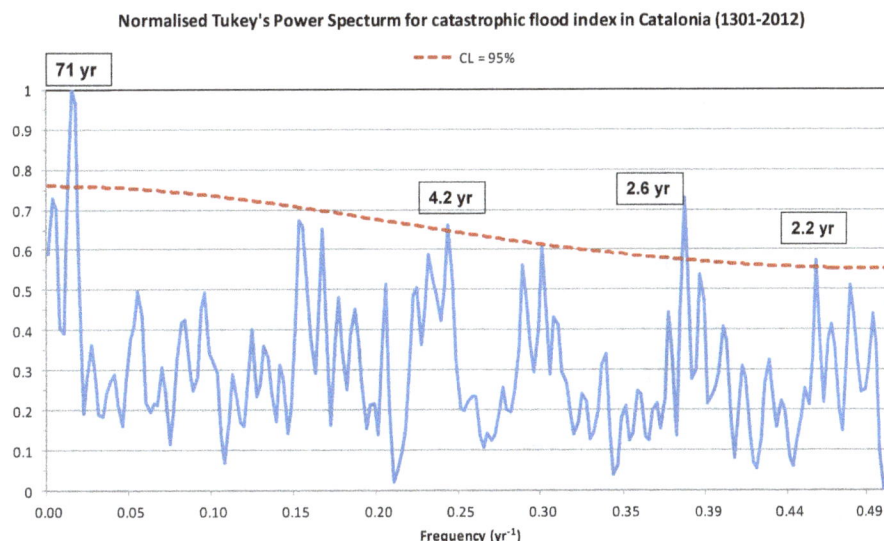

Figure 6. Spectral analysis applied to the catastrophic flood index series for Catalonia (1301–2012) by means of Tukey's power spectrum. The red dashed line represents the 95 % confidence level of the spectral analysis applied.

al. (2006) outlined the fact that urban growth in the city of Barcelona has had an impact on flood vulnerability and flood frequency from the fourteenth century onwards, especially from the late nineteenth and early twentieth centuries. This fact is corroborated when analysing the 31 year moving correlations for the above-mentioned variables for raw data (Fig. 8). Considering the total annual number of floods and the number of days above 50 mm day^{-1}, they reached values above +0.60 for 1936–1985, which could be considered to be a homogenous period because the city drainage system did not experience significant changes (Martín-Pascual, 2009). The construction of water tanks, from the 1990s on, diminished the correlation with the 50 mm threshold and improved the one with the 100 mm threshold, arriving at +0.61. On the contrary, after the wall demolition and initial urban occupation of flood-prone areas, the 20 mm threshold shows the best correlations. This fact corroborates the strong sensitivity of the rainfall threshold associated with floods to changes in vulnerability.

As a result, rainfall patterns alone cannot explain the changes in floods in the region. Extraordinary floods are more frequently related to flood vulnerability and land use, while catastrophic floods are associated with climatic factors.

4.3 Climatic factors

4.3.1 General circulation

While there is no single atmospheric synoptic pattern associated with floods in Catalonia, the analysis of recent and past floods suggests a predominant southern circulation for autumn floods. This implies a relation with a negative NAO phase (Trigo et al., 2004; Llasat et al., 2005), while summer events would be more closely associated with northern cir-

culation in low levels in the region, with a meso-low in eastern Catalonia. In this latter instance, the NAO phase would be more likely to be positive when taking into account the position of the Azores anticyclone during the summer. However, the correlation between NAO and precipitation in the Mediterranean region for the summer season is very low and not significant, due to the important role of local convective developments. For this reason, it does not make sense to analyse the potential relationship between NAO and summer floods.

Following the previous discussion, the influence of general circulation on catastrophic floods in Catalonia has only been analysed for the autumn season (accounting for 55 % of all catastrophic floods; see Fig. 2), by means of the NAO reconstruction set out by Luterbacher et al. (2002) for 1500–2000 (Fig. 9). The temporal correlation between these variables is fairly low, with a significant value of +0.09. The low level of correlation is not unusual, because some floods are isolated local events produced by short heavy rainfall that can take place under positive or negative NAO phases, depending more on mesoscale features than synoptic patterns. Besides this, the frequency of flood events is extremely low, and this affects the significance of any potential correlation. Seasonal shifts in flood distribution could also be produced as a consequence of changes in atmospheric conditions and their effect in precipitation features and snowmelt. On the other hand, the relationship between the NAO and precipitation has changed over time for the twentieth century (Knippertz et al., 2003; Trigo et al., 2004; Beranová and Huth, 2008). For the last 500 years, the temporal evolution of the 31 year moving correlations between floods and NAO (Fig. 9) also show a similar behaviour with a high variability. The highest correlations were found during the central part of the LIA:

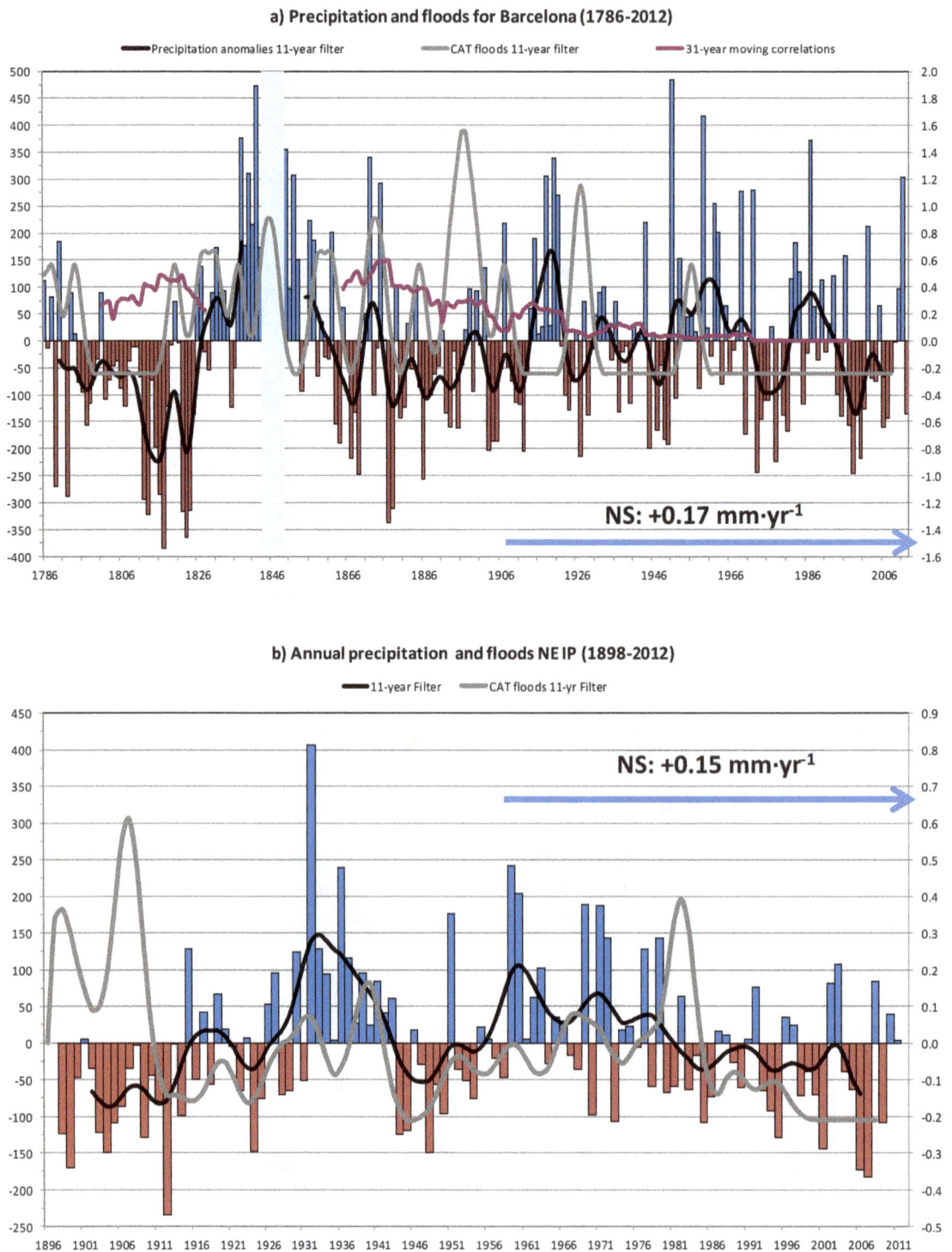

Figure 7. **(a)** Temporal evolution (1786–2012) of annual mean precipitation anomalies (coloured bars), catastrophic flood index series smoothed by an 11 year low-pass Gaussian filter (grey line) and the 31 year moving correlations between precipitation and catastrophic floods (pink line) for Barcelona. **(b)** Temporal evolution (1898–2012) of annual mean areal precipitation anomalies (coloured bars) for the north-eastern Iberian Peninsula (NEIP) and catastrophic flood index series smoothed by an 11 year low-pass Gaussian filter (grey line) for Catalonia. In both figures, black lines are the temporal evolution of annual mean precipitation anomalies smoothed by an 11 year low-pass Gaussian filter. The results of applying a trend analysis in the annual mean precipitation anomalies are also shown.

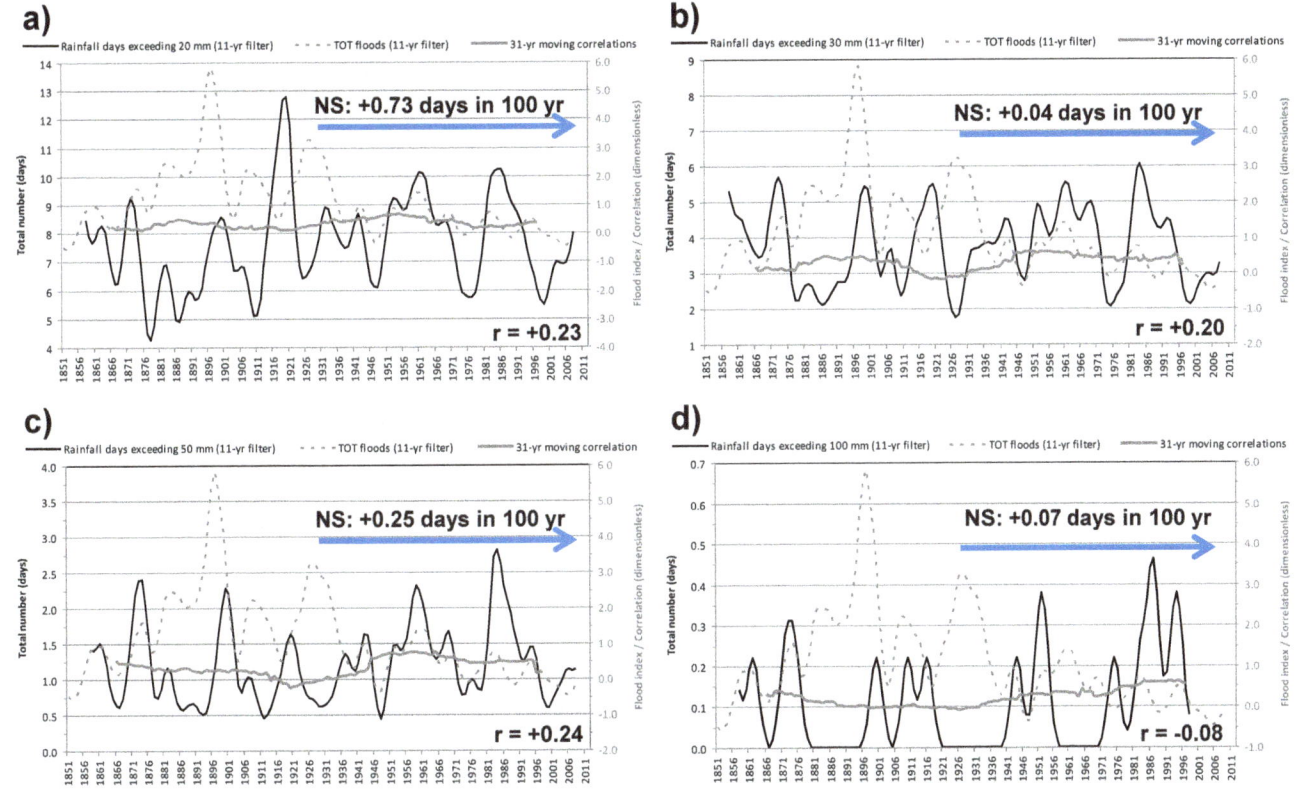

Figure 8. Temporal evolutions of total flood index series and the number of days exceeding a daily precipitation threshold for Barcelona (1854–2012): (**a**) 20 mm, (**b**) 30 mm, (**c**) 50 mm and (**d**) 100 mm. Data have been smoothed by an 11 year Gaussian low-pass filter. The 31 year moving correlations between floods and the different daily precipitation thresholds are also displayed in each panel. The results of applying a trend analysis in the number of days and the temporal correlations between them and floods for all the periods are also shown.

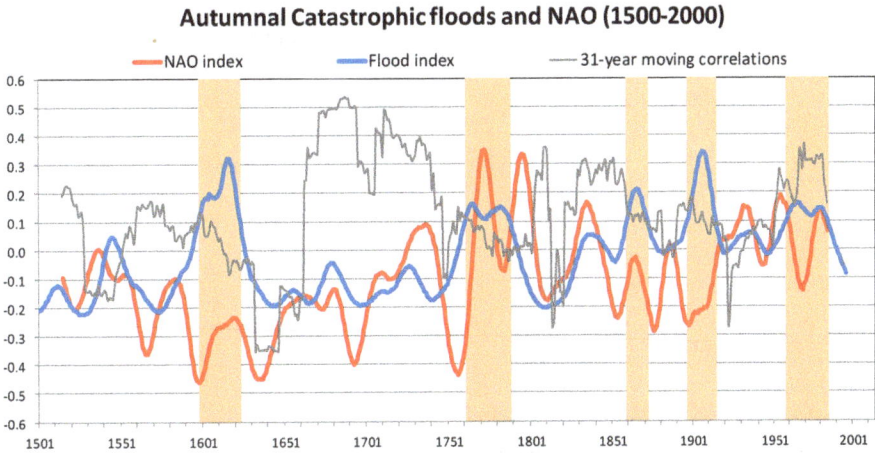

Figure 9. Temporal evolution (1500–2000) of autumnal NAO (Luterbacher et al., 2002; red line), autumnal catastrophic flood index series for Catalonia (blue line) and 31 year moving correlations between both variables (grey line). Data are smoothed by a 31 year low-pass Gaussian filter. The anomalous periods of high flood frequency are highlighted.

a maximum value of $+0.54$ for 1673–1703 and a minimum value of -0.35 for 1618–1648, both periods with low flood frequency. The most important peak in floods was at the beginning of the LIA, a period with the minimum NAO values. The end of the LIA, this so-called "Modernist" oscillation, and the 1970s–1980s, also correspond to a negative NAO phase. On the other hand, the Maldà oscillation was a special period, with both floods and droughts, mainly associated with a great NAO variability (but with a predominance of positive values). A major occurrence of unusual winter thunderstorms and heavy rainfalls related to zonal circulation could explain this anomalous period on an annual scale (Barriendos and Llasat, 2003). Finally, some periods of relatively high flood frequency occur within periods of strong changes to the NAO.

4.3.2   Solar variability

As previously mentioned, the most significant oscillations are shown in different basins in Europe, and are correlated with the main phases of the LIA (Camuffo and Enzi, 1996; Brázdil et al., 1999; Pfister, 1999; Benito et al., 2003; Llasat et al., 2005; Glaser et al., 2010). The first oscillation (LMA) was produced at the end of the Wolf minimum, while the second one (m16) corresponded to the end of Spörer minimum (Fig. 3a). On the contrary, the most significant period of high flood frequency (bLIA) was recorded near maximum solar activity levels that started in 1580 at the beginning of the LIA and, during the Maunder minimum, flooding activity decreased (Fig. 3a). The Wolf, Spörer and Maunder are considered the last "grand minima" (Usoskin et al., 2007), for which sunspot activity decreased considerably more than for the other minima. The first half of the Dalton minimum coincided with the Maldà oscillation, which was characterised by high climatic irregularity accompanied by hydrologic extremes. The maximum period of flood frequency for the end of the LIA (eLIA) corresponds to the highest levels of solar activity recorded between 1849 and 1875. Summarising, flood-rich periods are only related to maximum solar activity at the beginning and end of the LIA.

In central Europe (Brázdil et al., 1999), the periods with the most flooding were recorded in the mid-sixteenth century and in the late Maunder Minimum (1675–1715), corresponding to periods with less solar activity. On the contrary, Vaquero (2004), from a visual inspection, points to a major flood frequency in the Tagus River (Iberian Peninsula) associated with maxima solar activity. This suggests that the regional component is very important. This fact is not strange if we consider the different circulation patterns associated with heavy rainfalls and floods (including snowmelt) and their potential seasonal shift for different periods. Other authors (Borgmark, 2005; Versteegh, 2005; Wilhelm et al., 2012) also attribute the periodicities found in many geological flood records to extraterrestrial forcings, such as centennial and decadal solar cycles.

Figure 10a shows a comparison of the annual evolution in solar activity (1389–1994) taken from the [10]Be concentration and the catastrophic flood index series for Catalonia (1301–2012). [10]Be concentration is high for periods of low solar activity and low for periods of high solar activity. The temporal correlation between raw data is extremely low ($r = -0.06$), but its long-term correlation (with 31 year filtered data) arrive at $r = -0.33$. This last value implies that lower [10]Be concentrations (greater solar activity) would be related to periods with higher flooding activity. However, the 31 year moving correlations between them have changed over time and shown a great variability, with a minimum value of -0.34 for 1600–1630 (a period with high solar activity and the maximum flood frequency), and a maximum value of $+0.37$ for 1726–1756 (a period with a significant increase in solar activity and a decrease in flood frequency). Then, the most flood-rich period recorded at the beginning of the LIA period would be related to a maximum of solar activity and strongly negative NAO values. On the contrary, visual inspection shows that secondary flood peaks could be related to the Wolf, Spörer and Dalton minima, mainly characterised by positive NAO values (Fig. 9). If the analysis focuses solely on autumn (SON; Fig. 10b), the temporal correlation between solar activity and floods is higher, reaching values of -0.08 for raw data and -0.42 for smoothed data. The related 31 year moving correlations also show higher correlations, with a minimum value of -0.38 for 1616–1646 and a maximum value of $+0.62$ for 1726–1756. Finally, it is also interesting to note that a significant change in flood occurrence could be associated with transient periods between solar maxima and minima, and periods of solar maximum (Fig. 10). In addition to this observation, we should mention that the reconstruction of solar activity from [10]Be concentrations does not give the exact dates of the maxima, due to dating uncertainties, and the length of periods of minimum solar activity are not strictly delimitated (Berggren et al., 2009).

The possible link between floods and solar activity is somewhat controversial (Benito et al., 2004; Vaquero, 2004). If we consider the accepted hypothesis that flood-producing mechanisms in the past are similar to those in the present, then marked clusters of historical floods could be associated with changes to the climatic pattern on both a regional and global scale, and, in turn, to changes in solar activity. Although this potential relationship is still a challenge for the scientific community and merits further research, some studies have revealed the influence of solar activity in North Atlantic atmospheric and ocean circulations (Moffa-Sánchez et al., 2014). Following on from this, solar activity would have an impact on the development and trajectory of Atlantic perturbations that could arrive in Europe; low solar irradiance would promote the development of frequent and persistent atmospheric blocking events, with a quasi-stationary high-pressure system in the eastern North Atlantic, which would modify the flow of westerly winds. This kind of pattern could be identified by a positive NAO phase, and would make it

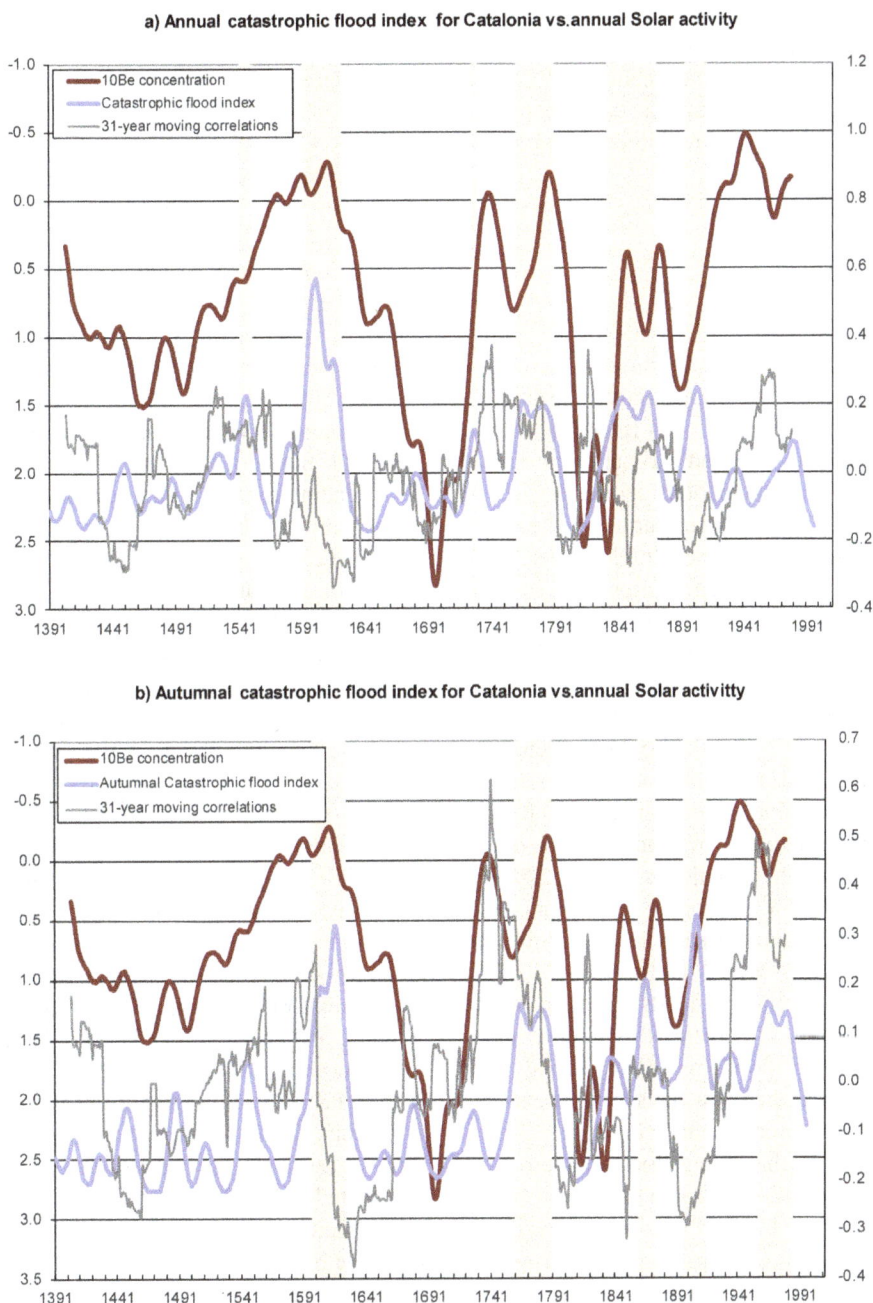

Figure 10. Temporal evolution (1389–2012) of solar activity taken from [10]Be (Berggren et al., 2009; brown lines) versus annual (**a**) and autumnal (**b**) catastrophic flood index series for Catalonia (blue lines). The temporal evolution of the 31 year moving correlations between solar activity and floods are also displayed (grey lines). Solar activity and flood data are smoothed by a 31 year low-pass Gaussian filter. The scale for [10]Be concentration is inverted because its concentration is high for periods of minimum solar activity, and low for periods of maximum solar activity. The anomalous periods of high flood frequency are highlighted in both figures.

more difficult for low-pressure systems and the associated perturbations to arrive in southern Europe. On the other hand, this would favour their arrival in central and northern Europe. Given that floods produced by heavy rainfall are associated with different circulation patterns, the effect of solar variability could change from one region to another. These expla-

nations are coherent with our previous results that correlate positive (negative) NAO with minimum (maximum) solar activity.

Furthermore, maximum solar activity is associated with low-intensity cosmic rays in the stratosphere, which results in greater ozone production in some regions and subsequent warming (Ermolli et al., 2013). Besides the influence of solar activity on general circulation and winds, this differential warming of the stratosphere might influence the development of potential vorticity and the dynamics near the tropopause. Therefore, it could have an important role in heavy rainfall.

5   Conclusions and discussion

The analysis of a reviewed flood index series for the northeastern Iberian Peninsula (1301-2012) shows that catastrophic floods (the most severe ones) are mainly concentrated in the autumn, while extraordinary floods (moderate ones) generally occur in the late summer and early autumn. These results corroborate those obtained for other shorter periods in Catalonia (i.e. Llasat et al., 2014), the similarities to the south of France, and the differences with the central Mediterranean region, where maximum flood frequency is recorded in winter (i.e. Llasat et al., 2013). This distribution is coherent with the bimodal precipitation distribution in Catalonia and with a principal maximum in autumn, usually related to convective events, while the secondary one recorded in spring proceeds from usually stratiform events with not very high intensities (Jansà, 1997; Llasat et al., 2007). On the contrary, summer is characterised by the contribution of convective precipitation to total precipitation, usually caused by thunderstorms or local heavy rainfall giving way to flash floods (Llasat, 2001).

Trend analysis does not show any notable trends for catastrophic floods, although there is a statistically significant trend for extraordinary floods, which implies that a significant increase in the total number of floods has been found since the fourteenth century. This increase is mainly associated with the extraordinary floods recorded in small and torrential basins located near the coast. This trend is exacerbated in the case of urban floods in Barcelona since 1351, and it can be explained by the strong flood increase that occurred in the mid-nineteenth century, which is probably rooted in both climatic (a slight trend in summer precipitation has been found) and human causes (wall destruction, urban development, etc.). Notwithstanding, the increase in extraordinary floods in small coastal basins is mainly related to the marked increase in developed land over the last century, and especially over the last 30 years, which implies a significant change in flood vulnerability and land use, as was recently stated by the IPCC (2012). On the contrary, the evolution of daily rainfall thresholds associated with urban floods in Barcelona corroborates the positive role of prevention measures, such as the improvement of the drainage system, including pluvial tanks or the early warning systems. In effect, this threshold has evolved from 20 mm for the late nineteenth and early twentieth centuries, to 50 mm for most of the twentieth century and 100 mm from the 1990s on. These results point to the necessity of considering floods from a holistic point of view, and support the remarks from

the IPCC (2012) or the recent review paper from Hall et al. (2014).

The temporal evolution shows seven different anomalous periods of high catastrophic flood for Catalonia: (1) 1325–1334 (late Middle Age oscillation), (2) 1541–1552 (mid-sixteenth century oscillation), (3) 1591–1623 (beginning of the LIA), (4) 1725–1729 (what we propose calling the "Enlightened" oscillation), (5) 1761–1790 (Maldà oscillation), (6) 1833–1871 (end of the LIA), and finally (7) 1895–1910 (what we propose calling the "Modernist" oscillation). The anomalous periods 2, 3, 5 and 6 coincide totally or partially with those common periods with high flood frequency for central and eastern Europe identified by Glaser et al. (2010). This fact points to the search for climatic causes that would affect the northern atmospheric circulation pattern.

Attending to the fact that extraordinary floods can be seriously affected by non-climatic factors, only the correlation between catastrophic floods and the NAO and solar variability has been analysed. Results have shown that the correlation with the NAO for autumn changes over time: in the second half of the seventeenth century, it reaches +0.54, while in the first half of the same century, it reaches just −0.35, both periods with low flood frequency. This is not unusual in itself, given that the present correlation between the NAO and precipitation in this specific region and season does not reach −0.50 (Barrera and Llasat, 2004; Martín-Vide and López-Bustins, 2006). The detailed analysis for all of Spain in the last 100 years also shows how negative correlation has increased in the last few decades (Barrera-Escoda, 2008). This complex relationship between the NAO and flooding has also been found for Britain (MacDonald, 2014). However, the most significant flood oscillations occur at the beginning and end of the LIA, and this so-called "Modernist" oscillation coincides with a strong negative NAO phase, as for the last anomalous period in the twentieth century.

Spectral analysis shows two key periodicities for the annual catastrophic flood index. The first one (71 years) could be related to the Gleissberg solar cycle, and the second one (2.6 years) to the quasi-biennial oscillation. Although the possible links between floods and solar activity are still controversial, the correlation between [10]Be concentration (maximum when solar activity reaches a minimum) and the catastrophic floods may provide some interesting information. Up to the present moment, approaches to this issue have been mainly qualitative (i.e. Vaquero, 2004). This correlation shows a significant value of −0.33 (−0.42 if only the autumn season is taken into consideration) and points to the fact that major solar activity is usually associated with periods of higher flooding activity. This is particularly important at the beginning and end of the LIA, and corroborates the positive significant relationship between solar magnetic activity and floods in the UK (MacDonald, 2014). Significant changes in flood occurrence could be associated with transient periods between solar maxima and minima. Recent studies (Martín-Puertas et al., 2012; Moffa-Sánchez et al.,

2014) have revealed the possible influence of solar activity through changes to solar irradiance (mainly in the ultraviolet) and changes in cosmic rays and solar particles as they arrive in the stratosphere. Although these studies mainly refer to winter–early spring North Atlantic circulation (both atmospheric and ocean circulations), and floods in Catalonia are mainly produced in autumn, they provide a departure point for future research on this possible "top-down" mechanism. Changes in solar irradiance over the North Atlantic would be amplified through atmospheric feedbacks including the Atlantic meridional overturning circulation, which would in turn affect the formation of persistent atmospheric blocking events. The latter factor would also affect the predominant circulation patterns (i.e. NAO), with the consequent differential regional influence for heavy precipitation. Less interaction between cosmic rays and the ozone in the stratosphere during periods of maximum solar activity would increase ozone presence, diminish UV radiation arriving on the Earth's surface, and increase the stratospheric temperature in some regions, with a consequent impact on the dynamics of the high atmosphere.

Acknowledgements. The authors would like to thank B. Sanahuja and COST Action TOSCA (ES1005) for their inputs and explanations of solar variability, and A. Sánchez-Cabeza for the information provided about the applications of ^{10}Be for climate analysis. We would also like to acknowledge G. Blösch and his FloodChange project for supporting our contribution.

Edited by: R. Brázdil

References

Adhikari, P., Hong, Y., Douglas, K. R., Kirschbaum, D. B., Gourley, J., Adler, R., and Brakenridge, G. R.: A digitized global flood inventory (1998–2008): compilation and preliminary results, Nat. Hazards, 55, 405–422, 2010.

Alpert, P., Ben-gai, T., Baharad, A., Benjamini, Y., Yekutieli, D., Colacino, M., Diodato, L., Ramis, C., Homar, V., Romero, R., Michaelides, S., and Manes, A.: The paradoxical increase of Mediterranean extreme daily rainfall in spite of decrease in total values, Geophys. Res. Lett., 29, 1536, doi:10.1029/2001GL013554, 2002.

Altava-Ortiz, V.: Caracterització i monitoratge de les sequeres a Catalunya i nord del País Valencià, Càlcul d'escenaris climàtics per al segle XXI (Characterising and monitoring of droughts in Catalonia and north of Valencian Country. Calculation of climate scenarios for the 21st century), PhD, Internal publication, Department of Astronomy and Meteorology, University of Barcelona, Barcelona, Spain, 296 pp., 2010.

Altava-Ortiz, V., Llasat, M. C., Ferrari, E., Atencia, A., and Sirangelo, B.: Monthly rainfall changes in central and western mediterranean basins, at the end of the 20th and beginning of the 21st centuries, Int. J. Climatol., 31, 1943–1958, 2010.

Baldwin, M. P., Gray, L. J., Dunkerton, T. J., Hamilton, K., Haynes, P. H., Randel, W. J., Holton, J. R., Alexander, M. J., Hirota, I.,

Horinouchi, T., Jones, D. B. A., Kinnersley, J. S., Marquardt, C., Sato, K., and Takahashi, M.: The Quasi-Biennial Oscillation, Rev. Geophys., 39, 179–229, 2001.

Barnolas, M. and Llasat, M. C.: A flood geodatabase and its climatological applications: the case of Catalonia for the last century, Nat. Hazards Earth Syst. Sci., 7, 271–281, doi:10.5194/nhess-7-271-2007, 2007.

Barnolas, M., Rigo, T., and Llasat, M. C.: Characteristics of 2-D convective structures in Catalonia (NE Spain): an analysis using radar data and GIS, Hydrol. Earth Syst. Sci., 14, 129–139, doi:10.5194/hess-14-129-2010, 2010.

Barredo, J. I., Saurí, D., and Llasat, M. C.: Assessing trends in insured losses from floods in Spain 1971–2008, Nat. Hazards Earth Syst. Sci., 12, 1723–1729, doi:10.5194/nhess-12-1723-2012, 2012.

Barrera, A. and Llasat, M. C.: Evolución regional de la precipitación en España en los últimos 100 años, Ingeniería Civil, 135, 105–113, 2004.

Barrera, A., Llasat, M. C., and Barriendos, M.: Estimation of extreme flash flood evolution in Barcelona County from 1351 to 2005, Nat. Hazards Earth Syst. Sci., 6, 505–518, doi:10.5194/nhess-6-505-2006, 2006.

Barrera-Escoda A.: Evolución de los extremos hídricos en Catalunya en los últimos 500 años y su modelización regional (Evolution of hydric extremes in Catalonia during the last 500 years and its regional modelling), PhD, Internal publication, Department of Astronomy and Meteorology, University of Barcelona, Barcelona, Spain, 319 pp., http://www.zucaina.net/Publicaciones/Barrera-Escoda-TESIS-2008.pdf (last access: 26 January 2015), 2008.

Barriendos, M. and Llasat, M. C.: The case of the 'Maldá' Anomaly in the Western Mediterranean basin (AD 1760–1800): An example of a strong climatic variability, Climatic Change, 61, 191–216, 2003.

Barriendos, M., Gómez, B., and Peña, J. C.: Old series of meteorological readings for Madrid and Barcelona (1780–1860), Documentary and observed characteristics, in: Advances in Historical Climatology in Spain, edited by: Martín-Vide, J., Oikos-Tau, Barcelona, 157-172, 1997.

Barriendos, M., Coeur, D., Lang, M., Llasat, M. C., Naulet, R., Lemaitre, F., and Barrera, A.: Stationarity analysis of historical flood series in France and Spain (14th–20th centuries), Nat. Hazards Earth Syst. Sci., 3, 583–592, doi:10.5194/nhess-3-583-2003, 2003.

Barriendos, M., Ruiz-Bellet, J. L., Tuset, J., Mazón, J., Balasch, J. C., Pino, D., and Ayala, J. L.: The "Prediflood" database of historical floods in Catalonia (NE Iberian Peninsula) AD 1035–2013, and its potential applications in flood analysis, Hydrol. Earth Syst. Sci., 18, 4807–4823, doi:10.5194/hess-18-4807-2014, 2014.

Benito, G., Díez-Herrero, A., and Fernández de Villalta, M.: Magnitude and frequency of flooding in the Tagus Basin (central Spain) over the last millennium, Climatic Change, 58, 171–192, 2003.

Benito, G., Díez-Herrero, A., and Fernández de Villalta, M.: Flood response to solar activity in the Tagus Basin (Central Spain) over the last millennium, Response to J. M. Vaquero 'Solar Signal in the Number of Floods Recorded for the Tagus River over the Last Millennium', Climatic Change, 66, 27–28, 2004.

Beranová, R. and Huth, R.: Time variations of the effects of circulation variability modes on European temperature and precipitation in winter, Int. J. Climatol., 28, 139–158, 2008.

Berggren, A. M., Beer, J., Possnert, G., Aldahan, A., Kubik, P., Christl, M., Johnsen, S. J., Abreu, J., and Vinther, B. M.: A 600-year annual [10]Be record from the NGRIP ice core, Greenland, Geophys. Res. Lett., 36, L11801, doi:10.1029/2009GL038004, 2009.

Borgmark, A.: Holocene climate variability and periodicities in south-central Sweden, as interpreted from peat humification analysis, Holocene, 15, 387–395, 2005.

Brázdil, R., Glaser, R., Pfister, C., Antoine, J. M., Barriendos, M., Camuffo, D., Deutsch, M., Enzi, S., Guidoboni, E., and Rodrigo, F. S.: Flood events of selected rivers of Europe in the Sixteenth Century, Climatic Change, 43, 239–285, 1999.

Brázdil, R., Kundzewicz, Z. W., and Benito, G.: Historical hydrology for studying flood risk in Europe, Hydrolog. Sci. J., 51, 739–764, 2006.

Brázdil, R., Chromá, K., Řezníčková, L., Valášek, H., Dolák, L., Stachoň, Z., Soukalová, E., and Dobrovolný, P.: The use of taxation records in assessing historical floods in South Moravia, Czech Republic, Hydrol. Earth Syst. Sci., 18, 3873–3889, doi:10.5194/hess-18-3873-2014, 2014.

Camuffo, D. and Enzi, S.: The analysis of two bi-millennial series: Tiber and Po river floods, in: Climatic Variations and Forcing Mechanisms of the Last 2000 Years, edited by: Jones, P. D., Bradley, R. S., and Jouzel, J., Springer, Berlin, 433–450, 1996.

Dankers, R. and Feyen, L.: Flood hazard in Europe in an ensemble of regional climate scenarios, J. Geophys. Res., 114, D16108, doi:10.1029/2008JD011523, 2009.

DGPC: INUNCAT, Pla Especial d'emergències per inundacions de Catalunya, Direcció General de Protecció Civil de Catalunya, Generalitat de Catalunya, 133 pp', 2012.

Di Baldassarre, G., Castellarin, A., and Brath, A.: Analysis on the effects of levee heightening on flood propagation: some thoughts on the River Po, Hydrolog. Sci. J., 54, 1007–1017, 2009.

Ermolli, I., Matthes, K., Dudok de Wit, T., Krivova, N. A., Tourpali, K., Weber, M., Unruh, Y. C., Gray, L., Langematz, U., Pilewskie, P., Rozanov, E., Schmutz, W., Shapiro, A., Solanki, S. K., and Woods, T. N.: Recent variability of the solar spectral irradiance and its impact on climate modelling, Atmos. Chem. Phys., 13, 3945–3977, doi:10.5194/acp-13-3945-2013, 2013.

García, J. A., Gallego, M. C., Serrano, A., and Vaquero, J. M.: Trends in Block-Seasonal Extreme Rainfall over the Iberian Peninsula in the Second Half of the Twentieth Century, J. Climate, 20, 113–130, 2007.

Glaser, R.: Data and methods of climatological evaluation in historical climatology, Hist. Soc. Res., 21, 56–88, 1996.

Glaser, R. and Stangl, H.: Climate and floods in Central Europe since AD 1000: Data, methods, results and consequences, Surv. Geophys., 25, 485–510, 2004.

Glaser, R., Riemann, D., Schönbein, J., Barriendos, M., Brázdil, R., Bertolin, C., Camuffo, D., Deutsch, M., Dobrovolný, P., van Engelen, A., Enzi, S., Halíčková, Koening, S. J., Kotyza, O., Limanówka, D., Macková, J., Sghedoni, M., Martin, B., and Himmelsbach, I.: The variability of European floods since AD 1500, Climatic Change, 101, 235–256, 2010.

González-Hidalgo, J. C., López-Bustins, J. A., Stepánek, P., Martín-Vide, J., and De Luis, M.: Monthly precipitation trends on the Mediterranean fringe of the Iberian Peninsula during the second-half of the twentieth century (1951–2000), Int. J. Climatol., 29, 1415–1429, 2009.

Goodess, C. M. and Jones, P. D.: Links between circulation and changes in the characteristics of Iberian rainfall, Int. J. Climatol., 22, 1593–1615, 2002.

Hall, J., Arheimer, B., Borga, M., Brázdil, R., Claps, P., Kiss, A., Kjeldsen, T. R., Kriaučiūnienė, J., Kundzewicz, Z. W., Lang, M., Llasat, M. C., Macdonald, N., McIntyre, N., Mediero, L., Merz, B., Merz, R., Molnar, P., Montanari, A., Neuhold, C., Parajka, J., Perdigão, R. A. P., Plavcová, L., Rogger, M., Salinas, J. L., Sauquet, E., Schär, C., Szolgay, J., Viglione, A., and Blöschl, G.: Understanding flood regime changes in Europe: a state-of-the-art assessment, Hydrol. Earth Syst. Sci., 18, 2735–2772, doi:10.5194/hess-18-2735-2014, 2014.

Heine, R. and Pinter, N.: Levee effects upon flood levels: an empirical assessment, Hydrol. Process., 26, 3225–3240, 2012.

Herget, J., Roggenkamp, T., and Krell, M.: Estimation of peak discharges of historical floods, Hydrol. Earth Syst. Sci., 18, 4029–4037, doi:10.5194/hess-18-4029-2014, 2014.

Hurrell, J.W., Kushnir, Y., Visbeck, M., and Ottersen, G.: An Overview of the North Atlantic Oscillation, in: The North Atlantic Oscillation: Climate Significance and Environmental Impact, edited by: Hurrell, J. W., Kushnir, Y., Ottersen, G., and Visbeck, M., American Geophysical Union, Geophys. Monogr. Ser., 134, 1–35, 2003.

IPCC: Managing the risks of extreme events and disasters to advance climate change adaption (SREX), Intergovernmental Panel on Climate Change, Cambridge University Press, Cambridge, 582 pp., 2012.

IPCC: Climate Change 2013: The Physical Science Basis, in: Contribution of Working Group I to the Fifth Assessment Report of the Intergovernmental Panel on Climate Change, Cambridge University Press, Cambridge, 1535 pp., 2013.

Jansà, A.: A general view about Mediterranean meteorology: cyclones and hazardous weather, In Proceedings of the INM/WMO International Symposium on Cyclones and Hazardous Weather in the Mediterranean, Instituto Nacional de Meteorología and Universitat de les Illes Balears, Palma de Mallorca, 33–42, 1997.

Jones, P. D., Osborn, T. J., and Briffa, K. R.: The evolution of climate over the last millennium, Science, 292, 662–667, 2001.

Kiss, A., Brázdil, R., and Blöschl, G. (Eds.): Floods and their changes in historical times – a European perspective, HESSD special issue, 2014.

Knippertz, P., Ulbrich, U., Marques, F., and Corte-Real, J.: Decadal changes in the link between El Niño and springtime North Atlantic oscillation and European-North African rainfall, Int. J. Climatol., 23, 1293–1311, 2003.

Lang, M., and Cœur, D.: Flood knowledge: history, hydraulics and hydrology, Case study on three French rivers, In Chinese-French Conference on water resources, Shanghai/Suzhou, 6–9 November, AFCRST, Paris, 96–102, 2002.

Lang, M., Ouarda, T., and Bobée, B.: Towards operational guidelines for over-threshold modeling, J. Hydrol., 225, 103–117, 1999.

Lang, M., Naulet, R., Brochot, S., and Cœur, D.: Historisque-Isere et torrents affluents. Utilisation de l'information historique pour une meilleure définition du risque d'inondation, Rapport Final, CEMAGREF, Lyon, 248 pp., 2000.

Lehner, B., Döll, P., Alcamo, J., Henrichs, T., and Kaspar, F.: Estimating the impact of global change on flood and drought risks in Europe: A continental, integrated analysis, Climatic Change, 75, 273–299, 2006.

Livezey, R. E. and Chen, W. Y.: Statistical field significance and its determination by Monte Carlo techniques, Mon. Weather Rev., 111, 46–59, 1983.

Llasat, M. C.: An objective classification of rainfall events on the basis of their convective features: Application to rainfall intensity in the North-East of Spain, Int. J. Climatol., 21, 1385–1400, 2001.

Llasat, M. C. and Corominas, J.: Riscos associats al clima, in: Segon informe sobre el canvi climàtic a Catalunya, edited by: Llebot, J. E., Institut d'Estudis Catalans and Generalitat de Catalunya, Barcelona, 243–307, 2010.

Llasat, M. C. and Quintas, L.: Stationarity of monthly rainfall series, since the middle of the XIXth century, Application to the case of peninsular Spain, Nat. Hazards, 31, 613–622, 2004.

Llasat, M. C., Barriendos, M., Barrera, A., and Rigo, T.: Floods in Catalonia (NE Spain) since the 14th Century, Climatological and meteorological aspects from historical documentary sources and old instrumental records, J. Hydrol., 313, 32–47, 2005.

Llasat, M. C., Ceperuelo, M., and Rigo, T.: Rainfall regionalization on the basis of the precipitation convective features using a rain-gauge network and weather radar observations, Atmos. Res., 83, 415–426, 2007.

Llasat, M. C., Llasat-Botija, M., Barnolas, M., López, L., and Altava-Ortiz, V.: An analysis of the evolution of hydrometeorological extremes in newspapers: the case of Catalonia, 1982–2006, Nat. Hazards Earth Syst. Sci., 9, 1201–1212, doi:10.5194/nhess-9-1201-2009, 2009.

Llasat, M. C., Llasat-Botija, M., Petrucci, O., Pasqua, A. A., Rosselló, J., Vinet, F., and Boissier, L.: Towards a database on societal impact of Mediterranean floods within the framework of the HYMEX project, Nat. Hazards Earth Syst. Sci., 13, 1337–1350, doi:10.5194/nhess-13-1337-2013, 2013.

Llasat, M. C., Marcos, R., Llasat-Botija, M., Gilabert, J., Turco, M., and Quintana, P.: Flash flood evolution in North-Western Mediterranean, Atmos. Res., 149, 230–243, 2014.

López-Moreno, J. I., Vicente-Serrano, S. M., Angulo-Martínez, M., Beguería, S., and Kenawy, A.: Trends in daily precipitation on the northeastern Iberian Peninsula, 1955–2006, Int. J. Climatol., 30, 1026–1041, 2010.

Luterbacher, J., Xoplaki, E., Dietrich, D., Jones, P. D., Davies, T. D., Portis, D., González-Rouco, J. F., von Storch, H., Gyalistras, D., Casty, C., and Wanner, H.: Extending North Atlantic Oscillation reconstructions back to 1500, Atmos. Sci. Lett., 2, 114–124, 2002.

MacDonald, N.: Millennial scale variability in high magnitude flooding across Britain, Hydrol. Earth Syst. Sci. Discuss., 11, 10157–10178, doi:10.5194/hessd-11-10157-2014, 2014.

Martín-Pascual, M.: Barcelona, Aigua i ciutat, Fundació AGBAR, Barcelona, 455 pp., 2009.

Martín-Puertas, C., Matthes, K., Brauer, A., Muscheler, R., Hansen, F., Petrick, Ch., Aldahan, A., Possnert, G., and van Geel, B.: Regional atmospheric circulation shifts induced by a gran solar minimum, Nat. Geosci., 5, 397–401, 2012.

Martín-Vide, J. and Lopez-Bustins, J.-A.: The Western Mediterranean Oscillation and rainfall in the Iberian Peninsula, Int. J. Climatol., 26, 1455–1475, 2006.

Merz, B., Vorogushyn, S., Uhlemann, S., Delgado, J., and Hundecha, Y.: HESS Opinions "More efforts and scientific rigour are needed to attribute trends in flood time series", Hydrol. Earth Syst. Sci., 16, 1379–1387, doi:10.5194/hess-16-1379-2012, 2012.

Moffa-Sánchez, P., Born, A., Hall, I. R., Thornalley, D. J. R., and Barker, S.: Solar forcing of North Atlantic surface temperature and salinity over the past millennium, Nat. Geosci., 7, 275–278, 2014.

Mouri, G., Kanae, S., and Oki, T.: Long-term changes in flood event patterns due to changes in hydrological distribution parameters in a rural-urban catchment, Shikoku, Japan, Atmos. Res., 10, 164–177, 2011.

Munich Re: Annual review: Natural catastrophes 2005, Topics Geo, Munich Reinsurance Group, Munich, 56 pp., http://www.preventionweb.net/files/1609_topics2005.pdf (last access: 26 January 2015), 2006.

Naef, F., Scherrer, S., and Weiler, M.: A process based assessment of the potential to reduce flood runoff by land use change, J. Hydrol., 267, 74–79, 2002.

Petrucci, O., Pasqua, A. A., and Polemio, M.: Flash flood occurrences since 17th century in steep drainage basins in southern Italy, Environ. Manage., 50, 807–818, 2012.

Pfister, C.: Wetternachhersage: 500 Jahre Klimavariationen und Naturkatastrophen, Verlag Paul Haupt, Bern, 1999.

Pryor, S. C., Howe, J. A., and Kunkel, K. E.: How spatially coherent and statistically robust are temporal changes in extreme precipitation in the contiguous USA?, Int. J. Climatol., 45, 31–45, 2009.

Remo, J., Megan, C., and Pinter, N.: Hydraulic and flood-loss modeling of levee, floodplain, and river management strategies, Middle Mississippi River, USA, Nat. Hazards, 61, 551–575, 2012.

Retsö, D.: Documentary evidence of historical floods and extreme rainfall events in Sweden 1400–1800, Hydrol. Earth Syst. Sci. Discuss., 11, 10085–10116, doi:10.5194/hessd-11-10085-2014, 2014.

Rigo, T. and Llasat, M. C.: A methodology for the classification of convective structures using meteorological radar: Application to heavy rainfall events on the Mediterranean coast of the Iberian Peninsula, Nat. Hazards Earth Syst. Sci., 4, 59–68, doi:10.5194/nhess-4-59-2004, 2004.

Rodrigo, F. S.: Changes in the probability of extreme daily precipitation observed from 1951 to 2002 in the Iberian Peninsula, Int. J. Climatol., 30, 1512–1525, 2010.

Rodrigo, F. S. and Trigo, R. M.: Trends in daily rainfall in the Iberian Peninsula from 1951 to 2002, Int. J. Climatol., 27, 513–529, 2007.

Sivapalan, M., Blöschl, G., Merz, R., and Gutknecht, D.: Linking flood frequency to long-term water balance: Incorporating effects of seasonality, Water Resour. Res., 41, W06012, doi:10.1029/2004WR003439, 2005.

Sturm, K., Glaser, R., Jacobeit, J., Deutsch, M., Brázdil, R., and Pfister, C.: Floods in Central Europe since AD 1500 and their relation to the atmospheric circulation, Petermanns Geogr. Mit., 148, 18–27, 2001.

Trigo, R. M., Pozo-Vázquez, D., Osborn, T. J., Castro-Díez, Y., Gámiz-Fortis, S., and Esteban-Parra, M. J.: North Atlantic oscillation influence on precipitation, river flow and water resources in the Iberian Peninsula, Int. J. Climatol., 24, 925–944, 2004.

Turco, M. and Llasat, M. C.: Trends in indices of daily precipitation extremes in Catalonia (NE Spain), 1951–2003, Nat. Hazards Earth Syst. Sci., 11, 3213–3226, doi:10.5194/nhess-11-3213-2011, 2011.

Usoskin, I. G., Solanki, S. K., and Kovaltsov, G. A.: Grand minima and maxima of solar activity: new observational constraints, Astron. Astrophys., 471, 301–309, 2007.

Vaquero, J. M.: Solar signal in the number of floods recorded for the Tagus river basin over the last millennium, Climatic Change, 66, 23–26, 2004.

Versteegh, G. J. M.: Solar Forcing of Climate, 2: Evidence from the Past, Space Sci. Rev., 120, 243–286, 2005.

Wilhelm, B., Arnaud, F., Sabatier, P., Crouzet, C., Brisset, E., Chaumillion, E., Disnar, J. R., Guiter, F., Malet, E., Reyss, J. L., Tachikawa, K., Bard, E., and Delannoy, J. J.: 1400 years of extreme precipitation patterns over the Mediterranean French Alps and possible forcing mechanisms, Quaternary Res., 78, 1–12, 2012.

Zhang, X., Alexander, L., Hegerl, G. C., Jones, P., Tank, A. K., Peterson, T. C., Trewin, B., and Zwiers, F. W.: Indices for monitoring changes in extremes based on daily temperature and precipitation data, WIREs Clim. Change, 2, 851–870, 2011.

Permissions

All chapters in this book were first published in HESS, by Copernicus Publications; hereby published with permission under the Creative Commons Attribution License or equivalent. Every chapter published in this book has been scrutinized by our experts. Their significance has been extensively debated. The topics covered herein carry significant findings which will fuel the growth of the discipline. They may even be implemented as practical applications or may be referred to as a beginning point for another development.

The contributors of this book come from diverse backgrounds, making this book a truly international effort. This book will bring forth new frontiers with its revolutionizing research information and detailed analysis of the nascent developments around the world.

We would like to thank all the contributing authors for lending their expertise to make the book truly unique. They have played a crucial role in the development of this book. Without their invaluable contributions this book wouldn't have been possible. They have made vital efforts to compile up to date information on the varied aspects of this subject to make this book a valuable addition to the collection of many professionals and students.

This book was conceptualized with the vision of imparting up-to-date information and advanced data in this field. To ensure the same, a matchless editorial board was set up. Every individual on the board went through rigorous rounds of assessment to prove their worth. After which they invested a large part of their time researching and compiling the most relevant data for our readers.

The editorial board has been involved in producing this book since its inception. They have spent rigorous hours researching and exploring the diverse topics which have resulted in the successful publishing of this book. They have passed on their knowledge of decades through this book. To expedite this challenging task, the publisher supported the team at every step. A small team of assistant editors was also appointed to further simplify the editing procedure and attain best results for the readers.

Apart from the editorial board, the designing team has also invested a significant amount of their time in understanding the subject and creating the most relevant covers. They scrutinized every image to scout for the most suitable representation of the subject and create an appropriate cover for the book.

The publishing team has been an ardent support to the editorial, designing and production team. Their endless efforts to recruit the best for this project, has resulted in the accomplishment of this book. They are a veteran in the field of academics and their pool of knowledge is as vast as their experience in printing. Their expertise and guidance has proved useful at every step. Their uncompromising quality standards have made this book an exceptional effort. Their encouragement from time to time has been an inspiration for everyone.

The publisher and the editorial board hope that this book will prove to be a valuable piece of knowledge for researchers, students, practitioners and scholars across the globe.

List of Contributors

A. J. Newman
National Center for Atmospheric Research, Boulder CO, USA

M. P. Clark
National Center for Atmospheric Research, Boulder CO, USA

K. Sampson
National Center for Atmospheric Research, Boulder CO, USA

A. Wood
National Center for Atmospheric Research, Boulder CO, USA

L. E. Hay
United States Geological Survey, Modeling of Watershed Systems, Lakewood CO, USA

A. Bock
United States Geological Survey, Modeling of Watershed Systems, Lakewood CO, USA

R. J. Viger
United States Geological Survey, Modeling of Watershed Systems, Lakewood CO, USA

D. Blodgett
United States Geological Survey, Center for Integrated Data Analytics, Middleton WI, USA

L. Brekke
US Department of Interior, Bureau of Reclamation, Denver CO, USA

J. R. Arnold
US Army Corps of Engineers, Institute for Water Resources, Seattle WA, USA

T. Hopson
National Center for Atmospheric Research, Boulder CO, USA

Q. Duan
Beijing Normal University, Beijing, China

T. A. McMahon
Department of Infrastructure Engineering, University of Melbourne, Victoria, 3010, Australia

M. C. Peel
Department of Infrastructure Engineering, University of Melbourne, Victoria, 3010, Australia

D. J. Karoly
School of Earth Sciences and ARC Centre of Excellence for Climate System Science, University of Melbourne, Victoria, 3010, Australia

P. Licznar
Faculty of Environmental Engineering, Wroclaw University of Technology, Wrocław, Poland

C. De Michele
Department of Civil and Environmental Engineering, Politecnico di Milano, Milan, Italy

W. Adamowski
Institute of Environmental Engineering, John Paul II Catholic University of Lublin, Stalowa Wola, Poland

K. Schröter
Helmholtz Centre Potsdam, GFZ German Research Centre for Geosciences, Section Hydrology, Potsdam, Germany
CEDIM – Center for Disaster Management and Risk Reduction Technology, Potsdam, Germany

M. Kunz
Karlsruhe Institute of Technology, Institute for Meteorology and Climate Research, Karlsruhe, Germany
CEDIM – Center for Disaster Management and Risk Reduction Technology, Potsdam, Germany

F. Elmer
Helmholtz Centre Potsdam, GFZ German Research Centre for Geosciences, Section Hydrology, Potsdam, Germany
CEDIM – Center for Disaster Management and Risk Reduction Technology, Potsdam, Germany

B. Mühr
Karlsruhe Institute of Technology, Institute for Meteorology and Climate Research, Karlsruhe, Germany
CEDIM – Center for Disaster Management and Risk Reduction Technology, Potsdam, Germany

B. Merz
Helmholtz Centre Potsdam, GFZ German Research Centre for Geosciences, Section Hydrology, Potsdam, Germany
CEDIM – Center for Disaster Management and Risk Reduction Technology, Potsdam, Germany

S. Wi
Department of Civil and Environmental Engineering, University of Massachusetts Amherst, USA

Y. C. E. Yang
Department of Civil and Environmental Engineering, University of Massachusetts Amherst, USA

S. Steinschneider
Department of Civil and Environmental Engineering, University of Massachusetts Amherst, USA

A. Khalil
The World Bank, Washington, DC, USA

C. M. Brown
Department of Civil and Environmental Engineering, University of Massachusetts Amherst, USA

G. Bruni
Department of Water Management, Faculty of Civil Engineering and Geosciences, Delft University of Technology, Delft, the Netherlands

R. Reinoso
Department of Geoscience and Remote Sensing, Faculty of Civil Engineering and Geosciences, Delft University of Technology, Delft, the Netherlands

N. C. van de Giesen
Department of Water Management, Faculty of Civil Engineering and Geosciences, Delft University of Technology, Delft, the Netherlands

. H. L. R. Clemens
Department of Water Management, Faculty of Civil Engineering and Geosciences, Delft University of Technology, Delft, the Netherlands
Deltares, Delft, the Netherlands

J. A. E. ten Veldhuis
Department of Water Management, Faculty of Civil Engineering and Geosciences, Delft University of Technology, Delft, the Netherlands

A. Md Ali
Department of Integrated Water System and Knowledge Management, UNESCO-IHE Institute for Water Education, Delft, the Netherlands
Department of Irrigation and Drainage, Kuala Lumpur, Malaysia

D. P. Solomatine
Department of Integrated Water System and Knowledge Management, UNESCO-IHE Institute for Water Education, Delft, the Netherlands
Water Resource Section, Delft University of Technology, the Netherlands

G. Di Baldassarre
Water Resource Section, Delft University of Technology, the Netherlands
Department of Earth Sciences, Uppsala University, Sweden

H. Vuollekoski
University of Helsinki, Department of Physics, Helsinki, Finland

M. Vogt
University of Helsinki, Department of Physics, Helsinki, Finland
Norwegian Institute for Air Research, Oslo, Norway

V. A. Sinclair
University of Helsinki, Department of Physics, Helsinki, Finland

J. Duplissy
University of Helsinki, Department of Physics, Helsinki, Finland

H. Järvinen
University of Helsinki, Department of Physics, Helsinki, Finland

E.-M. Kyrö
University of Helsinki, Department of Physics, Helsinki, Finland

R. Makkonen
University of Helsinki, Department of Physics, Helsinki, Finland

T. Petäjä
University of Helsinki, Department of Physics, Helsinki, Finland

N. L. Prisle
University of Helsinki, Department of Physics, Helsinki, Finland

P. Räisänen
Finnish Meteorological Institute, Helsinki, Finland

M. Sipilä
University of Helsinki, Department of Physics, Helsinki, Finland

J. Ylhäisi
University of Helsinki, Department of Physics, Helsinki, Finland

M. Kulmala
University of Helsinki, Department of Physics, Helsinki, Finland
Finnish Meteorological Institute, Helsinki, Finland

B. Asadieh
Civil Engineering Department and NOAA-CREST, The City College of New York, City University of New York, New York, USA

N. Y. Krakauer
Civil Engineering Department and NOAA-CREST, The City College of New York, City University of New York, New York, USA

J. Teng
CSIRO Land and Water Flagship, Canberra, Australia

N. J. Potter
CSIRO Land and Water Flagship, Canberra, Australia

F. H. S. Chiew
CSIRO Land and Water Flagship, Canberra, Australia

L. Zhang
CSIRO Land and Water Flagship, Canberra, Australia

B. Wang
CSIRO Land and Water Flagship, Canberra, Australia

J. Vaze
CSIRO Land and Water Flagship, Canberra, Australia

J. P. Evans
Climate Change Research Centre and ARC Centre of Excellence for Climate System Science, University of New South Wales, Sydney, Australia

A. Barrera-Escoda
Climate Change Unit, Meteorological Service of Catalonia, Barcelona, Spain

M. C. Llasat
Meteorological Hazard Analysis Team (GAMA), Department of Astronomy and Meteorology, University of Barcelona, Barcelona, Spain